This series reports on new developments in all areas of mathematics and their applications - quickly, informally and at a high level. Mathematical texts analysing new developments in modelling and numerical simulation are welcome. The type of material considered for publication includes:

1. Research monographs
2. Lectures on a new field or presentations of a new angle in a classical field
3. Summer schools and intensive courses on topics of current research.

Texts which are out of print but still in demand may also be considered if they fall within these categories. The timeliness of a manuscript is sometimes more important than its form, which may be preliminary or tentative. Please visit the LNM Editorial Policy (https://drive.google.com/file/d/19XzCzDXr0FyfcV-nwVojWYTIIhCeo2LN/view?usp=sharing)

Titles from this series are indexed by Scopus, Web of Science, Mathematical Reviews, and zbMATH.

Kôhei Uchiyama

Potential Functions of Random Walks in $\mathbb{Z}$ with Infinite Variance

Estimates and Applications

 Springer

Kôhei Uchiyama (ID)
Tokyo Institute of Technology
Tokyo, Japan

ISSN 0075-8434   ISSN 1617-9692   (electronic)
Lecture Notes in Mathematics
ISBN 978-3-031-41019-2   ISBN 978-3-031-41020-8   (eBook)
https://doi.org/10.1007/978-3-031-41020-8

Mathematics Subject Classification (2020): 60G50, 60G52, 60K05

This Springer imprint is published by the registered company Springer Nature Switzerland AG
The registered company address is: Gewerbestrasse 11, 6330 Cham, Switzerland

Paper in this product is recyclable.

Preface

This treatise consists of nine chapters and an Appendix. Chapters 3–5 are adaptations from the preprint [85]; Chapters 6–7 are edited versions of the preprints [86] and [87] and can be read independently of the preceding three chapters, except for occasional uses of the results of the latter. Chapters 8–9 and the Appendix are written for the present treatise. Chapter 1 is the introduction, where the main results of the treatise are summarised. In Chapter 2, some preliminary results used throughout the later chapters are presented.

We are concerned with a one-dimensional random walk (r.w.) in the integer lattice $\mathbb{Z}$ with a common step distribution function F. For the most part, the r.w. will be assumed to be recurrent and have infinite variance, and we are primarily concerned with the fluctuation theory of the r.w.

In the classical book [71], F. Spitzer derived many beautiful formulae, especially those that involve the potential function, $a(x)$, under very natural assumptions on F. If F has a finite variance when $a(x)$ has a simple asymptotic form, they give explicit asymptotic expressions involving $a(x)$ for objects like hitting probabilities, Green functions, occupation time distributions, etc., of some subsets of $\mathbb{Z}$. Even if the results do not explicitly involve $a(x)$, the asymptotics of $a(x)$ often play crucial roles in their derivation. If the variance of F is infinite, the behaviour of the function $a(x)$, possibly varying in diverse ways depending on the tails of F, is known in a few particular cases. In this treatise, we shall obtain some estimates of $a(x)$ under conditions valid for a broad class of F, and with the estimates obtained, we shall address typical problems of the fluctuation theory, where the potential function is at the centre of the analysis.

Many features of the r.w. in $\mathbb{Z}$ may be expected to be shared by more general real-valued stochastic processes with independent stationary increments, such as Lévy processes and real-valued random walks. Our principal concern in this treatise is the asymptotic form of the potential function of the r.w. and its applications. S.C. Port and C.J. Stone showed that there exists a potential operator that corresponds to $a(x)$ for recurrent r.w.'s on d-dimensional Euclidean space (necessarily $d = 1$ or 2) [59] and for Lévy processes [60], and thereby developed a potential theory for these processes and established analogues of some of the basic formulae obtained for arithmetic random walks. However, it is worth studying the problem in the simple setting of arithmetic r.w.'s. Some significant problems that require delicate analytic estimates remain open for general processes, yet they are solved for the arithmetic r.w.'s. Even when an extension to a broader class of processes is possible, one often needs to figure out additional problems that make the exposition much longer or more complicated – by nature, some problems for r.w.'s may not even exist for any Lévy process other than compound Poisson processes. By the functional limit theorem, many r.w. results yield those on a stable process and vice versa; the proof for the r.w. case is usually elementary, direct, and intuitively comprehensible, with the principal results established by Spitzer taken for granted, while the proof for the stable process often requires an advanced analytic theory.

Tokyo   *Kôhei Uchiyama*
April, 2023

Contents

Chapter 1
Introduction

The potential function for the recurrent random walk (r.w.) in $\mathbb{Z}^d$ ($d = 1, 2$), denoted by $a(x)$, was introduced by Frank Spitzer at the beginning of the 1960s. He established its existence, derived its fundamental properties, and developed a potential theory of the recurrent r.w. analogous to those for Brownian motion or transient Markov processes that had been studied by S. Kakutani, J. Doob, M. Kac, C. Hunt, and others. As in the classical case of transient Markov processes, the function $a(x)$ appears in various objects of probabilistic significance like the hitting distribution and Green function for a subset of $\mathbb{Z}$, and it determines their asymptotic forms. (The definition and fundamental properties of $a(x)$ and some of the previous results inherently involving $a(x)$ are presented in Sections 2.1 and 2.2, respectively.)

For $d = 1$, Spitzer distinguishes two cases: one with $\sigma^2 < \infty$ and the other with $\sigma^2 = \infty$, where σ^2 denotes the variance of the common distribution, F, of increments of the r.w., because of the sharp differences between them: $a(x)$ is a positive harmonic function for the recurrent r.w., absorbed as the r.w. visits the origin, and in the first case, there is a continuum of positive harmonic functions other than $a(x)$, while in the second case, there are no others.

If $\sigma^2 < \infty$ and $d = 1$, we have a simple asymptotic form of $a(x)$ as $|x| \to \infty$, asymptotically linear growth with slopes $\pm 1/\sigma^2$ as $x \to \pm\infty$; the ladder processes, playing a vital role in the fluctuation theory, are easy to deal with since the ladder height variables have finite expectations. These, sometimes combined with the central limit theorem, allow the theory of r.w.'s to be described in a unified way with rather explicit formulae under the single condition $\sigma^2 < \infty$.

On the other hand, if $\sigma^2 = \infty$ there are a variety of circumstances that are characterised by features different from one another; the exact asymptotic behaviour of $a(x)$ for large x has been given only in some special cases; $\lim_{x\to\infty} \bar{a}(x)/x = 0$ and $\lim_{x\to\infty} a(x) \in [0, \infty]$ always exist, but there is a wide range of ways in which $a(x)$ can behave as $x \to \pm\infty$, depending on the behaviour of the tails of F (it can behave very violently with $\limsup a(rx)/a(x) = \infty$ for each $r \in (0, 1)$; see Section 5.6), and it is desirable to find conditions under which $a(x)$ behaves in a specific way. The ladder height, at least the ascending or descending ladder height, has infinite mean.

© The Author(s), under exclusive license to Springer Nature Switzerland AG 2023
K. Uchiyama, *Potential Functions of Random Walks in* $\mathbb{Z}$ *with Infinite Variance*,
Lecture Notes in Mathematics 2338, https://doi.org/10.1007/978-3-031-41020-8_1

The central limit theorem (convergence to stable laws) requires a specific regularity of tails of F that depend on the exponent of a limit stable law.

At the time when Spitzer introduced $a(x)$, several works appeared which studied problems closely related to or involving the potential function, both in its application (H. Kesten, F. Spitzer, B. Belkin) and its extension to other processes (D. Ornstein, S.G. Port, C.J. Stone). For instance, Kesten and Spitzer [49] obtained certain ratio limit theorems for the distributions of the hitting times and sojourn times of a finite set. Kesten [41] gave refinements under mild additional assumptions. (An excellent exposition on the principal contents of [70], [49], [41] is given in Chapter VII of Spitzer's book [71]; extensions to non-lattice random walks are given by Ornstein [55], Port and Stone [59], and Stone [74].) However, in the last half-century, only a few works have appeared related to the potential function of one-dimensional r.w.'s. During the same period, in contrast, many papers have appeared that address subjects concerning r.w.'s, recurrent or transient, but not involving the potential function, like renewal theory, r.w.'s conditioned to stay positive, large deviations, issues related to Spitzer's condition, and so on.

In this treatise, we shall deal with one-dimensional arithmetic r.w.'s and obtain upper and lower bounds of the potential function of the r.w. under a widely applicable condition on the tails of F. We will then use these to solve several problems that are significant from both a probabilistic and an analytic viewpoint. It would be hard to obtain precise results without extra conditions in the case $\sigma^2 = \infty$, so in our applications we shall assume some specific conditions on the tails of F that, in particular, allow us to compute exact asymptotic forms of $a(x)$. Based on the estimates of $a(x)$ obtained, we shall delve into the detailed study of the fluctuation theory in the following cases:

- F is attracted to a strictly stable law.
- The r.w. is relatively stable, or one tail of F is negligible relative to the other in a certain average.

Many authors have studied various aspects of the r.w. for the first case and established remarkable results. However, important questions still need to be answered, some of which we shall resolve satisfactorily. On the other hand, the second case has been much less studied. With the help of results obtained in the last half century on the ladder processes associated with a random walk, especially those on renewal theory, overshoot estimates, and relatively stable r.w.'s, in addition to the classical theory of stochastic processes, we shall derive some precise asymptotic results on hitting probabilities of finite sets, Green functions on long finite intervals or a half-line of $\mathbb{Z}$, absorption probabilities of two-sided exit problems and the like.

Summary

We let $S_n = S_0 + X_1 + X_2 + \cdots + X_n$ be a random walk in $\mathbb{Z}$ where the starting position S_0 is an unspecified integer and the increments $X_1, X_2, \ldots$ are i.i.d. random variables. X denotes a random variable having the same law as X_1, F the distribution function of X, P_x the probability of the random walk with $S_0 = x$ and E_x the expectation with respect to P_x. We assume the r.w. S is *oscillatory* and *irreducible* (as a Markov chain with the state space $\mathbb{Z}$).

Chapter 2 collects previously obtained results that are used later in this treatise. In particular, the first section presents the fundamental results and notation used throughout that readers should read through and absorb. The other sections may be read when referred to or when the current problems become relevant to them.

In Chapter 3, we are concerned with the upper and lower bounds of $a(x)$ that are of crucial importance in its applications, but we previously only know $\lim_{x\to\infty}[a(x) + a(-x)] = \infty$ in general under the condition $EX^2 = \infty$. We shall show that whenever $EX = 0$ (which we tacitly understand is valid only if $E|X| < \infty$),

$$a(x) + a(-x) \geq c_* x/m(x) \quad (x > 0),$$

where $m(x) = \int_0^x dt \int_t^\infty P[|X| > s]\, ds$ and c_* is a universal positive constant, while we give a set of sufficient conditions – also necessary if X is in the domain of attraction – for the upper bound

$$a(x) + a(-x) \leq Cx/m(x)$$

to be true. Putting $m_+(x) = \int_0^x dt \int_t^\infty P[X > s]\, ds$ we shall also show that

$$a(-x)/a(x) \to 0 \quad \text{if} \quad m_+(x)/m(x) \to 0.$$

In Chapter 4, we compute a precise asymptotic form of $a(x)$ under some regularity assumptions on the tails of F. In the first section, we verify that if

$$(*) \qquad \frac{A(x)}{xP[|X| > x]} \to \infty \quad (x \to \infty),$$

where $A(x) = \int_0^x (P[X > t] - P[X < -t])\, dt$, then

$$a(x) \sim \int_0^x \frac{P[X < -t]}{A^2(t)} dt \quad \text{and} \quad a(x) - a(-x) \sim \frac{1}{A(x)}.$$

The condition $(*)$ is satisfied if $EX = 0$ and as $x \to \infty$

$$(\sharp) \qquad P[X < -x] \asymp \frac{L(x)}{x} \quad \text{and} \quad \limsup \frac{\int_x^\infty P[X > t]dt}{\int_x^\infty P[X < -t]dt} < 1,$$

where L is some slowly varying function. For X in the domain of attraction with exponent $1 < \alpha < 2$, an asymptotic form of $a(x)$ expressed in terms of F has been obtained, whereas for $\alpha = 1$, no result like it is found in the existing literature. In the second section, we compute exact asymptotic forms of $a(x)$ for all the cases $1 \leq \alpha \leq 2$ (under some additional side conditions in extreme cases). We also obtain some estimates of the increments of $a(x)$ in the third section. We shall consider the case $(*)$ in detail in Chapter 7, focusing on the two-sided exit problem and related issues.

We deal with the case when two tails of dF are not comparable in Chapter 5, where we mainly consider the case when the negative tail is dominant in the sense that $m_+(x)/m(x) \to 0$ and apply the estimates of $a(x)$ obtained in Chapters 3 and 4 to evaluate some hitting probabilities. Denote by σ_B the first hitting time of $B \subset \mathbb{R}$:

$$\sigma_B = \inf\{n \geq 1 : S_n \in B\},$$

and by $U_a(x)$ and $V_d(x)$, the renewal functions of the strictly ascending and weakly descending ladder height processes, respectively. Among other things, we verify in Section 5.3 that uniformly for $0 \leq x \leq R$, as $R \to \infty$,

$$(**) \qquad P_x(\Lambda_R) \sim V_d(x)/V_d(R),$$

provided $m_+(x)/m(x) \to 0$. Here

$$\Lambda_R = \left[\sigma_{[R+1,\infty)} < \sigma_{(-\infty,-1]} \right].$$

The formula $(**)$ does not seem to have appeared previously in the literature, except in [76], although its analogue is known to hold for spectrally negative Lévy processes [22]. In Sections 5.5–5.7, we study estimations of $P_x[\sigma_B < \sigma_0]$, the probability of the r.w. escaping from 0 into a set $B \subset \mathbb{Z}$ for, e.g., $B = [R, \infty)$, $B = (-\infty, -Q] \cup [R, \infty)$ (with R, Q positive integers). For $x = 0$, the asymptotic form of this probability is of particular interest since it determines the rate of increase of the sojourn time of S spent before visiting B in each finite set (outside B) as R (and Q) becomes large. The results obtained about them will be complemented by those given in subsequent chapters.

In Chapter 6, we provide a sufficient condition for $(**)$ to hold that is also necessary if X belongs to the domain of attraction of a stable law and is satisfied if either $(*)$ holds or $m_+(x)/m(x) \to 0$ (provided $EX = 0$). Suppose that X belongs to the domain of attraction of a stable law of exponent $0 < \alpha \leq 2$ and the limit $\rho := \lim P_0[S_n > 0]$ exists. Then we see that if $(\alpha \vee 1)\rho = 1$, the sufficient condition obtained holds; if $0 < (\alpha \vee 1)\rho < 1$, there exist positive constants $\theta_* < \theta^* < 1$ such that $\theta_* < P_x(\Lambda_R)V_d(R)/V_d(x) < \theta^*$ for $0 \leq x < \delta R$; and if $\rho = 0$, $P_x(\Lambda_R)V_d(R)/V_d(x) \to 0$ as $R \to \infty$ uniformly for $0 \leq x < \delta R$, where δ is an arbitrarily fixed constant less than 1. If $0 < \rho < 1$, the asymptotic form of $P_x(\Lambda_R)$

as $x/R \to \xi \in (0, 1)$ is known,[1] which, however, does not yield the uniform estimate near the endpoints of the interval, as given in our result mentioned above.

Chapter 7 addresses, as mentioned above, the two-sided exit problem under $(*)$ (which implies $P_{\lfloor R/2 \rfloor}(\Lambda_R) \to 1$). We verify, among other things, that if $(*)$ holds then

$$1 - P_x(\Lambda_R) \sim \frac{U_a(R - x)V_d(R)}{R - x - 1} \int_{x-1}^{R} P[X < -t]dt$$

(uniformly for $\varepsilon R < x \le R$) and

$$v_d(x) := V_d(x) - V_d(x - 1) \sim \frac{V_d(x)P[X < -x]}{A(x)},$$

under some additional regularity condition on $P[X < -x]$ for large x, which is satisfied at least if $EX = 0$ and $(\sharp)$ holds. Of the above two relations, the latter is especially interesting since in general we know very little about $v_d(x)$, except that $v_d(x) \to 0$, from the fact that V_d is slowly varying (the case under $(*)$).

Chapter 8 mainly concerns the evaluation of the Green function $g_{B(R)}(x, y)$, $B(R) := \mathbb{R} \setminus [0, R]$, of the r.w. killed as it exits a long interval $[0, R]$. When the law of X is symmetric, the asymptotics of $g_{B(R)}(x, y)$ as $R \to \infty$ were studied by Spitzer and Stone [72] in the case $EX^2 < \infty$ and by Kesten [39], [40] under the condition $E[1 - e^{i\theta X}]/|\theta|^\alpha \longrightarrow Q$ $(\theta \to 0)$ with $1 \le \alpha \le 2$ and $Q > 0$.[2] In particular, Kesten proved that under this condition, if $1 < \alpha < 2$, then for $(\xi, \eta) \in (0, 1)^2$, as $x/R \to \xi$ and $y/R \to \eta$,[3]

(♭)
$$\frac{g_{B(R)}(x, y)}{R^{\alpha-1}} \longrightarrow \frac{|\eta - \xi|^{\alpha-1}}{Q[\Gamma(\alpha/2)]^2} \int_0^{\min\left\{\frac{\xi(1-\eta)}{\eta(1-\xi)}, \frac{\eta(1-\xi)}{\xi(1-\eta)}\right\}} (1 - w)^{-\alpha} w^{\frac{1}{2}\alpha-1} dw$$

(Theorem 3 of [40]), and if $\alpha = 1$ and x/R is bounded away from 0 and 1,

$$g_{B(R)}(x, x)/\log R \longrightarrow (\pi Q)^{-1}$$

(Theorem 3 of [39]). Kesten [40] left the case $\alpha = 1$ with $\xi \ne \eta$ open, although he seemed to be convinced of the truth of the convergence in (♭), saying "There are good reasons to believe" that it is valid also for $\alpha = 1$. We shall generalise these results to the r.w. that is attracted to a stable law with exponent $1 \le \alpha \le 2$, and refine them, the asymptotic formulae obtained being valid uniformly for $0 \le x, y \le R$ if $\alpha > 1$ and for $|y - x|/(x \vee y) > \varepsilon$ $(\varepsilon > 0)$ if $\alpha = 1$ (Theorem 8.4.3, Remark 8.4.4).

Chapter 9 concerns the r.w. conditioned to avoid a finite set when X is in the domain of attraction. For the r.w. conditioned to avoid a finite set up to time n, Belkin [3] proved a functional limit theorem, under appropriate scaling, as $n \to \infty$. We shall prove an extension of it using a different approach from [3], and apply

---

[1] The asserted formula in the literature is wrong, although it and its derivation can be easily rectified (see Remark 6.1.4).

[2] In [39], the asymmetric case where $\Re Q > 0$ is also considered.

[3] If $\sigma^2 < \infty$, this reduces to the result of [72] (cf. (8.2)), though not explicitly stated in [40].

the same method to estimate an escape probability from the set. We also obtain a functional limit theorem of the r.w. conditioned to avoid the set forever. Some related results obtained earlier are briefly reviewed.

Finally, we provide an appendix where miscellaneous results are presented (either known, or easily derived from known facts). We shall give proofs of many of the results that are not currently available in textbooks.

Chapter 2
Preliminaries

Here we briefly review several previously obtained results – mostly when the variance is infinite – that are used or closely related to the topics treated in this book. In the Appendix, we shall give proofs of some of them, mainly those not found in textbooks, to make this treatise as self-contained as possible. The first section gives the basic notation, concepts, and facts used throughout.

2.1 Basic Facts, Notation, and Conventions

Throughout, we shall apply the following facts from Spitzer's book [71], and employ the functions and the random variables involved in them repeatedly, usually without explicit reference.

Let $S_n = S_0 + X_1 + X_2 + \cdots + X_n$ be a random walk in $\mathbb{Z}$, where the starting position S_0 is an unspecified integer and the increments $X_1, X_2, \ldots$ are independent and identically distributed random variables defined on some probability space $(\Omega, \mathcal{F}, P)$ and taking values in $\mathbb{Z}$. Let X be a random variable having the same law as X_1. F denotes the distribution function of X and E indicates integration with respect to P as usual. Denote by P_x the probability of the random walk with $S_0 = x$ and E_x the expectation with respect to P_x. Our basic assumption is

the r.w. $S = (S_n)_{n=0}^{\infty}$ is *oscillatory* and *irreducible* (as a Markov chain on $\mathbb{Z}$),

namely

$$P_0 \left[\limsup S_n = -\liminf S_n = \infty \right] = 1 \quad \text{and} \quad \forall y \in \mathbb{Z}, \quad \sup_n P_0[S_n = y] > 0,$$

which we always suppose, except on a few occasions when explicitly stated otherwise.

K. Uchiyama, *Potential Functions of Random Walks in $\mathbb{Z}$ with Infinite Variance*,
Lecture Notes in Mathematics 2338, https://doi.org/10.1007/978-3-031-41020-8_2

Definition of the potential function

For $x \in \mathbb{Z}$, put $p^n(x) = P_0[S_n = x]$, write $p(x)$ for $p^1(x)$ and define the function $a(x)$ by

$$a(x) = \sum_{n=0}^{\infty} (p^n(0) - p^n(-x)).$$

Here the series on the RHS is convergent for every r.w. (cf. Spitzer [71, P28.8]). When the r.w. is recurrent, $a(x)$, called the *potential function* of S, plays a fundamental role in the potential theory of recurrent random walks, especially for random walks which are killed as they hit the origin, $a(x)$ being a non-negative *harmonic* function,[1] i.e.,

$$a(x) = \sum_{y \in \mathbb{Z} \setminus \{0\}} p(y - x)a(y) \quad x \neq 0,$$

and the Green function of the killed r.w. being expressed in a simple form in terms of a, as described in the next paragraph.

The first hitting time and the Green function of a set B

For a subset B of the whole real line $\mathbb{R}$ such that $B \cap \mathbb{Z} \neq \emptyset$, put

$$\sigma_B = \inf\{n \geq 1 : S_n \in B\},$$

and define the Green function $g_B(x, y)$ of the r.w. killed as it hits B by

$$g_B(x, y) = \sum_{n=0}^{\infty} P_x[S_n = y, \sigma_B > n]. \tag{2.1}$$

Our definition of g_B is not standard: $g_B(x, y) = \delta_{x,y} + E_x[g_B(S_1, y); S_1 \notin B]$ not only for $x \notin B$ but for $x \in B$, while $g_B(x, y) = \delta_{x,y}$ for all $x \in \mathbb{Z}, y \in B$. Here $\delta_{x,y}$ denotes the Kronecker delta symbol. The function $g_B(\cdot, y)$ restricted to B equals the hitting distribution of B by $\hat{S}^{(y)} := y - X_1 - X_2 - \cdots - X_n$, the dual r.w. started at y, in particular

$$g_{\{0\}}(0, y) = 1 \quad (y \in \mathbb{Z}). \tag{2.2}$$

In [71] Spitzer defined the Green function of the r.w. killed on hitting 0 by $g(x, y) = g_{\{0\}}(x, y) - \delta_{0,x}$ (so that $g(0, \cdot) = g(\cdot, 0) \equiv 0$) and proved that

$$g(x, y) = a(x) + a(-y) - a(x - y) \quad (x, y \in \mathbb{Z}). \tag{2.3}$$

---

[1] The uniqueness holds if and only if $\sigma^2 = \infty$.

The ladder height variables and renewal functions

Let Z be the first strictly ascending ladder height that is defined by

$$Z = S_{\sigma_{[S_0+1,\infty)}} - S_0. \tag{2.4}$$

We also define $\hat{Z} = S_{\sigma_{(-\infty,S_0-1]}} - S_0$, the first strictly descending ladder height. As S is oscillatory, both Z and $-\hat{Z}$ are proper random variables whose distributions concentrate on positive integers $x = 1, 2, \ldots$. Let U_a (resp. V_d) be the renewal function of the strictly ascending (resp. weakly descending) ladder height process:

$$U_a(x) = \sum_{n=1}^{\infty} P[Z_1 + \cdots + Z_n \leq x]; \quad V_d(x) = v^\circ \sum_{n=1}^{\infty} P[\hat{Z}_1 + \cdots + \hat{Z}_n \geq -x]$$

for $x = 1, 2, \ldots$, where Z_k (resp $\hat{Z}_k$) are i.i.d. copies of Z (resp $\hat{Z}$) and

$$1/v^0 = P_0[S_{\sigma(-\infty,0]} < 0] = P_0[S_{\sigma[0,\infty)} > 0] = \exp\left\{-\sum k^{-1}p^k(0)\right\}.$$

We put $U_a(0) = 1$, $V_d(0) = v^\circ$ and $U_a(x) = V_d(x) = 0$ for $x < 0$. The product $EZE|\hat{Z}|$ equals $[2v^\circ]^{-1}\sigma^2 \leq \infty$ (cf. [71, Section 18]). Put

$$\ell^*(x) = \int_0^x P[Z > t]\, dt \quad \text{and} \quad \hat{\ell}^*(x) = \frac{1}{v^\circ} \int_0^x P[-\hat{Z} > t]\, dt.$$

U_a varies regularly with index 1 if and only if ℓ^* is slowly varying (s.v.) (Theorem A.2.3) and if this is the case

$$u_a(x) := U_a(x) - U_a(x-1) \sim 1/\ell^*(x) \quad (x \to \infty) \tag{2.5}$$

[83] (see Lemma A.2.1 of the Appendix for a proof applicable in the current setting).

Let $v_d(x) = V_d(x) - V_d(x-1)$ and $\Omega = (-\infty, -1]$. Together with (2.3), the following identity [71, Propositions 18.7, 19.3]

$$g_\Omega(x, y) = \sum_{k=0}^{x \wedge y} v_d(x-k)u_a(y-k) \quad \text{for } x, y \geq 0. \tag{2.6}$$

is fundamental to our investigation. We shall refer to (2.6) as *Spitzer's formula*.

Remark 2.1.1 If U_d denotes the renewal function of the strictly descending ladder height process, then $U_d(x) = V_d(x)/V_d(0)$ (cf., e.g., [31, Sections XII.9, XVIII.3]). It is natural to work with the pair (U_a, V_d) instead of (U_a, U_d) because of Spitzer's formula (2.6) (see also, e.g., Lemma 6.5.1).

Time reversal and Feller's duality

Duality for a random walk with step law $p(x)$ usually refers to the relation between properties of the r.w. and corresponding properties of its dual walk (i.e., the r.w. with step law $p(-x)$). Feller [31] (Section XII.2) discusses a different sort of 'duality', the duality between a random walk path with increments $(X_1, \ldots, X_n)$ and that obtained by taking X_k in reverse order. Clearly, a path has the same probability as its 'dual' path. A little reflection shows that time-reversed paths and Feller's dual paths are symmetric with respect to the time axis. In many cases, the two dualities have essentially the same effects. For instance, Feller's duality yields $g_B(x, y) = g_{-B}(-y, -x)$. On the other hand, the corresponding duality relation (in the sense of time-reversal) reads $g_B(x, y) = \hat{g}_B(y, x)$. Here and throughout the treatise, $\hat{g}_B$ denotes the dual of g_B, namely, the Green function of the dual walk killed as it enters B.

Additional notation and conventions

In the above, we have introduced the functions $F(x)$, $a(x)$, $V_d(x)$, $p^n(x)$ and $g(x, y)$ and the random variables Z, $\hat{Z}$ and σ_B. In addition to these, the following notation is used throughout.

• $F(t) = P[X \le t]$ $(t \in \mathbb{R})$, $\quad p(x) = P[X = x]$ $(x \in \mathbb{Z})$, and

$$\mu_-(t) = P[X < -t], \quad \mu_+(t) = P[X > t], \quad \mu(t) = \mu_-(t) + \mu_+(t) \quad (0 \le t < \infty).$$

• A positive Borel function f defined on a neighbourhood of $+\infty$ is said to be *slowly varying (s.v.)* if $\lim f(\lambda x)/f(x) = 1$ as $x \to \infty$ for each $\lambda > 1$, *regularly varying* with index $\alpha \in \mathbb{R}$ if $x^{-\alpha} f(x)$ is s.v., and *almost decreasing (increasing)* if there exist numbers $C \ge 1$ and x_0 such that $f(y)/f(x) \le C$ $(\ge C^{-1})$ whenever $y > x \ge x_0$. An almost decreasing positive function f is said to be *of dominated variation* if $\liminf_{x \to \infty} f(2x)/f(x) > 0$. An s.v. $f(x)$ is represented as $c(x)e^{\int_1^x \varepsilon(t)t^{-1}/dt}$ with $c(t) \to 1$ and $\varepsilon(t) \to 0$ as $t \to \infty$. If $c(t) \equiv 1$, f is called *normalised*. (See Feller [31] and/or Bingham, Goldie and Teugels [8] for properties of these functions.)[2]

• $x \sim y$ (resp. $x \asymp y$) designates that the ratio of x and y approaches 1 (resp. is bounded away from zero and infinity). A formula of the form $f(x) \sim Cg(x)$ with an extended real number $0 \le C \le \infty$ and $g(x) > 0$ is understood to mean $f(x)/g(x) \to C$. [We often write $f/g \to C$ for this if doing so causes no ambiguity.]

• For non-negative functions f and g of $x \ge 0$, we sometimes write

$$f(x) \ll g(x)$$

for $f(x) = o(g(x))$, i.e., $f(x)/g(x) \to 0$; and similarly for $g(x) \gg f(x)$. These expressions tacitly entail $g(x) > 0$ for all x large enough.

---

[2] In these books, functions of dominated variation are presumed monotone.

- $x \wedge y$ and $x \vee y$ respectively denote the minimum and maximum of x and y; for a real number x, $\lfloor x \rfloor$ is its integer part, $x_+ = x \vee 0$, $x_- = (-x) \vee 0$; $\mathrm{sgn}(x) = x/|x|$. For $x \neq 0$, $x/0 = +\infty$ $(-\infty)$ if $x > 0$ $(x < 0)$.
- We shall write $EX = 0$ with the understanding that $E|X| < \infty$ is implicitly assumed. We often omit '$x \to \infty$' and/or '$R \to \infty$' and the like when it is obvious. We shall write $0 < x, y < R$ for the conjunction not of $0 < x$ and $y < R$ but of $0 < x < R$ and $0 < y < R$.
- The letters x, y, z, w designate space variables that may be real numbers or integers. Their exact nature will be specified only when significant ambiguity arises. For real numbers a, b, the summation sign $\sum_{a \leq x \leq b}$ will often be written $\sum_a^b$ when the contributions of the boundary terms to the sum are negligible.
- By C, C', C_1, etc., we denote finite positive constants whose values may change from line to line. For a subset B of $\mathbb{R}$ we write $-B = \{-x : x \in B\}$ and $B + y = \{x + y : x \in B\}$ $(y \in \mathbb{R})$ and $\sharp B$ stands for the cardinality of B.
- $\mathbf{1}(S)$ equals 1 or 0 according as a statement S is true or false; $\delta_{x,y} = \mathbf{1}(x = y)$.
- ϑ_t denotes the shift operator of sample paths: $S \circ \vartheta_n = S_{n+\cdot}$ for $n = 0, 1, 2, \ldots$; $f \circ \vartheta_t = f(t + \cdot)$ for a real function $f(t)$, $t \geq 0$.
- '$\longrightarrow_P$' ('$\Rightarrow$') designates convergence in probability (distribution).

2.2 Known Results Involving $a(x)$

Among the known results, the following are remarkable: for $x, y \in \mathbb{Z}$,

$$\lim_{n \to \infty} \frac{P_x[\sigma_0 > n]}{P_0[\sigma_0 > n]} = a^\dagger(x) \tag{2.7}$$

(valid for every irreducible r.w., recurrent or transient, of dimension $d \geq 1$); and

$$\lim_{n \to \infty} \frac{P_x[S_n = y, \sigma_0 \geq n]}{P_0[\sigma_0 = n]} = a^\dagger(x) a^\dagger(-y) + \frac{xy}{\sigma^4}, \tag{2.8}$$

provided F is recurrent and strongly aperiodic.[3] Here $\sigma_x = \inf\{n \geq 1 : S_n = x\}$, xy/∞ is understood to be zero,

$$a^\dagger(0) = 1 \quad \text{and} \quad a^\dagger(x) = a(x) \quad \text{for } x \neq 0,$$

and F is called *strongly aperiodic* if

for every x there exists an $n_0 \geq 1$ such that $P_0[S_n = x] > 0$ for all $n \geq n_0$. (2.9)

The first identity (2.7) is Theorem 32.1 of [71]. The formula (2.8) follows from Theorem 6a $(\sigma^2 < \infty)$ and Theorem 7 $(\sigma^2 = \infty)$ of Kesten [41]. Some extensions and refinements of (2.8) are given in [76], [79] in the case when $\sigma^2 < \infty$ (see Section

---

[3] We call F recurrent (or transient) if the r.w. S is recurrent (resp. transient).

9.9.1) and in [82] for the stable walks with exponent $1 < \alpha < 2$ (see Section 9.9.3). (See [78] in the case $\sigma^2 < \infty$ for further detailed estimates of $P_x[\sigma_0 = n]$.)

For recurrent r.w.'s, the following basic properties of $a(x)$ hold [71, Sections 28, 29]:

$$a(x+1) - a(x) \to \pm 1/\sigma^2 \quad \text{as} \quad x \to \pm\infty, \tag{2.10}$$

$$\bar{a}(x) := \frac{1}{2}[a(x) + a(-x)] \to \infty \tag{2.11}$$

and

$$a(x) - \frac{x}{\sigma^2} \begin{cases} = 0 \text{ for all } x > 0 \quad \text{if} \quad P[X \le -2] = 0, \\ > 0 \text{ for all } x > 0 \quad \text{otherwise} \end{cases} \tag{2.12}$$

(for the strict positivity in the second case of (2.12), not given in [71] when $\sigma^2 < \infty$, see, e.g., [76, Eq(2.9)]). When $\sigma^2 < \infty$ (2.10) entails the exact asymptotic $a(x) \sim |x|/\sigma^2$, whereas if $\sigma^2 = \infty$ it gives only $a(x) = o(|x|)$ and sharper asymptotic estimates are highly desirable.

Below we state some results concerning the properties of $a(x)$ that the present author has recently obtained, and which will be relevant to the main focus of this treatise. Let $\sigma^2 = \infty$. In [84] it is shown that for every recurrent r.w.,

$$\text{the limits} \lim_{x\to\infty} a(x) \le \infty \text{ and } \lim_{x\to\infty} a(-x) \le \infty \text{ exist;}$$

$$\begin{cases} 0 < \lim_{x\to\infty} a(-x) < \infty & \text{if} \quad \begin{cases} E|X| < \infty, \ P[X \ge 2] > 0 \quad \text{and} \\ \int_0^\infty [t/m_-(t)]^2 [1 - F(t)] \, dt < \infty, \end{cases} \\ \lim_{x\to\infty} a(-x) = \infty & \text{otherwise;} \end{cases} \tag{2.13}$$

$$\lim_{x\to\infty} a(-x) < \infty \quad \text{if and only if} \quad 0 < \sum_{x=1}^\infty P[Z > x]a(x) < \infty; \tag{2.14}$$

(Theorem 2.4 and Corollary 2.6 of [84]), and analogously for $\lim_{x\to\infty} a(x)$; and

$$\begin{cases} \dfrac{a(x)}{V_d(x)} \to 1/EZ, \quad \dfrac{E_x[a(S_{\sigma(-\infty,0]})]}{a(x)} \to 0 \quad (x \to \infty) \text{ if } EZ < \infty, \\ \liminf_{x\to+\infty} \dfrac{a(x)}{V_d(x)} = 0, \quad a(x) = E_x[a(S_{\sigma(-\infty,0]})] \quad (x > 0) \text{ otherwise} \end{cases} \tag{2.15}$$

(Corollary 2.2 of [84]). [Here $m_-(x) = \int_0^x dt \int_t^\infty F(-s) \, ds$ (see the beginning of Section 3.1).] Note that if $E|X| = \infty$, then $EX_+ = EX_- = \infty$ because of the assumed oscillation of the walk. In (2.15), lim inf can be replaced by lim in various cases, and no counter-example contrary to it has been found so far.

2.3 Domain of Attraction

We shall be concerned with the asymptotic stable walks that satisfy the condition

$$(AS) \begin{cases} \text{(a) } X \text{ is attracted to a stable law of exponent } 0 < \alpha \leq 2, \\ \text{(b) there exists the limit } \rho := \lim P_0[S_n > 0]. \end{cases}$$

Note that the assumption of S being oscillating (in particular $EX = 0$ if $E|X| < \infty$) is always in force. Condition (a) holds if and only if

$$\begin{aligned} E[X^2; |X| < x] &\sim L(x) & \text{for } \alpha = 2, \\ \mu_+(x) \sim pL(x)x^{-\alpha} \quad \text{and} \quad \mu_-(x) &\sim qL(x)x^{-\alpha} & \text{for } \alpha < 2, \end{aligned} \tag{2.16}$$

for some s.v. function L and constants $p \in [0, 1]$, $q = 1 - p$ (cf. [31]).[4]

In (b), ρ ranges exactly over $[1 - \alpha^{-1}] \vee 0 \leq \rho \leq \alpha^{-1} \wedge 1$. For $\alpha \neq 1$, (b) follows from (a), and according to [91],

$$\rho = 2^{-1} + (\pi\alpha)^{-1} \arctan[(p - q)\tan(\alpha\pi/2)],$$

while for $\alpha = 1$ (b) holds with $\rho = 1$ at least if either $p < 1/2, EX = 0$ or $p > 1/2, E|X| = \infty$ and analogously for the case $\rho = 0$; for $p = 1/2, 0 < \rho < 1$ is equivalent to the existence of $\lim nE[\sin(X/c_n)] \in \mathbb{R}$ as well as to that of $\lim nc_n^{-1}E[X; |X| < c_n] \in \mathbb{R}$ with constants c_n as specified below. (See [31, Sections XVII.5 and IX.8]; also (4.29).)

Let (c_n) be a positive sequence chosen so that

$$nL(c_n)/c_n^\alpha \to c_\sharp \tag{2.17}$$

for some arbitrary positive constant $c_\sharp$. Then if $\alpha \neq 1$, S_n/c_n converges in law to a strictly stable variable. For $\alpha = 1$, with the centring constants given by $b_n = nE[\sin(X/c_n)]$, $S_n/c_n - b_n$ converges in law to a stable variable. Denote by $Y^\circ = (Y^\circ(t))_{t \geq 0}$ the limit stable process (assumed to be defined on the same probability space as X). The characteristic exponent $\Phi(\theta) := -\log E[e^{i\theta Y^\circ(1)} \mid Y^\circ(0) = 0]$ ($\theta \in \mathbb{R}$) is then given by

$$\Phi(\theta) = \begin{cases} C_\Phi|\theta|^\alpha \left\{1 - i(\operatorname{sgn}\theta)(p - q)\tan\frac{1}{2}\alpha\pi\right\} & \text{if } \alpha \neq 1, \\ C_\Phi|\theta| \left\{\frac{1}{2}\pi + i(\operatorname{sgn}\theta)(p - q)\log|\theta|\right\} & \text{if } \alpha = 1, \end{cases} \tag{2.18}$$

where $C_\Phi = c_\sharp\Gamma(1 - \alpha)\cos\frac{1}{2}\alpha\pi$ if $\alpha \notin \{1, 2\}$, $C_\Phi = c_\sharp/\alpha$ if $\alpha \in \{1, 2\}$, and $\operatorname{sgn}\theta = \theta/|\theta|$ (cf. [31, (XVII.3.18-19)]).[5] We shall often write $\hat{\rho}$ for $1 - \rho$.

---

[4] We have chosen the formulation (2.16), which differs from the standard one (as given in [31]) where one has the factor $(2 - \alpha)/\alpha$ on the RHS in the case $\alpha < 2$, since it makes many constants simpler.

[5] If $\alpha \notin \{1, 2\}$, then the difference in the choice of L from [31] means that the constant C_Φ must be replaced by $[(2 - \alpha)/\alpha]C_\Phi$ in the standard formulation.

For convenience, we shall assume the norming sequence c_n to be extended to a continuous function on $[0, \infty)$ by linear interpolation.

2.4 Relative Stability, Distribution of Z and Overshoots

The class of relatively stable r.w.'s contains the extreme case $\rho\hat{\rho} = 0$ of $\alpha = 1$ in (AS), which has been least studied when the problem is treated in relation to limit stable processes. Here we summarise some of the known results on the relatively stable r.w.'s directly related to the topics treated in this treatise; see [48] for more details.

2.4.1 Relative Stability

We say a random walk S is *relatively stable* (abbreviated as r.s.) if there exists a (non-random) sequence $B_n \neq 0$ such that $S_n/B_n \to 1$ in probability under P_0. Here B_n is necessarily either ultimately positive or ultimately negative (cf. [48, Theorem 2.3], [45, p. 1806]). In the positive (negative) case, after [48], we say that S is *positively (negatively) relatively stable* (abbreviated as p.r.s. (n.r.s.)). Put

$$A_{\pm}(x) = \int_0^x \mu_{\pm}(y)\,dy \quad \text{and} \quad A(x) = A_+(x) - A_-(x). \tag{2.19}$$

According to [53], [65] (see also the remark around Eq(1.15) of [48]), S is p.r.s. if and only if $E[X; |X| < x]/x\mu(x) \to \infty$ ($x \to \infty$), or, what amounts to the same,

$$\frac{A(x)}{x\mu(x)} \longrightarrow \infty \qquad \text{as } x \to \infty; \tag{2.20}$$

and in this case B_n can be chosen so that $B_n \sim nA(B_n)$ [for a proof, see the beginning of Section A.4]; clearly $A'(x) = o(A(x)/x)$ ($x \to \infty$), so A is slowly varying (s.v.) at infinity, and the same is true for A_+ and $A_+ + A_- = \int_0^{\cdot} \mu(y)\,dy$. If F belongs to a domain of attraction of a stable law of exponent $\alpha \in (0, 2]$, (2.20) holds if and only if $\alpha = 1$ and $P[S_n > 0] \to 1$. (See Remark 6.1.2 and Section 7.1 for more about related matters.) Under (2.20) the function A is positive from some point on and in the sequel x_0 is a positive integer such that $A(x) > 0$ for $x \geq x_0$.

According to [63] the relative stability of S implies that for any $\varepsilon > 0$

$$P_0\left[\sup_{0 \leq t < 1/\varepsilon} |S_{\lfloor nt \rfloor}/B_n - t|\right] \longrightarrow 1 \quad (n \to \infty) \tag{2.21}$$

(see Theorem A.4.1), which evidently entails the following

$$\text{(a) } P[S_n > 0] \longrightarrow 1; \quad \text{(b) } P_x(\Lambda_{2x}) \longrightarrow 1; \quad \text{(c) } S_n \longrightarrow_P +\infty, \tag{2.22}$$

where Λ_R stands for the event $\{\sigma_{(R,\infty)} < \sigma_{(-\infty,0)}\}$. The converse holds under (AS), but not in general. An exact solution is obtained by Kesten and Maller [46], where they do not assume the oscillation of S, nor any moment condition. From Theorem 2.1 and Lemma 4.3 of [46] it follows that if either $EX_-^2 = \infty$ or $\mu_-(x) > 0$ for all $x < 0$ and if $\sigma^2 = \infty$, then each of (a) to (c) above is equivalent to

$$\exists x_0 > 0, \; A(x) > 0 \text{ for } x \geq x_0 \text{ and } \frac{x\mu_-(x)}{A(x)} \to 0 \tag{2.23}$$

(see also [47, Theorem 3] for another criterion).[6] [If $EX = 0$, (2.23) is rephrased as $\lim[\eta_-(x) - \eta_+(x)]/x\mu_-(x) = \infty$ (where $\eta_\pm(x) = \int_x^\infty \mu_\pm(y)\,dy$), which implies that $\eta_-(x)/x\mu_-(x) \to \infty$, hence η_- is s.v. at infinity. Note that if $\mu_-(x) = 0$ for x large enough (so that $EX = 0$ under our basic hypothesis), (2.23) is impossible.]

2.4.2 Overshoots and Relative Stability of Z

We define the overshoot that the r.w. S makes beyond the level R by

$$Z(R) := S_{\sigma_{[R+1,\infty)}} - R. \tag{2.24}$$

Let $S_0 = 0$. Then $Z(R)$ coincides with the overshoot made by the ladder height process $H_n = Z_1 + \cdots + Z_n$, where Z_k are i.i.d. copies of Z. When $EZ < \infty$, the law of the overshoot itself accordingly converges weakly as $R \to \infty$ to a proper probability distribution by standard renewal theory [31, (XI.3.10)]. On the other hand, if $EZ = \infty$, according to Kesten [43] (or [44, Section 4]), $\limsup_{n\to\infty} Z_n/H_{n-1} = \infty$ a.s., or, what amounts to the same, $\limsup Z(R)/R = \infty$ a.s. In particular, it follows that

$$\lim_{R\to\infty} Z(R)/R = 0 \quad \text{a.s.} \quad \text{if and only if} \quad EZ < \infty.$$

It is important to find the corresponding criterion for convergence in probability, which holds under a much weaker condition. In this respect, the following result due to Rogozin [63] is relevant:

(i) *the following are equivalent*

$$\text{(a) } Z(R)/R \longrightarrow_P 0; \quad \text{(b) } Z \text{ is r.s.}; \quad \text{(c) } U_a(x)/x \text{ is s.v.} \tag{2.25}$$

Note that (c) holds if and only if u_a is s.v. (see (2.5)). In [63], it is also shown that

---

[6] Under $\sigma^2 = \infty$ Griffin and McConnell [35] obtain an analytic condition equivalent to (b) that is quite different in appearance from (2.23), but equivalent to it, which Kesten and Maller proved in (iii) of Remarks to their Theorem 2.1.

(ii) *the following are equivalent*

$$\text{(a)} \ \ Z(R)/R \longrightarrow_P \infty; \quad \text{(b)} \ \ U_a(x) \ \textit{is s.v.} \tag{2.26}$$

(See Theorem A.2.3 for a proof of (i) and (ii).) By (2.21) each of (a) to (c) in (2.25) holds if the r.w. S is p.r.s. and each of (a) and (b) of (2.25) holds if S is n.r.s. (see Remark 6.1.2 for the latter). In Section 5.2 we shall obtain a reasonably fine sufficient condition for (2.25) to hold that complements the sufficiency of S being p.r.s. stated above.

2.5 Overshoot Distributions Under (AS)

Suppose that (AS) holds and that $0 < \rho < 1$ if $\alpha = 1$ (so that $P_0[S_n/c_n \in \cdot]$ converges to a non-degenerate stable law for appropriate constants c_n; the results below do not depend on the choice of c_n). The distribution of the overshoot $Z(R)$ is a rather classical result as given in [31] (see (A.16) of Theorem A.2.3), and here we consider the overshoot made by the r.w. S when it exits the interval $B(R) = [0, R]$.

Let $Y = (Y(t))_{t \geq 0}$ be a limit stable process with probabilities $P_\xi^Y, \xi \in \mathbb{R}$ (i.e., the limit law for $S_{\lfloor nt \rfloor}/c_n$ as $S_0/c_n \to \xi$). Denote by Λ_r^Y $(r > 0)$ the event for Y defined by

$$\Lambda_r^Y = \{\text{the first exit of } Y \text{ from the interval } [0, r] \text{ is on the upper side}\}.$$

Rogozin [64] established the overshoot law given in (2.27) below.

Put $Z^Y(r) = Y_{\sigma^Y(r,\infty)} - r$, the overshoot of Y as Y crosses the level $r > 0$ from the left to the right of it (the analogue of $Z(R)$ given in (2.24)). The law of $Z^Y(r)$ is obtained by Blumenthal, Getoor and Ray [9], Widom [90], Watanabe [89] and Kesten [40] for the symmetric case, by Port [61] for the spectrally one-sided case and by Rogozin [64] for the general case. We have only to consider the problem in the case when Λ_r^Y occurs. The result may read as follows: if $0 < \alpha\rho \leq 1$, then for $0 \leq \xi \leq r$ and $\eta > 0$,

$$P_\xi^Y \left[Z^Y(r) > \eta, \Lambda_r^Y \right] = \frac{\sin \alpha\rho\pi}{\pi} (r - \xi)^{\alpha\rho} \xi^{\alpha\hat\rho} \int_\eta^\infty \frac{t^{-\alpha\rho}(t + r)^{-\alpha\hat\rho}}{t + r - \xi} \, dt. \tag{2.27}$$

By the functional limit theorem, it follows that as $y/c_n \to \eta > 0$

$$P_x[Z(R) > y, \Lambda_R] \longrightarrow \frac{\sin \alpha\rho\pi}{\pi} (1 - \xi)^{\alpha\rho} \xi^{\alpha\hat\rho} \int_\eta^\infty \frac{t^{-\alpha\rho}(t + 1)^{-\alpha\hat\rho}}{t + 1 - \xi} \, dt, \tag{2.28}$$

where $\Lambda_R = \{\sigma[R + 1, \infty) < \sigma_\Omega\}$. [Note that if $\alpha\rho = 1$, the RHS vanishes. The above convergence is true also for $\alpha = \rho = 1$, the case excluded in [64]. Indeed we shall obtain somewhat stronger results (see Lemma 5.2.4 and Proposition 7.6.4).]

2.6 Table for Modes of the Overshoot $Z(R)$ Under (AS)

Let $S_0 = 0$. The following table[7] summarises the modes of $Z(R)$ valid under (AS), in which $\theta_Z(x)$ stands for $P[Z > x]$, R_v stands for the set of regularly varying functions (at infinity) with index v, and $\zeta^{(s)}$ is a random variable having the density $(\pi^{-1} \sin \pi s)[x^s(1+x)]^{-1}$, $x > 0$ (cf. Theorems A.2.3 and A.3.2 and Lemmas 6.4.8 (for $\alpha = 1$) and A.2.1 (for $\alpha\rho = 1$)).

	$Z(R)/R \longrightarrow_P \infty$	$Z(R)/R \Longrightarrow \zeta^{(\alpha\rho)}$	$Z(R)/R \longrightarrow_P 0$
	$\theta_Z \in R_0$	$\theta_Z \in R_{-\alpha\rho}$	$\int_0^{\cdot} \theta_Z(u)\, du \in R_0$
	$\rho = 0$	$0 < \alpha\rho < 1$	$\alpha\rho = 1$
	$U_a(x) \sim 1/\hat{\ell}_\sharp(x)$	$U_a(x) \sim x^{\alpha\rho}/\ell(x)$	$u_a(x) \sim 1/\ell^*(x)$
$\alpha = 2$	$*$	$*$	$0 \le p \le 1,\ \rho = \tfrac{1}{2}$
$1 < \alpha < 2$	$*$	$p > 0,\ \alpha - 1 \le \alpha\rho < 1$	$p = 0,\ \alpha\rho = 1$
$\alpha = 1$ $EX = 0$	$p > \tfrac{1}{2}$ or $p = \tfrac{1}{2}$ with $\rho = 0$	$0 < \rho < 1$	$p < \tfrac{1}{2}$ or $p = \tfrac{1}{2}$ with $\rho = 1$
$\alpha = 1$ $E\lvert X\rvert = \infty$	$p < \tfrac{1}{2}$ or $p = \tfrac{1}{2}$ with $\rho = 0$	(necessarily $p = \tfrac{1}{2}$)	$p > \tfrac{1}{2}$ or $p = \tfrac{1}{2}$ with $\rho = 1$
$0 < \alpha < 1$	$p = \rho = 0$	$p > 0,\ 0 < \rho \le 1$	$*$

- If $\alpha = 1$ the second (resp. fourth) column is represented by $\rho = 0$ (resp. $\rho = 1$) whether $E\lvert X\rvert$ is finite or not.
- $V_d(x) \sim x^{\alpha\hat{\rho}}/\ell(x)$ in all cases, with $\ell = \hat{\ell}^*$ if $(\alpha \vee 1)\hat{\rho} = 1$ and $\ell = \ell_\sharp$ if $\hat{\rho} = 0$.
- See (6.48) for $\hat{\ell}^\sharp$ and (6.2) for $\ell^\sharp$ (both when $\alpha \le 1$).

Further properties of $Z, \hat{Z}, Z(R), U_a$ and V_d are given in Sections 6.2–6.4 in some particular cases, in Section 6.5 under (AS), and in Sections 6.7, 7.6 and 8.7 for $Z(R)$; see also Sections 7.1 and 8.1 for asymptotics of u_a and v_d.

---

[7] This table is taken from [81] with the fourth row concerning U_a added.

Chapter 3
Bounds of the Potential Function

In this chapter, we study asymptotic properties of the potential function $a(x)$ of a recurrent r.w. under our basic setting. We obtain a lower bound of $a(x)$ that is valid without any extra assumptions, and show that a constant multiple of this bound serves as an upper bound under some condition that is satisfied by a large class of r.w.'s. The exact asymptotic form of $a(x)$ will be computed in the next chapter under specific conditions on the tails of F. Throughout this chapter, we suppose that $EX = 0$ and $\sigma^2 = \infty$.

3.1 Statements of Results

To state the results of this chapter, we need the following functionals of F that also bear a great deal of relevance to our analysis: for $x \geq 0$, $t \geq 0$,

$$\mu_-(x) = P[X < -x], \quad \mu_+(x) = P[X > x], \quad \mu(x) = \mu_-(x) + \mu_+(x);$$

$$\eta_\pm(x) = \int_x^\infty \mu_\pm(y)\,dy, \qquad\qquad \eta(x) = \eta_-(x) + \eta_+(x);$$

$$c_\pm(x) = \int_0^x y\mu_\pm(y)\,dy \qquad\qquad c(x) = c_+(x) + c_-(x);$$

$$m_\pm(x) = \int_0^x \eta_\pm(t)\,dt, \qquad\qquad m(x) = m_+(x) + m_-(x);$$

$$\alpha_\pm(t) = \int_0^\infty \mu_\pm(y)\sin ty\,dy, \qquad\qquad \alpha(t) = \alpha_+(t) + \alpha_-(t);$$

$$\beta_\pm(t) = \int_0^\infty \mu_\pm(y)(1 - \cos ty)\,dy, \qquad \beta(t) = \beta_+(t) + \beta_-(t);$$

$$\gamma(t) = \int_0^\infty [\mu_-(y) - \mu_+(y)]\cos ty\,dy = \beta_+(t) - \beta_-(t);$$

and

$$\psi(t) = Ee^{itX}.$$

K. Uchiyama, *Potential Functions of Random Walks in Z with Infinite Variance*, Lecture Notes in Mathematics 2338, https://doi.org/10.1007/978-3-031-41020-8_3

Integrating by parts shows

$$1 - \psi(t) = t\alpha(t) + it\gamma(t)^{\,1} \quad \text{and} \quad m_{\pm}(x) = c_{\pm}(x) + x\eta_{\pm}(x).$$

In the above formulae, x designates a real number, whereas x will always be an integer in $a(x)$; we shall not mention which usage will be adopted if this duplicity will cause no confusion.

Theorem 3.1.1 *For some universal constant $c_* > 0$*

$$\bar{a}(x) \geq c_* x / m(x) \qquad \text{for all sufficiently large} \quad x \geq 1.$$

We bring in the following condition to obtain an upper bound of $\bar{a}(x)$:

$$\textbf{(H)} \qquad \delta_H := \liminf_{t \downarrow 0} \frac{\alpha(t) + |\gamma(t)|}{\eta(1/t)} > 0.$$

Note that $\alpha(t) + |\gamma(t)| \asymp |1 - \psi(t)|/t \ (0 < t \leq \pi)$.

Theorem 3.1.2 (i) *If condition (H) holds, then for some constant C_H^* depending only on δ_H,*

$$\bar{a}(x) \leq C_H^* x / m(x) \qquad \text{for all sufficiently large} \quad x \geq 1.$$

(ii) *If* $\dfrac{\alpha(t) + |\gamma(t)|}{t\,m(1/t)} \to 0 \, (t \downarrow 0)$, *then* $\dfrac{\bar{a}(x)}{x/m(x)} \to \infty \, (x \to \infty)$.

[The converse is not true; see the end of Section 3.6 for a counter-example.]

According to Theorems 3.1.1 and 3.1.2, if (H) holds, then $\bar{a}(x) \asymp x/m(x)$ as $x \to \infty$ (i.e., the ratio of the two sides is bounded away from zero and infinity), which entails some regularity of $\bar{a}(x)$ like its being almost increasing and of dominated variation, while in general, $\bar{a}(x)$ may behave very irregularly, as will be exhibited by an example in Section 3.6.

Theorem 3.1.1 plainly entails the implication

$$\sum \bar{a}(x)\mu_+(x) < \infty \implies \int_1^{\infty} \frac{x\mu_+(x)}{m(x)} \, dx < \infty \tag{3.1}$$

which bears immediate relevance to the criteria for the summability of the first ascending ladder height Z. The integrability condition on the RHS of (3.1), implying $x\eta_+(x)/m(x) \to 0$ and $\int_1^{\infty} c_+(y) d(-1/m(y)) < \infty$, and hence $m_+(x)/m(x) \to 0$, is equivalent to the necessary and sufficient condition for $EZ < \infty$ due to Chow [15] ((3.1) is verified in [77] and constitutes one of the key observations there that lead to the following equivalences

---

[1] One may also use another expression $1 - \psi(t) = [\alpha^{\circ}(t) + i\gamma^{\circ}(t)]\sin t$, where $\alpha^{\circ}(t) = \sum_0^{\infty} \mu(n)[\sin tn + \tan\frac{1}{2}t \cos tn]$, $\gamma^{\circ}(t) = \sum_0^{\infty}[\mu_-(n) - \mu_+(t)][\cos tn - \tan\frac{1}{2}t \sin tn]$, which, though preferable for some problems, is not applied in this treatise.

$$\sum \bar{a}(x)\mu_+(x) < \infty \iff EZ < \infty \iff \int_1^\infty \frac{x\mu_+(x)}{m_-(x)}\,dx < \infty \qquad (3.2)$$

without recourse to Chow's result (see Section 4 (especially Lemma 4.1) of [77]; cf. also [84, Remark 2.3(b)]). By the same token, it follows from the dual of (2.13) (see also (3.6)) that

$$\sum [\bar{a}(x)]^2 \mu_+(x) < \infty \iff \limsup a(-x) < \infty. \qquad (3.3)$$

For condition (H) to hold each of the following conditions (3.4) to (3.8) is sufficient:

$$\text{either } \limsup_{x\to\infty} \frac{\mu_-(x)}{\mu(x)} < \frac{1}{2} \text{ or } \limsup_{x\to\infty} \frac{\mu_+(x)}{\mu(x)} < \frac{1}{2}; \qquad (3.4)$$

$$\limsup_{x\to\infty} \frac{x\eta_+(x)}{m_+(x)} < 1 \text{ or } \limsup_{x\to\infty} \frac{x\eta_-(x)}{m_-(x)} < 1 \text{ or } \limsup_{x\to\infty} \frac{x\eta(x)}{m(x)} < 1; \qquad (3.5)$$

$$\frac{m_+(x)}{m(x)} \text{ converges as } x \to \infty \text{ to a number } \neq \frac{1}{2}; \qquad (3.6)$$

$$\lim_{x\to\infty} \frac{x\eta(x)}{m(x)} = 1 \text{ and either } \limsup_{x\to\infty} \frac{\eta_-(x)}{\eta(x)} < \frac{1}{2} \text{ or } \limsup_{x\to\infty} \frac{\eta_+(x)}{\eta(x)} < \frac{1}{2}; \qquad (3.7)$$

$$\limsup_{x\to\infty} \frac{x[\eta_+(x) \wedge \eta_-(x)]}{m_+(x) \vee m_-(x)} \leq \frac{1}{9}. \qquad (3.8)$$

That (3.4) is sufficient for (H) follows from $\liminf_{t\downarrow 0} \beta(t)/\eta(1/t) > 0$ (cf. (3.12b)) since $\liminf_{t\downarrow 0} |\gamma(t)|/\beta(t) > 0$ under (3.4). The sufficiency of the conditions (3.5) and (3.8) will be verified in Section 3.2 (Lemmas 3.2.9 and 3.2.5 for (3.5) and Lemma 3.2.10 for (3.8)); as for (3.6) and (3.7), see Remark 3.3.7 and the comment given immediately after Theorem 4.1.1, respectively. In (3.7), the first condition is equivalent to the slow variation of η, which together with the second one implies either η_+ or η_- is s.v. (so that S is r.s. under (3.7); see Theorem 4.1.1 for consequences on $a(x)$).

We do not know whether (H) is necessary for $\bar{a}(x)m(x)/x$ to be bounded. The next result entails that the necessity of (H) follows when the distribution of X is nearly symmetric in the sense that the limit in (3.6) equals $1/2$, namely

$$m_+(x)/m(x) \to 1/2 \quad \text{as} \quad x \to \infty. \qquad (3.9)$$

Theorem 3.1.3 *Suppose (3.9) holds. Then*

(i) $\lim_{t\downarrow 0} \gamma(t)/[t\,m(1/t)] = 0$; *and*

(ii) *each of the three inequalities in the disjunction (3.5) is necessary (as well as sufficient) in order that* $\limsup_{x\to\infty} \bar{a}(x)m(x)/x < \infty$.

Corollary 3.1.4 *Under (3.9),* $\limsup \bar{a}(x)m(x)/x < \infty$ *if and only if (H) holds.*

Proof If $\limsup \bar{a}(x)m(x)/x < \infty$, then by Theorem 3.1.3(ii) $\limsup x\eta(x)/m(x) <$ 1, or equivalently, $\liminf c(x)/m(x) > 0$, which implies $\liminf \alpha(t)/\eta(1/t) > 0$, as we shall see (Lemma 3.2.5), whence (H) holds. The converse follows from Theorem 3.1.2.   □

In practice, in most cases, Theorem 3.1.3 together with (3.4) to (3.8) provides the criterion, expressed in terms of $\mu_\pm$, $m_\pm$ and/or $\eta_\pm$, to judge whether the condition (H) holds. When F is attracted to a stable law, the result is simplified so that $\lim \bar{a}(x)m(x)/x = \infty$ if $x\eta(x)/m(x) \to 1$ and $m_+(x)/m(x) \to 1/2$; otherwise $\bar{a}(x) \sim Cx/m(x)$ (see Proposition 4.2.1(i, iv) and Remark 4.2.2(ii)).

Proposition 3.1.5 *If (H) holds, then for* $0 < x < 2R$,

$$\left| 1 - \frac{\bar{a}(x)}{\bar{a}(R)} \right| \le C \left| 1 - \frac{x}{R} \right|^{1/4}$$

for some constant C that depends only on δ_H.

From Proposition 3.1.5 it easily follows that $\bar{a}(x)/\bar{a}(R) \to 1$ as $x/R \to 1$ if (H) holds, which fact, not trivial, is what we need for some of our applications. [Proposition 4.3.1 in Section 4.3 entails that the exponent $1/4$ can be replaced by 1 if $\liminf c(x)/m(x) > 0$, but does not ensure $\bar{a}(x)/\bar{a}(R) \to 1$ otherwise.]

In the next result, as well as the applications in Chapter 5, we consider the r.w. S under the condition $m_+(x)/m(x) \to 0$ as $x \to +\infty$, which for simplicity we abbreviate as $m_+/m \to 0$. (Similar conventions will also apply to η_+/η, c/m, etc.)

Theorem 3.1.6 *Suppose* $m_+/m \to 0$. *Then* (H) *holds with* $\delta_H = 1$, *and*

$$\frac{a(-x)}{a(x)} \to 0 \quad as \quad x \to +\infty;$$

in particular $a(x) \sim 2\bar{a}(x) \asymp x/m(x)$.

In the proof of Theorem 3.1.6 we shall see that if $m_+/m \to 0$ and

$$b_\pm(x) = \frac{1}{\pi} \int_0^\pi \frac{t\beta_\pm(t) \sin xt}{|1 - \psi(t)|^2} \, dt,$$

then $\bar{a}(x) \sim b_-(x)$ and $b_+(x) = o(b_-(x))$, which together indeed imply that $a(-x)/a(x) \to 0$ since $\frac{1}{2}[a(x) - a(-x)] = b_-(x) - b_+(x)$.

Applications of the above theorems are made mostly in the case $a(-x)/a(x) \to 0$ in this treatise. In general, however, it is significant to consider the r.w. S under the condition

$$\delta_* := \liminf_{x\to\infty} a(-x)/\bar{a}(x) > 0 \quad \text{and/or} \quad \delta^* := \limsup_{x\to\infty} a(-x)/\bar{a}(x) < 1,$$

but we do not know when this condition holds except for some restricted classes of r.w.'s (as studied in Chapter 4; see also Section 7.7 for $\delta^* < 1$) and for the case $\delta_* = \delta^* = 1$ dealt with in the following

Proposition 3.1.7 *If (H) holds and $m_+/m \to \frac{1}{2}$, then $\lim a(\pm x)/\bar{a}(x) \to 1$.*

In the next section we derive some fundamental facts about $a(x)$ and the functionals introduced above, which incidentally yield (i) of Theorem 3.1.2 (see Lemmas 3.2.3 and 3.2.6) and the sufficiency for (H) of (3.5) and (3.8) (Lemmas 3.2.5, 3.2.9 and 3.2.10); also the lower bound of Theorem 3.1.1 is verified under a certain side condition. The proof of Theorem 3.1.1 is more involved and given in Section 3.3, in which we also prove Theorems 3.1.2(ii) and 3.1.3. Proposition 3.1.5 is proved in Section 3.4, and Theorem 3.1.6 and Proposition 3.1.7 are in Section 3.5. In the last section we present an example of F for which $\bar{a}(x)$ is not almost increasing and $c(x)/m(x)$ oscillates between ε and δ for any $0 < \varepsilon < \delta < 1$.

3.2 Auxiliary Results

In this section we first present some known or easily derived facts and then give several lemmas, in particular, Lemmas 3.2.3 and 3.2.7, which together assert that $\bar{a}(x) \asymp x/m(x)$ under the last inequality in (3.5) and whose proofs involve typical arguments that are implicitly used in Sections 3.3 to 3.5.

As in [77], we bring in the following functionals of F in addition to those introduced in Section 3.1:

$$\tilde{c}(x) = \frac{1}{x} \int_0^x y^2 \mu(y)\, dy, \quad \tilde{m}(x) = \frac{2}{x} \int_0^x y\eta(y)\, dy,$$

$$h_\varepsilon(x) = \int_0^{\varepsilon x} y\, [\mu(y) - \mu(\pi x + y)]\, dy \quad (0 < \varepsilon \le \pi/2);$$

also $\tilde{c}_\pm(x)$ and $\tilde{m}_\pm(x)$ are defined with $\mu_\pm$ in place of μ, so that $\tilde{c}(x) = \tilde{c}_-(x) + \tilde{c}_+(x)$, etc. Since $\sigma^2 = \infty$, $c(x)$ and $h_\varepsilon(x)$ tend to infinity as $x \to \infty$; η, c and h_ε are monotone and $\Im\psi(t) = -t\gamma(t)$. Here and throughout the rest of this section and Sections 3.3 to 3.5, $x \ge 0$ and $t \ge 0$. We shall be concerned with the behaviour of these functions as $x \to \infty$ or $t \downarrow 0$ and omit "$x \to \infty$" or "$t \downarrow 0$" when it is obvious.

As noted previously, the function m admits the decomposition

$$m(x) = x\eta(x) + c(x).$$

The function m is relatively tractable: it is increasing and concave, hence subadditive, and

for any $k > 1$, $\quad m(kx) \le km(x)$ and $m(x)/k \le m(x/k)$.

However c, though increasing, may vary quite differently. The ratio $c(x)/m(x)$ may converges to 0 or to 1 as $x \to \infty$ depending on μ and possibly oscillates asymptotically between 0 and 1; and

$$c(kx) = k^2 \int_0^x \mu(ku)u\, du \le k^2 c(x) \quad (k > 1),$$

where the factor k^2 cannot be replaced by $o(k^2)$ for the upper bound to be valid (cf. Section 3.6). It also follows that $\tilde{m}$ is increasing and concave and that

$$\tilde{m}(x) = x\eta(x) + \tilde{c}(x).$$

For $x > 0$, $c(x) = \tilde{c}(x) + x^{-1}\int_0^x c(y)\,dy$, in particular $\tilde{m}(x) < m(x)$.
   It holds that

$$\bar{a}(x) = \frac{1}{2\pi}\int_{-\pi}^{\pi}\Re\frac{1}{1-\psi(t)}(1-\cos xt)\,dt$$

(cf. [71, Eq(28.2)]). Recalling $1 - \psi(t) = t\alpha(t) + it\gamma(t)$, we have

$$\frac{1}{1-\psi(t)} = \frac{\alpha(t) - i\gamma(t)}{\alpha^2(t) + \gamma^2(t)} \cdot \frac{1}{t}.$$

Hence

$$\bar{a}(x) = \frac{1}{\pi}\int_0^{\pi}\frac{\alpha(t)}{[\alpha^2(t) + \gamma^2(t)]t}(1 - \cos xt)\,dt. \qquad (3.10)$$

Note that $\alpha_{\pm}(t)$ and $\beta_{\pm}(t)$ are all positive (for $t > 0$); by Fatou's lemma

$$\liminf \alpha(t)/t = \liminf \int_0^{\infty}\frac{1-\cos tx}{t^2}d(-\mu(x)) \geq \frac{1}{2}\sigma^2,$$

so that $\alpha(t)/t \to \infty$ under the present setting.
   To find the asymptotics of $\bar{a}$ we need to know those of $\alpha(t)$ and $\beta(t)$ as $t \downarrow 0$ (which entail those of $\alpha_{\pm}$ and $\beta_{\pm}$ as functionals of $\mu_{\pm}$). We present some of them in the following two lemmas. Although the arguments made therein are essentially the same as in [77], we provide complete proofs since some constants in [77] are wrong or inadequate for the present need and must be rectified – the values of the constants involved are not significant in [77], but some of them turn out to be of crucial importance for our proof of Theorem 3.1.1.

Lemma 3.2.1 *For $0 < \varepsilon \leq \pi/2$ and $0 < t \leq \pi$,*

$$[\varepsilon^{-1}\sin\varepsilon]h_{\varepsilon}(1/t) < \alpha(t)/t < [5c(1/t)] \wedge m(1/t). \qquad (3.11)$$

Proof By monotonicity of μ it follows that

$$\alpha(t) > \left(\int_0^{\varepsilon/t} + \int_{\pi/t}^{(\pi+\varepsilon)/t}\right)\mu(z)\sin tz\,dz = \int_0^{\varepsilon/t}[\mu(z) - \mu(\pi/t + z)]\sin tz\,dz,$$

which by $\sin tz \geq \varepsilon^{-1}(\sin\varepsilon)tz$ ($tz \leq \varepsilon$) shows the first inequality of the lemma. Splitting the defining integral at $1/t$ we see $\alpha(t) < tc(1/t) + \eta(1/t) = tm(1/t)$. Similarly we see $\alpha(t) < \int_0^{\pi/t}\mu(z)\sin tz\,dz \leq tc(\frac{1}{2}\pi/t) + \mu(1/t)\int_{\frac{1}{2}\pi/t}^{\pi/t}\sin tz\,dz \leq \frac{1}{4}\pi^2 tc(1/t) + t^{-1}\mu(1/t) < 5tc(1/t)$. Thus the second inequality follows. $\square$

Lemma 3.2.2 *For $0 < t \le \pi$,*

$$\begin{cases} (a) & \frac{1}{2}\tilde{m}(1/t)t \le \beta(t) \le 2\tilde{m}(1/t)t, \\ (a') & |\beta(t) - \eta(1/t)| \le 4tc(1/t), \\ (b) & \frac{1}{3}m(1/t)t \le \alpha(t) + \beta(t) \le 3m(1/t)t. \end{cases} \quad (3.12)$$

Proof Integrating by parts and using the inequality $\sin u \ge (2/\pi)u \ (u < \frac{1}{2}\pi)$ in turn we see

$$\beta(t) = t \int_0^\infty \eta(x) \sin tx \, dx \ge \frac{2t^2}{\pi} \int_0^{\pi/2t} \eta(x)x \, dx + t \int_{\pi/2t}^\infty \eta(x) \sin tx \, dx.$$

Observing that the first term of the expression of the rightmost member equals $\frac{1}{2}t\tilde{m}(\pi/2t) \ge \frac{1}{2}t\tilde{m}(1/t)$ and the second one equals $-\int_{\pi/2t}^\infty \mu(x) \cos tx \, dx > 0$, we obtain the left-hand inequality of (3.12a). As for the right-hand inequality of (3.12a), we split the defining integral of β at $1/t$ to see that $\beta(t) \le \frac{1}{2}t\tilde{c}(1/t) + 2\eta(1/t) \le 2t\tilde{m}(1/t)$. The proof of (a') is similar (rather simpler): one has only to note that $\int_{1/t}^\infty \mu(x) \cos tx \, dx$ is less than $t^{-1}\mu(1/t)(1 - \sin 1)$ and larger than $-2t^{-1}\mu(\pi/2t)$ and that $t\mu(1/t) \le 2tc(1/t)$.

The upper bound of (3.12b) is immediate from those in (3.11) and (a) proved above since $\tilde{m}(x) \le m(x)$. To verify the lower bound, use (3.11) and the inequalities $h_1(x) \ge c(x) - \frac{1}{2}x^2\mu(x)$ and $\sin 1 > 5/6$ to obtain

$$[\alpha(t) + \beta(t)]/t > (5/6) \left[c(1/t) - \mu(1/t)/2t^2 \right] + 2^{-1} \left[\tilde{c}(1/t) + \eta(t)/t \right].$$

Since $x^2\mu(x) \le [2c(x)] \wedge [3\tilde{c}(x)]$ it follows that $\frac{5}{6}\mu(1/t)/2t^2 \le \frac{1}{2}c(1/t) + \frac{1}{2}\tilde{c}(1/t)$ and hence $\alpha(t) + \beta(t) > \left[\frac{2}{6}c(1/t) + \frac{1}{2}\eta(1/t)/t \right] t > \frac{1}{3}m(1/t)t$, as desired. $\qquad\square$

For $t > 0$ define

$$f_m(t) = \frac{1}{t^2 m^2(1/t)} \quad \text{and} \quad f^\circ(t) = \frac{1}{\alpha^2(t) + \gamma^2(t)}. \quad (3.13)$$

Observe that

$$\left(\frac{x}{m(x)} \right)' = \frac{c(x)}{m^2(x)}, \quad (3.14)$$

which, in particular, entails $x/m(x)$ is increasing and $f_m(t)$ is decreasing.

Lemma 3.2.3 *For some universal constant C*

$$\int_0^\pi \frac{f_m(t)\alpha(t)}{t} (1 - \cos xt) \, dt \le C \frac{x}{m(x)}, \quad (3.15)$$

in particular if $\liminf f_m(t)[\alpha^2(t) + \gamma^2(t)] > \delta$ for some $\delta > 0$, then for all sufficiently large x, $\bar{a}(x) < C[\pi\delta]^{-1}x/m(x)$.

Proof Break the integral on the LHS of (3.15) into two parts

$$J(x) = \int_0^{\pi/2x} \frac{f_m(t)\alpha(t)}{t}(1 - \cos xt)\,dt \quad \text{and}$$

$$K(x) = \int_{\pi/2x}^{\pi} \frac{f_m(t)\alpha(t)}{t}(1 - \cos xt)\,dt.$$

Using $\alpha(t) \le 5c(1/t)t$ we then observe

$$0 \le \frac{K(x)}{5} \le \int_{\pi/2x}^{\pi} \frac{2c(1/t)}{t^2m^2(1/t)}\,dt \tag{3.16}$$

$$= \int_{1/\pi}^{2x/\pi} \frac{2c(y)}{m^2(y)}\,dy$$

$$= \left[\frac{2y}{m(y)}\right]_{y=1/\pi}^{2x/\pi} < \frac{2x}{m(x)}.$$

Similarly

$$\frac{J(x)}{5} \le \frac{x^2}{2}\int_0^{\pi/2x} f_m(t)c(1/t)t^2\,dt = \frac{x^2}{2}\int_{2x/\pi}^{\infty} \frac{c(y)}{y^2m^2(y)}\,dy$$

and, observing

$$\int_x^{\infty} \frac{c(y)}{y^2m^2(y)}\,dy \le \int_x^{\infty} \frac{dy}{y^2m(y)} \le \frac{1}{xm(x)}, \tag{3.17}$$

we have $J(x) \le \frac{5}{2}x/m(x)$, finishing the proof.   □

If there exists a constant $B_0 > 0$ such that for all t small enough,

$$\alpha(t) \ge B_0 c(1/t)t, \tag{3.18}$$

then the estimation of $\bar{a}$ becomes much easier. Unfortunately, condition (3.18) may fail to hold in general: in fact, the ratio $\alpha(t)/[c(1/t)t]$ may oscillate between $1 - \varepsilon$ and ε as $t \downarrow 0$ for any $0 < \varepsilon < 1$ (cf. Section 3.6). The following lemma will be used crucially to cope with such a situation.

Lemma 3.2.4 *Let* $0 \le \varepsilon \le \pi/2$. *Then for all* $x > 0$,

$$h_\varepsilon(x) \ge c(\varepsilon x) - (2\pi)^{-1}\varepsilon^2 x\,[\eta(\varepsilon x) - \eta(\pi x + \varepsilon x)].$$

Proof On writing $h_\varepsilon(x) = c(\varepsilon x) - \int_0^{\varepsilon x} u\mu(\pi x + u)\,du$, integration by parts yields

$$h_\varepsilon(x) - c(\varepsilon x) = -\int_0^{\varepsilon x} [\eta(\pi x + u) - \eta(\pi x + \varepsilon x)]\,du.$$

By monotonicity and convexity of η it follows that if $0 < u \leq \varepsilon x$,

$$0 \leq \eta(\pi x + u) - \eta(\pi x + \varepsilon x) \leq \frac{\varepsilon x - u}{\pi x} \left[\eta(\varepsilon x) - \eta(\pi x + \varepsilon x) \right],$$

and substitution leads to the inequality of the lemma. □

Lemma 3.2.5 *Let* $0 < \delta \leq 1$, $0 < t < \pi$ *and put* $s = [1 \wedge (\delta\pi)] t$. *Then*

(i) *if* $c(1/t) \geq \delta\eta(1/t)/t$, *then* $\alpha(s) > \lambda(\pi^{-1} \wedge \delta)^2 \eta(1/s)$ *with* $\lambda := \frac{1}{2}\pi \sin 1 > 1$, *in particular if* $\delta := \liminf c(x)/x\eta(x) > 0$, *then* $\alpha(t) > (\pi^{-1} \wedge \delta)^2 \eta(1/t)$ *for all sufficiently small* $t > 0$ – *so that (H) holds;*
(ii) *if* $c(1/t) \geq \delta m(1/t)$, *then* $\alpha(s) > \lambda(\pi^{-1} \wedge \delta)^2 sm(1/s)$ $(\lambda = \frac{1}{2}\pi \sin 1)$.

Proof Suppose $\delta\pi \leq 1$ and take $\varepsilon = \delta\pi$ in Lemma 3.2.4. Then $\varepsilon^{-1} \sin \varepsilon > \sin 1 = 2\lambda/\pi$, while the premise of the first statement of the lemma implies

$$h_\varepsilon(1/\varepsilon t) > c(1/t) - \frac{\varepsilon\eta(1/t)}{2\pi t} \geq \frac{\delta\eta(1/t)}{2t} \geq \frac{\delta\varepsilon\eta(1/\varepsilon t)}{2t} = \frac{\pi\delta^2\eta(1/\delta\pi t)}{2\delta\pi t},$$

whence by Lemma 3.2.1 $\alpha(s)/s \geq [\varepsilon^{-1} \sin \varepsilon] h_\varepsilon(1/s) > \lambda\delta^2\eta(1/s)/s$ for $s = \delta\pi t$. Similarly, if $\delta\pi > 1$, then taking $\varepsilon = 1$ we have

$$\alpha(t)/t > (\sin 1)(\delta - (2\pi)^{-1})\eta(1/t)/t > \lambda\pi^{-2}\eta(1/t)/t,$$

showing (i). The same proof applies to (ii). □

Lemma 3.2.6 *If (H) holds, then*

(H') $\quad \alpha(t) + |\gamma(t)| > [\pi^{-2} \wedge \frac{1}{3}\delta_H] tm(1/t) \quad$ *for all sufficiently small* $t > 0$,

and

$$\int_0^1 \frac{\beta(t)}{\alpha^2(t) + \gamma^2(t)} \, dt < \infty. \tag{3.19}$$

[By Lemma 3.2.2, (H') ensures that $f^\circ(t) \asymp f_m(t)$.]

Proof If $\frac{1}{2}m(1/t) \leq c(1/t)$, then by Lemma 3.2.5(ii), $\alpha(t) > \frac{1}{4}tm(1/t)$, entailing (H') (for this t), while if $\frac{1}{2}m(1/t) > c(1/t)$, then $\eta(1/t)/t > 2^{-1}m(1/t)$. Hence (H) implies (H').

By Lemma 3.2.2 (H') implies that the integral in (3.19) is at most a constant multiple of

$$\int_0^1 \frac{\tilde{m}(1/t)}{tm^2(1/t)} \, dt = \int_1^\infty \frac{\tilde{m}(x)}{xm^2(x)} \, dx$$

$$= \int_1^\infty \frac{2 \int_0^x y\eta(y) \, dy}{x^2 m^2(x)} \, dx.$$

Interchanging the order of integration, one deduces that the last integral above equals

$$\int_1^\infty \frac{2\int_0^1 y\eta(y)\,dy}{x^2 m^2(x)}\,dx + \int_1^\infty 2y\eta(y)\,dy \int_y^\infty \frac{dx}{x^2 m^2(x)} < C + \int_1^\infty \frac{2\eta(y)}{m^2(y)}\,dy < \infty,$$

Thus (3.19) is verified. □

Lemma 3.2.7 *If* $\delta := \liminf c(x)/m(x) > 0$, *then for some constant* $C > 0$ *that depends only on* δ,

$$C^{-1}x/m(x) \le \bar{a}(x) \le Cx/m(x) \quad \text{for all } x \text{ large enough.}$$

Proof By Lemmas 3.2.5(ii) and 3.2.6 condition (H′) is satisfied, provided $\delta > 0$. Hence the upper bound follows from Lemma 3.2.3.

Although the lower bound follows from Theorem 3.1.1, which will be shown independently of Lemma 3.2.7 in the next section, here we provide a direct proof. Let $K(x)$ be as in the proof of Lemma 3.2.3. By Lemma 3.2.5(ii) we may suppose that $\alpha(t)/t \ge B_1 c(1/t)$ with a constant $B_1 > 0$. Since both $c(1/t)$ and $f_m(t)$ are decreasing and hence so is their product, we see that

$$\frac{K(x)}{B_1} \ge \int_{\pi/2x}^\pi f_m(t)c(1/t)(1 - \cos xt)\,dt \ge \int_{\pi/2x}^\pi \frac{c(1/t)}{t^2 m^2(1/t)}\,dt,$$

from which we deduce, as in (3.16), that

$$\frac{K(x)}{B_1} \ge \left[\frac{y}{m(y)}\right]_{y=1/\pi}^{2x/\pi} \ge \frac{2}{\pi} \cdot \frac{x}{m(x)} - \frac{1/\pi}{m(1/\pi)}. \tag{3.20}$$

Thus the desired lower bound obtains. □

Lemma 3.2.8 *Suppose that* $0 = \liminf c(x)/m(x) < \limsup c(x)/m(x)$. *Then for any* $\varepsilon > 0$ *small enough there exists an unbounded sequence* $x_n > 0$ *such that*

$$c(x_n) = \varepsilon m(x_n) \quad \text{and} \quad \alpha(t)/t > 2^{-1}\varepsilon^2 m(x_n) \quad \text{for} \quad 0 < t \le 1/x_n.$$

Proof Put $\lambda(x) = c(x)/m(x)$ and $\delta = \frac{1}{2}\limsup \lambda(x)$. Let $0 < \varepsilon < \delta^2$. Then there exists two sequences x_n and x'_n such that $x_n \to \infty$, $x'_n < x_n$,

$$\lambda(x_n) = \varepsilon < \lambda(x) \quad \text{for } x'_n < x < x_n \quad \text{and} \quad \delta = \lambda(x'_n), \tag{3.21}$$

provided $\liminf \lambda(x) = 0$. Observing $x'_n/x_n \le m(x'_n)/m(x_n) < \lambda(x_n)/\lambda(x'_n) = \varepsilon/\delta < \delta$, we see that $1/x_n < \delta/x'_n$ and then by using (3.21),

$$\frac{\lambda(x_n)}{x_n} < \frac{1}{x_n} < \frac{\lambda(x'_n)}{x'_n}.$$

Hence the intermediate value theorem ensures that there exists a solution of the equation $\lambda(x) = x/x_n$ in the interval $x'_n < x < x_n$. Let y_n be the largest solution and put $\varepsilon_n = y_n/x_n$. Then

$$\varepsilon < \lambda(y_n) = c(y_n)/m(y_n) = \varepsilon_n < 1 \quad \text{and} \quad y_n = \varepsilon_n x_n.$$

Hence by Lemma 3.2.4

$$h_{\varepsilon_n}(x_n) \geq \left[c(y_n) - \frac{\varepsilon_n}{2\pi} m(y_n) \right] = \frac{2\pi - 1}{2\pi} \varepsilon_n m(\varepsilon_n x_n) > \frac{5}{6}\varepsilon^2 m(x_n).$$

Since h_ε is non-decreasing, we have for $0 < t \leq 1/x_n$,

$$\alpha(t)/t \geq (\sin 1) h_{\varepsilon_n}(1/t) > \frac{5}{6} h_{\varepsilon_n}(x_n) \geq 2^{-1}\varepsilon^2 m(x_n). \tag{3.22}$$

Thus the lemma is verified. □

Lemma 3.2.9 *For any $0 < \delta \leq 1/\pi$ and $0 < t < \pi$, one has the implication*

$$c_+(1/t)/m_+(1/t) \geq \delta \implies \alpha_+(s) + |\gamma(s)| \geq 3^{-1}\delta^2\beta(s) \text{ for } s = \delta\pi t. \tag{3.23}$$

In particular, if $\liminf c_+(x)/m_+(x) > 0$, *then (H) holds.*

Proof For any positive numbers $\beta_\pm$ and $\delta < 2$, we have

$$\delta\beta_+ + |\beta_+ - \beta_-| \geq \frac{1}{2}\delta(\beta_+ + \beta_-). \tag{3.24}$$

If the LHS inequality of (3.23) holds, Lemma 3.2.5(ii) applied with μ_+ in place of μ shows that for $s = \delta\pi t$, $\alpha_+(s)/s \geq \frac{1}{2}(5/6)\pi\delta^2 m_+(1/s)$. In conjunction with Lemma 3.2.2(a) this entails $\alpha_+(s) \geq \frac{2}{3}\delta^2\beta_+(s)$, so that, by (3.24), the RHS inequality of (3.23) follows. The second assertion is immediate from (3.23) given Lemma 3.2.2(b). □

Lemma 3.2.10 *For (H) to hold, it is sufficient that*

$$\limsup \frac{x[\eta_+(x) \wedge \eta_-(x)]}{m_+(x) \vee m_-(x)} < \frac{1}{4}\left(1 - \frac{1}{\pi}\right)^2. \tag{3.25}$$

Proof We apply Lemma 3.2.9 with $\delta = 1/\pi$ so that $s = t$ in (3.23). Write $x = 1/s$. Then, using the identity $m_\pm(x) = c_\pm(x) + x\eta_\pm(x)$, we see that if $\delta_* = 1/(\pi - 1)$,

$$(*) \begin{cases} \text{if either } c_+(x) \geq \delta_* x\eta_+(x) \text{ or } c_-(x) \geq \delta_* x\eta_-(x), \\ \text{then } \alpha(s) + |\gamma(s)| \geq \frac{1}{3}\pi^{-2}\beta(s). \end{cases}$$

Put

$$\lambda(x) := \frac{\tilde{m}_+(x) \wedge \tilde{m}_-(x)}{\tilde{m}_+(x) \vee \tilde{m}_-(x)}, \quad \omega(s) = \frac{\beta_+(s) \wedge \beta_-(s)}{\beta_+(s) \vee \beta_-(s)}. \tag{3.26}$$

Then applying Lemma 3.2.2(a) to obtain $\omega(s) \leq 4\lambda(x)$, we infer that

$$|\gamma(s)| = [1 - \omega(s)] (\beta_+(s) \vee \beta_-(s)) \geq 2^{-1} [1 - 4\lambda(x)] \beta(s).$$

Suppose $c_+(x) < \delta_* x \eta_+(x)$ and $c_-(x) < \delta_* x \eta_-(x)$, complementarily to the case of (∗). Then $m_\pm(x)/(1 + \delta_*) < x \eta_\pm(x) < \tilde{m}_\pm(x)$, so that the ratio under the lim sup on the LHS of (3.25) is larger than $(1+\delta_*)^{-2} \lambda(x)$. It therefore follows that (3.25) implies lim sup $\lambda(x) < 1/4$. Hence $|\gamma(s)| > \varepsilon \beta(s)$ with some $\varepsilon > 0$ for all sufficiently small $s > 0$. Combined with (∗) this shows (H) to be valid, since $\beta(s) > \eta(1/s)$.   □

The following lemma is used in order to handle the oscillating part of the integrals defining $\alpha_\pm(t)$ and/or $\beta_\pm(t)$ in Sections 3.4, 3.5 and 4.3 (see (3.54)).

Lemma 3.2.11 *Let* $0 < s < t \le \pi$. *Then*

$$|\alpha(t) - \alpha(s)| \vee |\beta(t) - \beta(s)| \le \begin{cases} (t + 7s)c(1/s), \\ 5(t - s)m(1/s). \end{cases}$$

If (H) *holds, then by Lemma 3.2.6 the second bound entails that for some constant* C *depending only on* δ_H,

$$|f^\circ(t) - f^\circ(s)| \le C[(t - s)/t]f^\circ(t) \quad (s < t < 2s). \tag{3.27}$$

Proof By definition $\beta(s) - \beta(t) = \int_0^\infty \mu(z)(\cos tz - \cos sz)\,dz$, and we deduce

$$\left| \int_0^{1/s} \mu(z)\,(\cos tz - \cos sz)\,dz \right| \le (t - s) \int_0^{1/s} z\mu(z)\,dz$$

$$= (t - s)c(1/s),$$

and

$$\left| \int_{1/s}^{\infty} \mu(z)\,(\cos tz - \cos sz)\,dz \right| \le \left(\frac{2}{s} + \frac{2}{t} \right) \mu(1/s) \tag{3.28}$$

$$\le 8s \cdot c(1/s).$$

Evidently we have the corresponding bound for $\alpha(t) - \alpha(s)$, hence the first bound of the lemma. For the second one, substituting $\cos tz - \cos sz = \int_s^t z \sin z\tau\,d\tau$ and interchanging the order of integration in turn result in

$$\int_{1/s}^{\infty} \mu(z)\,(\cos tz - \cos sz)\,dz = \int_s^t d\tau \int_{1/s}^{\infty} z\mu(z)\sin \tau z\,dz. \tag{3.29}$$

According to (A.38) of Section A.5.1, the inner integral on the RHS is at most $4\tau^{-1}s \cdot m(1/s)$, so that the RHS of (3.28) may be replaced by $4[s\log(t/s)]m(1/s)$. The same bound is valid if the cos's and sin's are interchanged. Since $s\log t/s \le t - s$, we now obtain the second bound of the lemma.   □

3.3 Proofs of Theorems 3.1.1 to 3.1.3

The main content of this section consists of the proof of Theorem 3.1.1. Theorem 3.1.2 is verified after it: part (i) of Theorem 3.1.2 was virtually proved in the preceding section, whereas part (ii) is essentially a corollary of the proof of Theorem 3.1.1. The proof of Theorem 3.1.3, which partly uses Theorem 3.1.2(ii), is given at the end of the section.

3.3.1 Proof of Theorem 3.1.1

By virtue of the right-hand inequality of (3.12b), $f^{\circ}(t) \geq \frac{1}{9} f_m(t)$, and for the present purpose, it suffices to bound the integral in (3.15) from below by a positive multiple of $x/m(x)$. As a lower bound we take the contribution to the integral from the interval $\pi/2x < t < 1$. We also employ the lower bound $\alpha(t) \geq \int_0^{2\pi/t} \mu(z) \sin tz \, dz$ and write down the resulting inequality as follows: with the constant $B = (9\pi)^{-1}$,

$$\frac{\bar{a}(x)}{B} \geq \int_{\pi/2x}^{1} \frac{f_m(t)\alpha(t)}{t} (1 - \cos xt) \, dt$$

$$\geq \int_{\pi/2x}^{1} \frac{f_m(t)}{t} (1 - \cos xt) \, dt \int_0^{2\pi/t} \mu(z) \sin tz \, dz$$

$$= K_I(x) + K_{II}(x) + K_{III}(x),$$

where

$$K_I(x) = \int_{\pi/2x}^{1} f_m(t) \frac{dt}{t} \int_0^{\pi/2t} \mu(z) \sin tz \, dz,$$

$$K_{II}(x) = \int_{\pi/2x}^{1} f_m(t) \frac{dt}{t} \int_{\pi/2t}^{2\pi/t} \mu(z) \sin tz \, dz$$

and

$$K_{III}(x) = \int_{\pi/2x}^{1} f_m(t)(-\cos xt) \frac{dt}{t} \int_0^{2\pi/t} \mu(z) \sin tz \, dz.$$

Lemma 3.3.1 $K_I(x) \geq \dfrac{5}{3\pi} \cdot \dfrac{x}{m(x)} - \dfrac{1}{m(1)}.$

Proof Since $\sin 1 \geq 5/6$ it follows that

$$\int_0^{\pi/2t} \mu(z) \sin tz \, dz > \int_0^{1/t} \mu(z) \sin tz \, dz \geq \frac{5}{6} tc(1/t),$$

and hence

$$K_I(x) > \frac{5}{6} \int_{\pi/2x}^{1} \frac{c(1/t)}{t^2 m^2(1/t)} \, dt \geq \frac{5}{6} \int_{1}^{2x/\pi} \frac{c(z)}{m^2(z)} \, dz$$

$$\geq \frac{5}{3\pi} \cdot \frac{x}{m(2x/\pi)} - \frac{5}{6m(1)},$$

implying the inequality of the lemma because of the monotonicity of m.   □

Lemma 3.3.2 $K_{III}(x) \geq -4\pi f_m(1/2)/x$ for all sufficiently large x.

Proof Put $g(t) = t^{-1} \int_0^{2\pi/t} \mu(z) \sin tz \, dz$. One sees that $0 \leq g(1) < 2$. We claim that g is decreasing. Observe that

$$\frac{d}{dt} t^{-1} \int_0^{2\pi/t} \sin tz \, dz = t^{-2} \int_0^{2\pi/t} (tz \cos tz - \sin tz) \, dz = 0$$

and

$$g'(t) = \frac{1}{t^2} \int_0^{2\pi/t} \mu(z)(tz \cos tz - \sin tz) \, dz,$$

and that $u \cos u - \sin u < 0$ for $0 < u < \pi$ and the integrand of the last integral has a unique zero in the open interval $(0, 2\pi/t)$. Then the monotonicity of μ leads to $g'(t) < 0$, as claimed. Now f being decreasing, it follows that $K_{III}(x) \geq -\int_{(2n+\frac{1}{2})\pi/x}^{1} f_m(t) g(t) \, dt$ for any positive integer n such that $(2n+\frac{1}{2})\pi/x \leq 1$. Since one can choose n so that $0 \leq 1 - (2n+\frac{1}{2})\pi/x \leq 2\pi/x$, the inequality of the lemma obtains.   □

Lemma 3.3.3 $K_{II}(x) \geq -\dfrac{4}{3\pi} \cdot \dfrac{x}{m(x)}.$

Proof Since μ is non-increasing, we have

$$\int_{\pi/2t}^{2\pi/t} \mu(z) \sin tz \, dz \geq \int_{3\pi/2t}^{2\pi/t} \mu(z) \sin tz \, dz,$$

so that

$$K_{II}(x) \geq \int_{\pi/2x}^{1} f_m(t) \frac{dt}{t} \int_{3\pi/2t}^{2\pi/t} \mu(z) \sin tz \, dz.$$

We wish to integrate with respect to t first. Observe that the region of the double integral is included in

$$\{3\pi/2 \leq z \leq 4x; \, 3\pi/2z < t < 2\pi/z\},$$

where the integrand of the inner integral is negative. Hence

$$K_{II}(x) \geq \int_{3\pi/2}^{4x} \mu(z) \, dz \int_{3\pi/2z}^{2\pi/z} f_m(t) \frac{\sin tz}{t} \, dt.$$

Put

$$\lambda = 3\pi/2,$$

so that $t \geq \lambda/z$. Then, since f_m is decreasing, the RHS is further bounded below by

$$\int_\lambda^{4x} \mu(z) f_m(\lambda/z)\, dz \int_{3\pi/2z}^{2\pi/z} \frac{\sin tz}{t}\, dt.$$

The inner integral being equal to $\int_{3\pi/2}^{2\pi} \sin u\, du/u$, which is larger than $-1/\lambda$, after a change of variable we obtain

$$K_{II}(x) \geq -\int_1^{4x/\lambda} \mu(\lambda z) f_m(1/z)\, dz \geq -\int_1^x \mu(\lambda z) f_m(1/z)\, dz.$$

Recall $f_m(1/z) = z^2/m^2(z)$. Since $\int_x^\infty \mu(\lambda z)\, dz = \lambda^{-1}\eta(\lambda x)$, by integration by parts,

$$-\int_1^x \mu(\lambda z) f_m(1/z)\, dz = \frac{1}{\lambda}\left[\frac{\eta(\lambda z)z^2}{m^2(z)}\right]_{z=1}^x - \frac{2}{\lambda}\int_1^x \frac{z\eta(\lambda z)c(z)}{m^3(z)}\, dz$$

$$\geq -\frac{2}{\lambda}\int_1^x \frac{z\eta(z)c(z)}{m^3(z)}\, dz - \frac{\eta(1)}{\lambda m^2(1)}.$$

Noting that $z\eta(z) = m(z) - c(z)$, we have

$$\frac{z\eta(z)c(z)}{m^3(z)} = \frac{c(z)}{m^2(z)} - \frac{c^2(z)}{m^3(z)}.$$

Since $m(1) > \eta(1)$, we conclude

$$K_{II}(x) \geq -\frac{2}{\lambda}\int_1^x \frac{c(z)}{m^2(z)}\, dz - \frac{1}{\lambda m(1)} = -\frac{4}{3\pi}\frac{x}{m(x)} + \frac{1}{\lambda m(1)},$$

hence the inequality of the lemma.   □

Completion of the proof of Theorem 3.1.1. Combining Lemmas 3.3.1 to 3.3.3 we obtain

$$\frac{\bar{a}(x)}{B} \geq \int_{\pi/2x}^1 f_m(t)\alpha(t)\frac{1 - \cos xt}{t}\, dt \geq \frac{1}{3\pi}\cdot\frac{x}{m(x)} - \frac{1}{m(1)} + O(1/x), \quad (3.30)$$

showing Theorem 3.1.1.   □

Proof (of Theorem 3.1.2) The first part (i) of Theorem 3.1.2 is obtained by combining Lemmas 3.2.3 and 3.2.6. As for (ii) its premise implies that for any $\varepsilon > 0$ there exists a $\delta > 0$ such that $f^\circ(t) > \varepsilon^{-1} f_m(t)$ for $0 < t < \delta$, which concludes the result in view of the second inequality of (3.30), since $\int_\delta^1 f_m(t)\alpha(t)(1 - \cos xt)\, dt/t$ is bounded for each $\delta > 0$.   □

3.3.2 Proof of Theorem 3.1.3

Lemma 3.3.4 *Suppose that there exists* $p := \lim m_+(x)/m(x)$. *Then as* $t \downarrow 0$

$$\beta_+(t) = p\beta(t) + o\,(tm(1/t)) \quad and \quad \alpha_+(t) = p\,\alpha(t) + o\,(tm(1/t)).$$

Proof Take a non-increasing summable function $\tilde{\mu}(x)$, $x \geq 0$, denote by $\tilde{m}, \tilde{\eta}, \tilde{\alpha}$ and $\tilde{\beta}$ the corresponding functions and suppose $\tilde{m}/m \to 1$. It suffices to show that

$$\beta(t) - \tilde{\beta}(t) = o(tm(1/t)) \quad and \quad \alpha(t) - \tilde{\alpha}(t) = o(tm(1/t)) \tag{3.31}$$

as $t \downarrow 0$.[2] Pick a positive number M such that $\sin M = 1$. Then

$$\left| \int_{M/t}^{\infty} \mu(y) \cos ty\, dy \right| \leq 2\frac{\mu(M/t)}{t} \leq \frac{4t}{M^2}c(M/t) \leq \frac{4}{M}tm(1/t),$$

while by integrating by parts,

$$\int_0^{M/t} (\mu - \tilde{\mu})(y)(1 - \cos ty)\, dy = -(\eta - \tilde{\eta})(M/t) + t \int_0^{M/t} (\eta - \tilde{\eta})(y) \sin ty\, dy.$$

Since $\int_{M/t}^{\infty}(\mu - \tilde{\mu})(y)\, dy = (\eta - \tilde{\eta})(M/t)$ (which will cancel out the first term on the RHS above), these together yield

$$\beta(t) - \tilde{\beta}(t) = t \int_0^{M/t} (\eta - \tilde{\eta})(y) \sin ty\, dy + tm(1/t) \times O(1/M)$$

and integrating by parts the last integral leads to

$$\frac{\beta(t) - \tilde{\beta}(t)}{t} = (m - \tilde{m})(M/t) - t \int_0^{M/t} (m - \tilde{m})(y) \cos ty\, dy + m(1/t) \times O(1/M).$$

If $\tilde{m}/m \to 1$, the first two terms on the RHS are $o(m(1/t))$ for each M fixed, and we conclude the first relation of (3.31) since M can be made arbitrarily large.

The second relation is verified in the same way but with M taken from $\pi\mathbb{Z}$.   □

Lemma 3.3.5 *Suppose* $\lim |\gamma(t)|/tm(1/t) = 0$. *If* $c/m \to 0$, *then* $\bar{a}(x)m(x)/x$ *diverges to infinity, and if* $\lim\inf c(x)/m(x) = 0$, *then* $\lim\sup \bar{a}(x)m(x)/x = \infty$.

Proof The first half follows from Theorem 3.1.2(ii) and Lemma 3.2.1. For the proof of the second one, we apply the trivial lower bound

$$\pi\bar{a}(\pi x_n) = \int_0^{\pi} \frac{\alpha(t)(1 - \cos \pi x_n t)}{[\alpha^2(t) + \gamma^2(t)]t}\, dt \geq \int_{1/2x_n}^{1/x_n} \frac{\alpha(t)\, dt}{[\alpha^2(t) + \gamma^2(t)]t}, \tag{3.32}$$

---

[2] We may let $p > 0$, for if $p = 0$ the result follows from Lemma 3.2.2.

valid for any sequence $x_n \in \mathbb{Z}/\pi$. Suppose that $\limsup c(x)/m(x) > 0$ in addition to $\liminf c(x)/m(x) = 0$, so that we can take a constant $\varepsilon > 0$ and an unbounded sequence x_n as in Lemma 3.2.8, according to which

$$c(x_n) = \varepsilon m(x_n) \quad \text{and} \quad \alpha(t)/t \geq 4^{-1}\varepsilon^2 m(1/t) \quad \text{if} \quad 1/2x_n < t \leq 1/x_n. \qquad (3.33)$$

Because of the assumption of the lemma, the second relation above leads to

$$|\gamma(t)| = o(m(1/t)t) = o(\alpha(t)) \qquad (n \to \infty), \qquad (3.34)$$

while, in conjunction with (3.21) and Lemma 3.2.1, the first one shows that for $1/2x_n \leq t \leq 1/x_n$,

$$\alpha(t)/t \leq 5c(1/t) \leq 5c(2x_n) \leq 4 \cdot 5c(x_n) = 20\varepsilon m(x_n).$$

These together yield

$$\frac{\alpha(t)}{\alpha^2(t) + \gamma^2(t)} \sim \frac{1}{\alpha(t)} \geq \frac{1}{t} \cdot \frac{1}{20\varepsilon m(x_n)},$$

and substitution into (3.32) shows that for all n large enough

$$\pi\bar{a}(\pi x_n) \geq \frac{1}{30\varepsilon m(x_n)} \int_{1/2x_n}^{1/x_n} \frac{dt}{t^2} = \frac{x_n}{30\varepsilon m(x_n)}.$$

Thus $\bar{a}(x)x/m(x)$ is unbounded, ε being made arbitrarily small. $\square$

Remark 3.3.6 Suppose $m_+/m \to 1/2$. Then for the second assertion of Lemma 3.3.5, the condition $\liminf c(x)/m(x) = 0$ can be replaced by

$$\liminf c_+(x)/m_+(x) = 0. \qquad (3.35)$$

Indeed, if $c_+/m_+ \to 0$, then $\alpha(t) = o(tm(1/t))$ owing to Lemma 3.3.4, and by Theorem 3.1.2(ii) we have only to consider the case $\limsup c_+(x)/m_+(x) > 0$. Take x_n as above but with c_+, m_+, α_+ in place of c, m, α. Then, for $1/2x_n \leq t \leq 1/x_n$, the same argument as leading to (3.33) verifies $m_+(1/t)t = O(\alpha_+(t))$, so that we have (3.34) and, by Lemma 3.3.4 again, $\alpha_+(t) \sim \alpha_-(t)$, and hence $\alpha(t)/[\alpha^2 + \gamma^2](t) \sim 1/2\alpha_+(t)$. We can follow the proof of Lemma 3.3.5 for the rest.

Remark 3.3.7 Let $p < 1/2$ in Lemma 3.3.4. Then we have the following.

(i) $\gamma(t) = (2p - 1)\beta(t) + o(tm(1/t))$, so that (H) holds owing to Lemma 3.2.2(a).
(ii) If $c/m \to 0$, or equivalently $m(x) \sim x\eta(x)$, then, by $c(x) \geq \frac{1}{2}x^2\mu(x)$, $x\mu(x)/\eta(x) \to 0$ so that η is s.v. due to Karamata's theorem. Hence $m_-(x) \sim (1 - p)x\eta(x) \sim x\eta_-(x)$. These properties of η_- and η entail that F is positively relatively stable, so $a(x)$ admits a simple explicit expression for its asymptotic form (see Section 4.1).

Completion of the proof of Theorem 3.1.3. The assertion (i) follows from Lemma 3.3.4; the necessity of the condition $\limsup x\eta(x)/m(x) < 1$ asserted in (ii) is immediate from Lemma 3.3.5. The necessity of the conditions $\limsup x\eta_\pm(x)/m_\pm(x) < 1$ is verified in Remark 3.3.6.   □

3.4 Proof of Proposition 3.1.5

Suppose (H) holds. Since then $\bar{a}(x) \asymp x/m(x)$ by Theorems 3.1.1 and 3.1.2, we may suppose $R/2 < x < R$ and on writing $\delta = (R - x)/x$ the assertion to be shown can be rephrased as

$$|\bar{a}(R) - \bar{a}(x)| \leq C\delta^{1/4}[x/m(x)] \qquad (R/2 < x < R). \qquad (3.36)$$

For $M > 1$ and $R/2 < x \leq R$ we make the decomposition $\bar{a}(x) = u^M(x) + v_M(x)$, where

$$u^M(x) = \int_0^{M/x} \frac{\alpha(t)f^\circ(t)}{t}(1 - \cos xt)\,dt,$$

$$v_M(x) = \int_{M/x}^{\pi} \frac{\alpha(t)f^\circ(t)}{t}(1 - \cos xt)\,dt.$$

By the inequality $|\cos xt - \cos Rt| \leq |Rt - xt|$ it is then easy to show

$$|u^M(x) - u^M(R)| < CM\delta[x/m(x)] \qquad (3.37)$$

for a (universal) constant C (as we shall see shortly), whereas to obtain a similar estimate for v_M we cannot help exploiting the oscillation of $\cos xt$. To the latter purpose one may seek some appropriate smoothness of $\alpha(t)f^\circ(t)$, which, however, is a property difficult to verify because of the intractable part $\int_{1/t}^\infty \mu(y)\sin ty\,dy$ involved in the integral defining $\alpha(t)$. In order to circumvent it, for each positive integer n we bring in the function

$$\alpha_n(t) := \int_0^{n\pi/t} \mu(y)\sin ty\,dy$$

and make use of the inequalities

$$\alpha_{2n}(t) < \alpha(t) < \alpha_{2n+1}(t).$$

If $v_M(x) \leq v_M(R)$, then

$$
\begin{aligned}
0 \leq v_M(R) - v_M(x) \leq &\int_{M/x}^{\pi} [\alpha_{2n+1}(t) - \alpha_{2n}(t)] \frac{f^\circ(t)(1 - \cos Rt)}{t} \, dt \\
&+ \int_{M/x}^{\pi} \frac{\alpha_{2n}(t) f^\circ(t)}{t} (\cos xt - \cos Rt) \, dt \qquad (3.38) \\
&+ \int_{M/R}^{M/x} \frac{\alpha_{2n+1}(t) f^\circ(t)}{t} (1 - \cos Rt) \, dt \\
= &\; I_n + II_n + III_n \quad \text{(say)};
\end{aligned}
$$

and if $v_M(x) > v_M(R)$, we have an analogous inequality. We consider the first case only, the other one being similar. Since $0 \leq \alpha_n(t) \leq \alpha(t) \leq \pi^2 t c(1/t)$ and by Lemma 3.2.6, which may read $f^\circ(t) \leq C_1/[m^2(1/t)t^2]$ under (H), we see that for any n,

$$
\int_{M/R}^{M/x} \frac{\alpha_n(t) f^\circ(t)}{t} \, dt \leq C \int_{x/M}^{R/M} \frac{c(y) \, dy}{m^2(y)} \leq C \frac{(R-x)/M}{m(x/M)} \leq C \frac{\delta x}{m(x)}, \quad (3.39)
$$

so that III_n admits a bound small enough for the present purpose.

Now we are able to verify (3.37). Noting $-(1/xm(x))' > 1/x^2 m(x)$, we have

$$
\int_x^\infty \frac{1}{y^2 m(y)} \, dy < \frac{1}{xm(x)}. \qquad (3.40)
$$

Since $|\cos xt - \cos Rt| = |[1 - \cos(R-x)t] \cos xt + \sin xt \, \sin(R-x)t| \leq 2\delta(xt)^2$ it therefore follows that

$$
\left| \int_0^{1/x} \frac{\alpha(t) f^\circ(t)}{t} (\cos xt - \cos Rt) \, dt \right| \leq C\delta x^2 \int_x^\infty \frac{c(y) \, dy}{m^2(y) y^2} \leq \frac{C\delta x}{m(x)}.
$$

Similarly, using $|\cos xt - \cos Rt| \leq |R-x|t$ one infers that the integral over $[1/x, M/x]$ is dominated in absolute value by

$$
C(R-x) \int_{x/M}^x \frac{c(y) \, dy}{m^2(y) y} \leq CM\delta \int_{x/M}^x \frac{c(y) \, dy}{m^2(y)} \leq \frac{CM\delta x}{m(x)},
$$

so that

$$
\left| \int_0^{M/x} \frac{\alpha(t) f^\circ(t)}{t} (\cos Rt - \cos xt) \, dt \right| \leq \frac{C'M\delta x}{m(x)},
$$

which concludes the proof of (3.37).

Lemma 3.4.1 *Under (H), $|I_n| \leq C[x/m(x)]/n$ whenever $M > 2n\pi$.*

Proof The integrand of the integral defining I_n is less than

$$
2 [\alpha_{2n+1}(t) - \alpha_{2n}(t)] f^\circ(t)/t \quad (\geq 0).
$$

Noting that $\alpha_{2n+1}(t) - \alpha_{2n}(t) = \int_{2n}^{(2n+1)\pi x/M} \mu(y) \sin yt \, dy$ and interchanging the order of integration, we infer that

$$I_n \le 2 \int_{2n}^{(2n+1)\pi x/M} \mu(y) \, dy \int_{2n\pi/y}^{[(2n+1)\pi/y]\wedge\pi} \frac{f^\circ(t)}{t} \sin yt \, dt.$$

Let (H) be satisfied. Then, by Lemma 3.2.6 again, in the range of the inner integral, where $y/2n\pi \ge 1/t \ge y/(2n+1)\pi$, we have

$$f^\circ(t)/t \le C_2(y/n)^3/m^2(y/n) \le C_2 n^{-1} y^3/m^2(y),$$

whereas the integral of $\sin yt$ over $0 < t < \pi/y$ equals $2/y$. Thus

$$I_n \le \frac{4C_2}{n} \int_{2n}^{(2n+1)\pi x/M} \mu(y) \frac{y^2}{m^2(y)} \, dy.$$

By $\mu(y)y^2 \le 2c(y)$

$$\int_0^z \frac{\mu(y)y^2}{m^2(y)} \, dy \le 2 \int_0^z \frac{c(y)}{m^2(y)} \, dy = \frac{2z}{m(z)}.$$

Hence we obtain the bound of the lemma because of the monotonicity of $x/m(x)$. $\square$

Lemma 3.4.2 *Under (H) it holds that if $\frac{1}{2}R < x \le R$ and $1 < M < x$,*

$$\left| \int_{M/x}^\pi \frac{\alpha_{2n}(t)f^\circ(t)}{t} \cos xt \, dt \right| \le \frac{Cn^2}{M} \cdot \frac{x}{m(x)}.$$

Proof Put $g(t) = f^\circ(t)/t$. Since $\alpha_n'(t) = \int_0^{n\pi/t} y\mu(y) \cos ty \, dy$, we have $|\alpha_n'(t)| \le c(n\pi/t) \le \pi^2 n^2 c(1/t)$. From this and Lemmas 3.2.11 and 3.2.6 (see (3.27)) we infer that if $s < t < 2s$,

$$\begin{aligned}
|\alpha_{2n}(t)g(t) - \alpha_{2n}(s)g(s)| &\le \alpha_{2n}(t)|g(t) - g(s)| + |\alpha_{2n}(t) - \alpha_{2n}(s)|g(s) \\
&\le C_1 \alpha_{2n}(t)g(t)|t - s|/s + C_2 n^2 c(1/t)g(t)|t - s| \\
&\le C\left[n^2|t - s|/s\right] c(1/t)g(t)t.
\end{aligned}$$

Put $s_k = (M + 2\pi k)/x$ for $k = 0, 1, 2, \ldots$. Then

$$|\alpha_{2n}(t)g(t) - \alpha_{2n}(s_k)g(s_k)| \le C\left[1/\sqrt{M} + n^2/M\right] c(1/t)g(t)t$$

for $s_{k-1} \le t \le s_k$ ($k \ge 1$) and if $N = \lfloor x/2 - M/2\pi \rfloor$, then on noting $\int_{s_{k-1}}^{s_k} \cos xt \, dt = 0$,

$$\left| \int_{M/x}^{s_N} \frac{\alpha_{2n}(t) f^\circ(t)}{t} \cos xt \, dt \right| = \left| \sum_{k=1}^N \int_{s_{k-1}}^{s_k} [\alpha_{2n}(t) g(t) - \alpha_{2n}(s_k) g(s_k)] \cos xt \, dt \right|$$

$$\le C \left[2\pi n^2 / M \right] \sum_{k=1}^N \int_{s_{k-1}}^{s_k} c(1/t) g(t) t \, dt$$

$$\le C' \left[n^2 / M \right] [x/m(x)],$$

where condition (H') is used for the last inequality. Since $0 \le \pi - s_N = O(1/x)$, this gives the bound of the lemma. □

Proof (of Proposition 3.1.5) By Lemmas 3.4.1 and 3.4.2

$$I_n + II_n \le C' M^{-1/3} [x/m(x)] \quad \text{for } \lfloor n = M^{1/3} \rfloor.$$

Hence taking $M = \delta^{-3/4}$ and recalling (3.39), we have $|v_M(x) - v_M(R)| \le C'' \delta^{1/4} [x/m(x)]$, which together with (3.37) yields the required bound (3.36). □

3.5 Proof of Theorem 3.1.6 and Proposition 3.1.7

If (H) holds, then by (3.19) of Lemma 3.2.6 $\Re\left\{ (1 - e^{ixt})/[1 - \psi(t)] \right\}$ is integrable over $|t| < \pi$, which ensures

$$a(x) = \frac{1}{2\pi} \int_{-\pi}^{\pi} \Re \frac{1 - e^{ixt}}{1 - \psi(t)} \, dt \tag{3.41}$$

since $\int_{-\pi}^{\pi} \Re \left[(1 - e^{ixy})/(1 - s\psi(t)) \right] dt = 2\pi \sum_{n=0}^\infty s^n [p^n(0) - p^n(-x)] \to a(x)$ $(s \uparrow 1)$ by virtue of Abel's lemma (alternatively see the proof of ([71, P28.4(a)]), and it follows that

$$a(x) = \frac{1}{\pi} \int_0^\pi \frac{\alpha(t)(1 - \cos xt) - \gamma(t) \sin xt}{[\alpha^2(t) + \gamma^2(t)]t} \, dt. \tag{3.42}$$

Recalling $\gamma(t) = \beta_+(t) - \beta_-(t)$, we put

$$b_\pm(x) = \frac{1}{\pi} \int_0^\pi \frac{\beta_\pm(t) \sin xt}{[\alpha^2(t) + \gamma^2(t)]t} \, dt \tag{3.43}$$

so that

$$a(x) = \bar{a}(x) + b_-(x) - b_+(x). \tag{3.44}$$

Proof (of Proposition 3.1.7) Suppose (H) holds. By (3.44) and Theorem 3.1.1 $a(x) \sim a(-x)$ if and only if $b_-(x) - b_+(x) = o(x/m(x))$. Let $m_+/m \to 1/2$. Since then $\gamma(t) = o(tm(1/t))$ $(t \downarrow 0)$ according to Theorem 3.1.3(i), it suffices to show

$$\int_0^\pi f^\circ(t) m(1/t) |\sin xt| \, dt \le C \frac{x}{m(x)}. \tag{3.45}$$

By virtue of Theorem 3.1.3(ii) and Lemma 3.2.6, $m(x) \le C_1 c(x)$ and $f^\circ \le C_2 f_m$. It therefore follows that

$$\int_{1/x}^{\pi} f^\circ(t) m(1/t) |\sin xt| \, dt \le C' \int_{1/x}^{\pi} f_m(t) c(1/t) \, dt \le C'' \int_{1/\pi}^{x} \frac{c(y)}{m^2(y)} \, dy \le \frac{C'' x}{m(x)}.$$

Noting that $\int_x^\infty \eta(y) \, dy / m^2(y) = 1/m(x)$, we have the same bound for the integral over $t \in [0, 1/x]$ in view of (3.46) below, which we take for granted to conclude (3.45).   □

The following inequalities hold whenever $EX = 0$:

$$\frac{1}{2} \int_0^{1/x} f_m(t) \beta_+(t) \, dt \le \int_x^\infty \frac{\tilde{m}_+(y)}{ym^2(y)} \, dy \le \frac{\tilde{m}_+(x)}{m^2(x)} + \int_x^\infty \frac{2\eta_+(y)}{m^2(y)} \, dy. \quad (3.46)$$

The first inequality of (3.46) is immediate from (3.12(a)) by changing the variable of integration. Putting $g(x) = \int_x^\infty dy / [y^2 m(y)]$ and integrating by parts we have

$$\int_x^\infty \frac{\tilde{m}_+(y)}{ym^2(y)} \, dy = - \left[g(y) \frac{y\tilde{m}_+(y)}{m(y)} \right]_{y=x}^\infty + \int_x^\infty g(y) \left(\frac{y\tilde{m}_+(y)}{m(y)} \right)' \, dy$$

as well as

$$g(y) = \frac{1}{ym(y)} - \int_y^\infty \frac{\eta(u)}{um^2(u)} \, du.$$

On observing

$$\left(\frac{y\tilde{m}_+(y)}{m(y)} \right)' = \frac{2y\eta_+}{m} - \frac{y\tilde{m}_+\eta}{m^2} \le \frac{2y\eta_+(y)}{m(y)}$$

substitution leads to the second inequality of (3.46).

Lemma 3.5.1 If $m_+/m \to 0$, then $\int_0^{1/x} f^\circ(t) \beta_+(t) \, dt = o(1/m(x))$.

Proof Let $m_+/m \to 0$. Then (H) holds so that we may replace f° by f_m owing to Lemma 3.2.6. The first term on the right of (3.46) is $o(1/m(x))$ since $\tilde{m}_+ \le m_+$. On the other hand, integrating by parts yields

$$\int_x^\infty \frac{\eta_+(y)}{m^2(y)} \, dy = -\frac{m_+(x)}{m^2(x)} + \int_x^\infty \frac{m_+(y)\eta(y)}{m^3(y)} \, dy = o(1/m(x)).$$

Thus the lemma is verified.   □

In preparation for the proof of Theorem 3.1.6, choose a positive integer N such that $E[|X|; |X| > N] \le P[|X| \le N]$ and define a function $p_*(x)$ on $\mathbb{Z}$ by $p_*(1) = E[|X|; |X| > N]$, $p_*(0) = P[|X| \le N] - p_*(1)$ and

$$p_*(k) = \begin{cases} p(k) + p(-k) & \text{if } k < -N, \\ 0 & \text{if } -N \le k \le -1 \text{ or } k \ge 2, \end{cases}$$

where $p(k) = P[X = k]$. Then p_* is a probability distribution on $\mathbb{Z}$ with zero mean. Denote the corresponding functions by $a_*, b_{*\pm}, \alpha_*, \alpha_{*\pm}$, etc. Since $p_*(z) = 0$ for $z \geq 2$ and $\sigma_*^2 = \infty$, we have for $x > 0$, $a_*(-x) = 0$, so that

$$a_*(x) = a_*(x)/2 + b_{*-}(x) - b_{*+}(x),$$

hence $\bar{a}_*(x) = 2^{-1} a_*(x) = b_{*-}(x) - b_{*+}(x)$.

We shall show that if $m_+/m \to 0$, then

$$\bar{a}(x) \sim \bar{a}_*(x), \quad |b_+(x)| + |b_{*+}(x)| = o(\bar{a}(x)) \tag{3.47}$$

and

$$b_{*-}(x) = b_-(x) + o(\bar{a}(x)). \tag{3.48}$$

These together with (3.44) yield

$$b_-(x) = \bar{a}_*(x)\{1 + o(1)\} = \bar{a}(x)\{1 + o(1)\}, \tag{3.49}$$

and hence $a(x) = \bar{a}(x)\{1 + o(1)\} + b_-(x) \sim 2\bar{a}(x)$, which shows $a(-x)/a(x) \to 0$.

The rest of this section is devoted to the proof of (3.47) and (3.48). It is easy to see that

$$\alpha_*(t) = \alpha(t) + O(t),$$
$$\beta_{*-}(t) = \beta_-(t) + \beta_+(t) + O(t^2), \tag{3.50}$$
$$\beta_{*+}(t) = p_*(1) \int_0^1 (1 - \cos tx)\, dx = O(t^2)$$

(as $t \downarrow 0$). Let $\Delta(t), t > 0$, denote the difference

$$\Delta(t) := f^\circ(t) - f_*^\circ(t) = \frac{1}{\alpha^2(t) + \gamma^2(t)} - \frac{1}{\alpha_*^2(t) + \gamma_*^2(t)}$$
$$= \{(\alpha_*^2 - \alpha^2)(t) + (\gamma_*^2 - \gamma^2)(t)\} f^\circ(t) f_*^\circ(t).$$

Observe that

$$(\gamma_* + \gamma)(t) = -2\beta_-(t) + o(t^2), \quad (\gamma_* - \gamma)(t) = -2\beta_+(t) + o(t^2);$$

$$(\alpha_*^2 - \alpha^2)(t) = 2\alpha(t) \times O(t) \quad \text{and} \quad (\gamma_*^2 - \gamma^2)(t) = 4\beta_-(t)\beta_+(t) + o(t^2). \tag{3.51}$$

Now we suppose $m_+/m \to 0$. Then it follows that $f_m(t) \asymp f^\circ(t) \asymp f_*^\circ(t)$ and $\beta_-(t)\beta_+(t)/[\alpha^2(t) + \beta^2(t)] \to 0$, and hence that $\Delta(t) = o(f_m(t))$, which implies that $\bar{a}(x) \sim \bar{a}_*(x)$, the first relation of (3.47).

The proofs of the second relation in (3.47) and of (3.48) are somewhat involved since we need to take advantage of the oscillating nature of the integrals defining $\beta_\pm(t)$. First, we dispose of the non-oscillatory parts of these integrals. By (3.12b) applied to $\alpha_\pm + \beta_\pm$ in place of $\alpha + \beta$ it follows that if $m_+/m \to 0$, then $\lim_{t \downarrow 0} \frac{\alpha_+(t) + \beta_+(t)}{\alpha_-(t) + \beta_-(t)} = 0$ (the converse is also true), which entails

$$\alpha(t) - \gamma(t) \sim \alpha(t) + \beta(t) \asymp m(1/t)t. \tag{3.52}$$

Lemma 3.5.2 *If $m_+/m \to 0$, then*

$$\int_1^x \frac{c_+(y)}{m^2(y)} \, dx = o\left(\frac{x}{m(x)}\right).$$

Proof The assertion of the lemma follows from the following identity for primitive functions

$$\int \frac{c_+(x)}{m^2(x)} \, dx = 2 \int \frac{c(x)}{m^2(x)} \cdot \frac{m_+(x)}{m(x)} \, dx - \frac{xm_+(x)}{m^2(x)}. \tag{3.53}$$

This identity may be verified by differentiation or by integration by parts, the latter giving

$$\int \frac{c_+(x)}{m^2(x)} \, dx = \frac{x}{m_+(x)} \cdot \frac{m_+^2(x)}{m^2(x)} - 2 \int \frac{x}{m_+(x)} \cdot \frac{(\eta_+ c - c_+\eta)m_+(x)}{m^3} \, dx,$$

from which we deduce (3.53) by an easy algebraic manipulation.   □

From Lemma 3.2.11 it follows that for $x \geq 4$,

$$|\alpha(t) - \alpha(s)| \vee |\beta(t) - \beta(s)| \leq 9c(1/t)t \quad \text{if} \quad t \geq \pi/x \text{ and } s = t + \pi/x. \tag{3.54}$$

We shall apply Lemma 3.2.11 only in this form in this section.

Proof (of (3.47)) We prove $b_+(x) = o(\bar{a}(x))$ only, b_{*+} being dealt with in the same way. In view of Theorem 3.1.1 and Lemma 3.5.1 it suffices to show that

$$\int_{\pi/x}^\pi \frac{f^\circ(t)\beta_+(t)}{t} \sin xt \, dt = o\left(\frac{x}{m(x)}\right). \tag{3.55}$$

We make the decomposition

$$2\int_{\pi/x}^\pi \frac{(f^\circ\beta_+)(t)}{t} \sin xt \, dt$$

$$= \int_{\pi/x}^\pi \frac{(f^\circ\beta_+)(t)}{t} \sin xt \, dt - \int_0^{\pi-\frac{\pi}{x}} \frac{(f^\circ\beta_+)(t+\pi/x)}{t+\pi/x} \sin xt \, dt \tag{3.56}$$

$$= I(x) + II(x) + III(x) + r(x),$$

where

$$I(x) = \int_{\pi/x}^\pi \frac{f^\circ(t) - f^\circ(t+\pi/x)}{t} \beta_+(t) \sin xt \, dt,$$

$$II(x) = \int_{\pi/x}^\pi f^\circ(t+\pi/x) \frac{\beta_+(t) - \beta_+(t+\pi/x)}{t} \sin xt \, dt,$$

$$III(x) = \int_{\pi/x}^\pi f^\circ(t+\pi/x)\beta_+(t+\pi/x)\left(\frac{1}{t} - \frac{1}{t+\pi/x}\right) \sin xt \, dt,$$

and

$$r(x) = -\int_0^{\pi/x} \frac{(f^\circ \beta_+)(t + \pi/x)}{t + \pi/x} \sin xt \, dt + \int_{\pi-\pi/x}^{\pi} \frac{(f^\circ \beta_+)(t + \pi/x)}{t + \pi/x} \sin xt \, dt.$$

From (3.54) (applied not only with μ but with $\mu_\pm$ in place of μ) we obtain

$$|\beta_+(t + \pi/x) - \beta_+(t)| \leq \kappa_\alpha^\circ c_+(1/t)t$$

and

$$|f^\circ(t + \pi/x) - f^\circ(t)| \leq C_1 c(1/t)[f_m(t)]^{3/2} t$$

for $t > \pi/x$. From the last inequality together with $f^{3/2}(t)t = 1/t^2 m^3(t)$, $\beta_+(t) \leq C_3 \tilde{m}_+(1/t)t$ and $\tilde{m}_+(x) \leq m_+(x) = o(m(x))$ we infer that

$$|I(x)| \leq C \int_{1/2}^{x/2} \frac{\tilde{m}_+(y)}{m(y)} \cdot \frac{c(y)}{m^2(y)} \, dy = o\left(\frac{x}{m(x)}\right).$$

Similarly

$$|II(x)| \leq C \int_{1/2}^{x/2} \frac{c_+(y)}{m^2(y)} \, dy = o\left(\frac{x}{m(x)}\right)$$

and

$$|III(x)| \leq \frac{C}{x} \int_{1/\pi}^{x/\pi} \frac{\tilde{m}_+(y)y}{m^2(y)} \, dy = o\left(\frac{x}{m(x)}\right),$$

where the equalities follow from Lemma 3.5.2 and the monotonicity of $y/m(y)$ in the bounds of $|II(x)|$ and $|III(x)|$, respectively. Finally

$$|r(x)| \leq C \int_{x/2\pi}^{x/\pi} \frac{\tilde{m}_+(y)}{m^2(y)} \, dy + O(1/x) = o\left(\frac{x}{m(x)}\right).$$

Thus we have verified (3.55) and accordingly (3.47). □

Proof (of (3.48)) Recalling $\Delta(t) = f^\circ(t) - f_*^\circ(t)$ we have

$$\pi[b_-(x) - b_{*-}(x)]$$
$$= \int_0^\pi \Delta(t) \frac{\beta_-(t)}{t} \sin xt \, dt + \int_0^\pi f_*^\circ(t) \frac{\beta_-(t) - \beta_{*-}(t)}{t} \sin xt \, dt$$
$$= J(x) + K(x) \quad \text{(say)}.$$

Suppose $m_+/m \to 0$. Since $\beta_{*-}(t) - \beta_-(t) = \beta_+(t) + O(t^2)$ and f_*° is essentially of the same regularity as f°, the proof of (3.55) and Lemma 3.5.1 applies to $K(x)$ on the RHS above to yield $K(x) = o(x/m(x))$. As for $J(x)$, we first observe that in view of (3.51),

$$|\Delta(t)| \leq C[f_m(t)]^{3/2}(t + \beta_+(t)), \tag{3.57}$$

so that the integral defining $J(x)$ restricted to $[0, \pi/x]$ is $o(x/m(x))$ in view of Lemma 3.5.1. It remains to show that

$$\int_{\pi/x}^{\pi} \Delta(t) \frac{\beta_-(t)}{t} \sin xt \, dt = o\left(\frac{x}{m(x)}\right). \tag{3.58}$$

We decompose $\Delta(t) = D_1(t) + D_2(t)$, where

$$D_1(t) = \left[(\alpha_*^2 - \alpha^2)(t) + (\gamma_*^2 - \gamma^2)(t) - 4(\beta_-\beta_+)(t)\right] f^\circ(t) f_*^\circ(t),$$
$$D_2(t) = 4(\beta_-\beta_+)(t) f^\circ(t) f_*^\circ(t).$$

By (3.51) $|D_1(t)\beta_-(t)/t| \le C_1 (\alpha(t) + t) [f_m(t)]^{3/2} \le C_2 c(1/t)/[t^2 m^3(1/t)]$ and

$$\left|\int_{\pi/x}^{\pi} D_1(t) \frac{\beta_-(t)}{t} \sin xt \, dt\right| \le C_2 \int_{1/\pi}^{x/\pi} \frac{c(y)}{m^3(y)} \, dy = o\left(\frac{x}{m(x)}\right),$$

as is easily verified. For the integral involving D_2 we proceed as in the proof of (3.55). To this end it suffices to evaluate the integrals corresponding to $I(x)$ and $II(x)$, namely,

$$J_I(x) := \int_{\pi/x}^{\pi} \frac{D_2(t) - D_2(t + \pi/x)}{t} \beta_-(t) \sin xt \, dt$$

and

$$J_{II}(x) := \int_{\pi/x}^{\pi} D_2(t + \pi/x) \frac{\beta_-(t) - \beta_-(t + \pi/x)}{t} \sin xt \, dt,$$

the other integrals being easily dealt with as before. By (3.54) the integrand for $J_{II}(x)$ is dominated in absolute value by a constant multiple of $[f_m(t)]^{3/2}\beta_+(t)c(1/t)$, from which it follows immediately that $J_{II}(x) = o(x/m(x))$. For the evaluation of $J_I(x)$, observe that

$$|(\beta_-\beta_+)(t) - (\beta_-\beta_+)(t + \pi/x)| \le C_1 \{\beta_+(t)c(1/t)t + \beta_-(t)c_+(1/t)t\}$$

and

$$|(f^\circ f_m)(t) - (f^\circ f_m)(t + \pi/x)| \beta_-(t) \le C_1 c(1/t)t[f_m(t)]^2$$

so that

$$\frac{|D_2(t) - D_2(t + \pi/x)|\beta_-(t)}{t} \le C \{\beta_+(t)c(1/t) + \beta_-(t)c_+(1/t)\} [f_m(t)]^{3/2}$$

$$\le C' \frac{\tilde{m}_+(1/t)c(1/t)}{t^2 m^3 1/t)} + C' \frac{c_+(1/t)}{t^2 m^2(1/t)}.$$

The integral of the first term of the last expression is immediately evaluated and that of the second by Lemma 3.5.2, showing

$$\left|\int_{\pi/x}^{\pi} D_2(t) \frac{\beta_-(t)}{t} \sin xt \, dt\right| \le C_2 \int_{1/\pi}^{x/\pi} \left[\frac{\tilde{m}_+(y)c(y)}{m^3(y)} + \frac{c_+(y)}{m^2(y)}\right] dy = o\left(\frac{x}{m(x)}\right).$$

The proof of (3.48) is complete. $\square$

3.6 An Example Exhibiting Irregular Behaviour of $a(x)$

We give a recurrent symmetric r.w. such that $E[|X|^\alpha/\log(|X| + 2)] < \infty$, and $P[|X| \geq x] = O(|x|^{-\alpha})$ for some $1 < \alpha < 2$ and

$$0 < {}^\forall r < 1, \quad \bar{a}(rx_n)/\bar{a}(x_n) \longrightarrow \infty \quad \text{as } n \to \infty \qquad (3.59)$$

for some sequence $x_n \uparrow \infty$, $x_n \in 2\mathbb{Z}$, which provides an example of an irregularly behaving $\bar{a}(x)$, as forewarned in the paragraph immediately after Theorem 3.1.2. In fact for the law F defined below it holds that for any $0 < \delta < 2 - \alpha$ there exists positive constants c_* such that for all sufficiently large n,

$$\bar{a}(x)/\bar{a}(x_n) \geq c_* n \quad \text{for all integers } x \text{ satisfying } 2^{-\delta n} < x/x_n \leq 1 - 2^{-\delta n} \quad (3.60)$$

(showing $\bar{a}$ diverges to ∞, fluctuating with valleys at the points x_n – having relatively steep (gentle) slopes on their left (right) – and very wide plateaus in between).

Put

$$x_n = 2^{n^2}, \quad \lambda_n = x_n^{-\alpha} = 2^{-\alpha n^2} \quad (n = 0, 1, 2, \ldots),$$

$$p(x) = \begin{cases} A\lambda_n & \text{if } x = \pm x_n \ (n = 0, 1, 2, \ldots), \\ 0 & \text{otherwise,} \end{cases}$$

where A is the constant chosen so as to make $p(\cdot)$ a probability.

Denote by $\eta_{n,k,t}^{(1)}$ the value of $1 - 2(1 - \cos u)/u^2$ at $u = x_{n-k}t$ so that uniformly for $|t| < 1/x_{n-1}$ and $k = 1, 2, \ldots, n$,

$$\frac{\lambda_{n-k}[1 - \cos(x_{n-k}t)]}{1 - \eta_{n,k,t}^{(1)}} = \frac{\lambda_{n-k}x_{n-k}^2}{2x_n^2}(x_n t)^2 = \frac{1}{2}2^{-(2-\alpha)(2n-k)k}\lambda_n(x_n t)^2$$

and

$$\eta_{n,k,t}^{(1)} = o(1) \quad \text{for} \quad k \neq 1 \quad \text{and} \quad 0 \leq \eta_{n,1,t}^{(1)} < (x_{n-1}t)^2/12.$$

Then for $|t| < 1/x_{n-1}$,

$$\sum_{x=1}^{x_{n-1}} p(x)(1 - \cos xt) = \sum_{k=1}^{n} p(x_{n-k})(1 - \cos x_{n-k}t)$$

$$= A\varepsilon_n \lambda_n(x_n t)^2\{1 - \eta_{n,1,t}^{(1)} + o(1)\}$$

with

$$\varepsilon_n = 2^{1-\alpha}2^{-2(2-\alpha)n},$$

since $p(x_{n-1})(1 - \cos x_{n-1}t)$, the last term of the series, is dominant over the rest. On the other hand

$$\mu_+(x) = \sum_{y>x} p(y) = \begin{cases} A\lambda_n + O(\lambda_{n+1}) & (x_{n-1} \leq x < x_n), \\ o(\lambda_n \varepsilon_n) & (x \geq x_n). \end{cases} \qquad (3.61)$$

On writing $1 - \psi(t) = 2 \sum_{x=1}^{x_n} p(x)(1 - \cos tx) + O(\mu(x_n))$, observe that uniformly for $|t| < 1/x_{n-1}$,

$$1 - \psi(t) = 2A\lambda_n \left[1 - \cos x_n t + \varepsilon_n (x_n t)^2 (1 - \eta_{n,1,t}^{(1)}) \right] + o(\lambda_n \varepsilon_n). \qquad (3.62)$$

Also, $2A\lambda_n (1 - \cos x_n t) = A\lambda_n (x_n t)^2 (1 - \eta^{(2)}(n, t))$, with $0 \le \eta^{(2)}(n, t) \le 1/12$ for $|t| < 1/x_n$. Recall that

$$\bar{a}(x) = \frac{1}{\pi} \int_0^\pi \frac{1 - \cos xt}{1 - \psi(t)} \, dt.$$

We break the integral into three parts

$$(\pi A)\bar{a}(x) = \left(\int_0^{1/x_n} + \int_{1/x_n}^{1/x_{n-1}} + \int_{1/x_{n-1}}^\pi \right) \frac{1 - \cos xt}{[1 - \psi(t)]/A} \, dt$$
$$= I(x) + II(x) + III(x) \quad \text{(say)}.$$

UPPER BOUND OF $\bar{a}(x_n)$. By the trivial inequality $1 - \psi(t) \ge 2p(x_n)(1 - \cos x_n t)$ one sees

$$I(x) \le 1/(2\lambda_n x_n) = x_n^{\alpha-1}/2.$$

Put $r = r(n, x) = x/x_n$. Then by (3.62) it follows that for sufficiently large n,

$$II(x) \le \int_{1/x_n}^{1/x_{n-1}} \frac{1 - \cos xt}{2\lambda_n (1 - \cos x_n t) + \varepsilon_n \lambda_n (x_n t)^2} \, dt$$
$$= \frac{1}{\lambda_n x_n} \int_1^{x_n/x_{n-1}} \frac{1 - \cos ru}{2(1 - \cos u) + \varepsilon_n u^2} \, du. \qquad (3.63)$$

If $x = x_n$, so that $r = 1$, this leads to

$$II(x_n) \le x_n^{\alpha-1} \sum_{k=0}^{2^{2n}} \int_{-\pi}^\pi \frac{1 - \cos u}{2(1 - \cos u) + \varepsilon_n (u + 2\pi k)^2} \, du.$$

Noting that $u + 2\pi k > 2\pi k - \pi$ and $2(u/\pi)^2 < 1 - \cos u < \frac{1}{2}u^2$ for $-\pi < u < \pi$, one observes

$$\int_0^\infty dk \int_0^\pi \frac{u^2}{u^2 + \varepsilon_n k^2} \, du = \frac{\pi/2}{\sqrt{\varepsilon_n}} \int_0^\pi u \, du = \frac{\pi^3/4}{\sqrt{\varepsilon_n}}$$

to conclude

$$II(x_n) \le \frac{C_1 x_n^{\alpha-1}}{\sqrt{\varepsilon_n}}.$$

For the evaluation of *III* we apply (3.62) with $n - 1$ in place of n to deduce that

$$\int_{1/x_{n-1}}^{1/x_{n-2}} \frac{1 - \cos x_n t}{[1 - \psi(t)]/A} \, dt \leq \int_{1/x_{n-1}}^{1/x_{n-2}} \frac{2}{2\lambda_{n-1}(1 - \cos x_{n-1}t) + \varepsilon_{n-1}\lambda_{n-1}(x_{n-1}t)^2} \, dt$$

$$= \frac{2}{\lambda_{n-1}x_{n-1}} \int_1^{x_{n-1}/x_{n-2}} \frac{1}{2(1 - \cos u) + \varepsilon_{n-1}u^2} \, du.$$

The last integral is less than

$$\int_1^\pi \frac{1}{u^2/3 + \varepsilon_{n-1}u^2} \, du + \sum_{k=1}^{2^{2n}} \int_{-\pi}^\pi \frac{1}{u^2/3 + \varepsilon_{n-1}(-\pi + 2\pi k)^2} \, du$$

$$\leq C_1 \sum_{k=1}^{2^{2n}} \frac{1}{\sqrt{\varepsilon_n} \, k} \leq C_2 \frac{n}{\sqrt{\varepsilon_n}}.$$

Since $\lambda_n x_n / \lambda_{n-1} x_{n-1} = 2^{-(\alpha-1)(2n-1)}$ and the remaining part of *III* is smaller,

$$III(x_n) \leq C_3 n 2^{-2(\alpha-1)n} x_n^{\alpha-1} / \sqrt{\varepsilon_n}.$$

Consequently

$$\bar{a}(x_n) \leq C x_n^{\alpha-1} / \sqrt{\varepsilon_n}. \tag{3.64}$$

LOWER BOUND OF $\bar{a}(x)$. By (3.62) it follows that for n sufficiently large,

$$[1 - \psi(t)]/A \leq 2\lambda_n(1 - \cos x_n t) + 3\varepsilon_n\lambda_n(x_n t)^2 \quad \text{for} \quad 1/x_n \leq |t| < 1/x_{n-1}.$$

Let $\frac{1}{2} < x/x_n \leq 1$ and put $b = b(n, x) = y/x_n = 1 - x/x_n \ (< 1/2)$. Then

$$\frac{II(x)}{x_n^{\alpha-1}} \geq \int_1^{x_n/x_{n-1}} \frac{1 - \cos(1 - b)u}{2(1 - \cos u) + 3\varepsilon_n u^2} \, du. \tag{3.65}$$

The RHS is bounded from below by

$$\sum_{k=1}^{2^{2n-1}/2\pi} \int_{-\pi}^\pi \frac{1 - \cos\left[(1 - b)u - 2\pi bk\right]}{u^2 + 3\pi^2\varepsilon_n(2k + 1)^2} \, du.$$

If $b \leq \sqrt{\varepsilon_n} = 2^{(1-\alpha)/2}2^{-(2-\alpha)n}$ then, restricting the summation to $k < 1/\sqrt{\varepsilon_n}$, one sees that

$$\frac{II(x)}{x_n^{\alpha-1}} \geq C \sum_{k=1}^{1/\sqrt{\varepsilon_n}} \int_0^\pi \frac{u^2 + b^2k^2}{u^2 + \varepsilon_n k^2} \, du \geq \frac{C\pi/2}{\sqrt{\varepsilon_n}} \quad \text{if} \quad 1 - \sqrt{\varepsilon_n} \leq \frac{x}{x_n} \leq 1. \tag{3.66}$$

For $b \geq \sqrt{\varepsilon_n}$, we restrict the summation to the intervals $(v + \frac{1}{4})/b \leq k \leq (v + \frac{3}{4})/b$, $v = 0, 1, 2, \ldots, \lfloor 2^{(2-\alpha)n}b/2 \rfloor$ to see that

$$\frac{\Pi(x)}{x_n^{\alpha-1}} \geq C \sum_{k=1/4b}^{1/\sqrt{\varepsilon_n}} \int_{-\pi/2}^{\pi/2} \frac{du}{u^2 + \varepsilon_n k^2} \geq C \frac{\log\left[4b/\sqrt{\varepsilon_n}\right]}{\sqrt{\varepsilon_n}} \quad \text{if} \quad \frac{1}{2} \leq \frac{x}{x_n} \leq 1 - \sqrt{\varepsilon_n}.$$

$$(3.67)$$

In particular this shows that if $0 < \delta < 2 - \alpha$, then

$$\bar{a}(x) \geq Cnx_n^{\alpha-1}/\sqrt{\varepsilon_n} \quad \text{for} \quad 2^{-1} < x/x_n < 1 - 2^{-\delta n}.$$

Let $1 \leq j < (2-\alpha)n + 1$ and $2^{-j-1} < x/x_n \leq 2^{-j}$. Put

$$b = y/x_n^{(j)} = 1 - 2^j x/x_n.$$

Then we have (3.65) with $(1-b)$ replaced by $2^{-j}(1-b)$ and making the changes of variable $u = s + 2\pi(2^{j-1} + k)$ in (3.65) with $k = 0, 1, \ldots$, we infer that

$$\frac{\Pi(x)}{x_n^{\alpha-1}} \geq c_1 \sum_{k=0}^{[2^{2n} - 2^j]/4\pi} \int_{-\pi}^{\pi} \frac{1 - \cos\theta_{j,x}(s,k)}{2(1 - \cos s) + 3\varepsilon_n\left[2\pi(2^j + k)\right]^2} \, ds,$$

where $\theta_{j,x}(s,k) = 2^{-j}(1-b)s + \pi + 2\pi\left[2^{-j}(1-b)k - \frac{1}{2}b\right]$. Noting $b < \frac{1}{2}$, we see that for $\nu = 0, 1, 2, \ldots$

$$\cos\theta_{j,x}(s,k) \leq \frac{1}{\sqrt{2}} \quad \text{if} \quad |s| < \frac{2^j\pi}{4} \quad \text{and} \quad \nu - \frac{1}{4} \leq \frac{(1-b)}{2^j}k - \frac{1}{2}b \leq \nu + \frac{1}{4}$$

and then find that the sum above is larger than a constant multiple of

$$\sum_{k=1}^{2^{2n}/5\pi} \int_0^1 \frac{ds}{s^2 + \varepsilon_n(2^j + k)^2} \sim \sum_{k=2^{j-1}}^{2^{2n}/5\pi} \frac{1}{\sqrt{\varepsilon_n}\,k} \arctan\frac{1}{\sqrt{\varepsilon_n}\,k}$$

$$\geq \sum_{k=2^{j-1}}^{1/\sqrt{\varepsilon_n}} \frac{1/2}{\sqrt{\varepsilon_n}\,k} \sim \frac{(2-\alpha)n - j + 1}{[2/\log 2]\sqrt{\varepsilon_n}}.$$

This together with (3.66) and (3.67) entails that for any $0 < \delta < 1/2$

$$\frac{\bar{a}(x)\sqrt{\varepsilon_n}}{x_n^{\alpha-1}} \geq \begin{cases} c_0 & \text{for } 1 - \sqrt{\varepsilon_n} \leq x/x_n \leq 1, \\ c_\delta n & \text{for } \varepsilon_n^\delta < x/x_n < 1 - \sqrt{\varepsilon_n}, \end{cases}$$

which combined with (3.64) shows (3.60).

In the present example, $c(x)/m(x)$ oscillates between 0 and 1, where $c(x) = \int_0^x t\mu(t)\,dt$. Indeed, without difficulty one can see that for $x_{n-1} \leq x \leq x_n$,

$$c(x)/2A \sim \tfrac{1}{2}(x^2 - x_{n-1}^2)\lambda_n + \tfrac{1}{2}x_{n-1}^2\lambda_{n-1},$$

$$x\eta(x)/2A \sim x(x_n - x)\lambda_n + xx_{n+1}\lambda_{n+1},$$

and

$$m(x)/2A \sim x(x_n - \tfrac{1}{2}x)\lambda_n + \tfrac{1}{2}x_{n-1}^2\lambda_{n-1} \asymp \begin{cases} xx_n^{1-\alpha} & x > \varepsilon_n x_n, \\ (x_{n-1})^{2-\alpha} & x < \varepsilon_n x_n, \end{cases} \tag{3.68}$$

and then that

$$\frac{x\eta(x)}{c(x)} \longrightarrow \begin{cases} 0 & \text{for } x \text{ with } 1 - o(1) < x/x_{n-1} < 1 + o(2^{2(\alpha-1)n}), \\ \infty & \text{for } x \text{ with } 2^{2(\alpha-1)n}x_{n-1} \ll x \ll x_n. \end{cases}$$

From (3.62) one also infers that $\alpha(t)/[tc(1/t)]$ oscillates between 8 and $4\varepsilon_n t^2 x_n^2\{1 - o(1)\}$ about $\lfloor M/2\pi \rfloor$ times as t ranges over the interval $[1/x_n, M/x_n]$ and that $\alpha(2\pi/x_n)/\alpha(\pi/x_n) = O(\varepsilon_n)$. Thus $\alpha(t)$ behaves quite irregularly.

This example also shows that the converse of Theorem 3.1.2(ii) is untrue. From what we observed above, it follows that

$$\liminf \frac{\alpha(t)}{\eta(1/t)} = 0 \quad \text{and} \quad \limsup \frac{\alpha(t)}{\eta(1/t)} \geq \limsup \frac{\alpha(t)}{tm(1/t)} > 0.$$

On the other hand

$$\lim \frac{\bar{a}(x)}{x/m(x)} = \infty \quad \text{if } \alpha < 4/3, \tag{3.69}$$

as is verified below. Thus the condition $\lim \alpha(t)/tm(1/t) = 0$ is not necessary for the ratio $\bar{a}(x)/[x/m(x)]$ to diverge to infinity. [For $\alpha > 4/3$ the ratio $\bar{a}(x)/[x/m(x)]$ is bounded along the sequence $x_n^* = \varepsilon_n x_n$ – as one may easily show.] To verify (3.69) we prove

$$\bar{a}(x) \geq c_1[x_{n-1}]^{\alpha-1}/\sqrt{\varepsilon_n} \quad \text{for } x \geq x_{n-1} \tag{3.70}$$

with a constant $c_1 > 0$. Observe that $\alpha < 4/3$ entails – in fact, is equivalent to – each of the relations $x_n^{\alpha-1} = o(x_{n-1}^{\alpha-1}/\sqrt{\varepsilon_n})$ and $\varepsilon_n x_n = o(x_{n-1}/\sqrt{\varepsilon_n})$. Then, comparing the above lower bound of $\bar{a}(x)$ to the upper bound of $x/m(x)$ obtained from (3.68) ensures the truth of (3.69). For the proof of (3.70) we evaluate $III(x)$. Changing the variable $t = u/x$, we have

$$III(x) \geq \frac{1}{3\lambda_{n-1}x} \int_{x/x_{n-1}}^{x/x_{n-2}} \frac{(1 - \cos u)\, du}{1 - \cos(x_{n-1}u/x) + \varepsilon_{n-1}(x_{n-1}u/x)^2}.$$

The integral on the RHS is easily evaluated to be larger than

$$\frac{1}{4} \int_{x/x_{n-1}}^{x/x_{n-2}} \frac{du}{1 + [\sqrt{\varepsilon_n}(x_{n-1}/x)u]^2} = \frac{x/x_{n-1}}{4\sqrt{\varepsilon_n}} \int_{\sqrt{\varepsilon_n}}^{2^{2n-3}\sqrt{\varepsilon_n}} \frac{ds}{1 + s^2} \sim \frac{\pi x}{8x_{n-1}\sqrt{\varepsilon_n}},$$

showing (3.70) since $\bar{a}(x) \geq III(x)/(\pi A)$.

Chapter 4
Some Explicit Asymptotic Forms of $a(x)$

Here we compute asymptotic forms of $a(x)$, explicit in terms of F, in two cases, first when F is relatively stable and second when F belongs to the domain of attraction of a stable law, in Sections 4.1 and 4.2, respectively. In Section 4.3, we give some estimates of the increments $a(x + 1) - a(x)$. Throughout this chapter S will be *recurrent* and $\sigma^2 = \infty$ except in (b) of Proposition 4.3.5. We continue to use the notation introduced in Chapter 3.

4.1 Relatively Stable Distributions

When the r.w. S is r.s. – we shall also call F r.s. – one can obtain some exact asymptotic forms of $a(x)$. Recall that S is p.r.s. if and only if

Eq(2.20) $A(x)/x\mu(x) \longrightarrow \infty$ as $x \to \infty$,

where $A(x) = A_+(x) - A_-(x)$, $A_\pm(x) = \int_0^x \mu_\pm(y)\,dy$; and in this case B_n can be chosen so that $B_n \sim nA(B_n)$; clearly $A'(x) = o(A(x)/x)$ $(x \to \infty)$, entailing A and A_+ are s.v. at infinity; we have chosen an integer $x_0 > 0$ so that $A(x) > 0$ for $x \geq x_0$ (see Section 2.4). Let $M(x)$ be a function of $x \geq x_0$ defined by

$$M_\pm(x) = \int_{x_0}^x \frac{\mu_\pm(y)}{A^2(y)}\,dy \quad \text{and} \quad M(x) = M_+(x) + M_-(x). \qquad (4.1)$$

Theorem 4.1.1 *Suppose that F is recurrent and p.r.s. Then as $x \to \infty$*

(i) $a(x) - a(-x) \sim 1/A(x)$;

(ii) $2\bar{a}(x) \sim M(x)$;

(iii) $a(-x)/a(x) \to 1$ *if and only if* $M_+(x)/M(x) \to 1/2$; *and*

(iv) *if* $EX = 0$ *and* $m_+/m \to 1/2$, *then* $\dfrac{a(x) - a(-x)}{x/m(x)} \to \infty$.

K. Uchiyama, *Potential Functions of Random Walks in $\mathbb{Z}$ with Infinite Variance*, Lecture Notes in Mathematics 2338, https://doi.org/10.1007/978-3-031-41020-8_4

Suppose that $EX = 0$ and (2.20) holds. Then $A(x) = \eta_-(x) - \eta_+(x)$ and $\alpha(t) \ll |\gamma(t)| \sim |A(1/|t|)|$, $(t \to 0)$ (see (P2) below), hence condition (H) holds if and only if $\limsup \eta_+(x)/\eta(x) < 1/2$, and if this is the case, by Theorems 3.1.1 and 3.1.2, $\bar{a}(x) \asymp x/m(x)$ [which also follows from (ii) above].

Let (2.20) hold and put $K(x) = \mu_+(x) - \mu_-(x)$. Then $\log[A(x)/A(x_0)] = \int_{x_0}^x \varepsilon(t)\,dt/t$ with $\varepsilon(t) = tK(t)/A(t)$, which approaches zero as $t \to \infty$, hence $A(x)$ is a normalised s.v. function. By $(1/A)'(x) = -K(x)/A^2(x)$, we have

$$\frac{1}{A(x)} - \frac{1}{A(x_0)} = \int_{x_0}^x \frac{-K(y)}{A^2(y)}\,dy = M_-(x) - M_+(x), \tag{4.2}$$

in particular $M_-(x) \geq M_+(x) \vee [1/A(x)] - 1/A(x_0)$. Because of the recurrence of the r.w., we have $M_-(x) \to \infty$ (see (P1) below), and from (i) and (ii) of Theorem 4.1.1 one infers that, as $x \to \infty$,

$$a(x) \sim M_-(x) \quad \text{and} \quad a(-x) = M_+(x) + o(M_-(x)); \tag{4.3}$$

it also follows that $x\mu(x)/A^2(x) = o\,(1/A(x)) = o(M_-(x))$, so that both M_- and M are s.v. Thus Theorem 4.1.1 yields the following

Corollary 4.1.2 *Suppose that (2.20) holds. Then both $a(x)$ and $\bar{a}(x)$ are s.v. at infinity, and (4.3) holds.*

The proof of Theorem 4.1.1 rests on the following results obtained in [83]:

If (2.20) holds, then

(P1) *S is recurrent if and only if $\int_{x_0}^\infty \mu(x)\,dx/A^2(x) = \infty$,*
(P2) $\alpha(\theta) = o(\gamma(\theta))$ *and* $\gamma(\theta) \sim -A(1/\theta)$ $(\theta \downarrow 0)$,[1]
(P3) *there exists a constant C such that for any $\varepsilon > 0$ and all x large enough,*

$$\left| \int_{1/\varepsilon x}^\pi \frac{f^\circ(t)\alpha(t)\cos xt}{t}\,dt \right| + \left| \int_{1/\varepsilon x}^\pi \frac{f^\circ(t)\gamma(t)\sin xt}{t}\,dt \right| \leq \frac{C\varepsilon}{A(x)}, \tag{4.4}$$

where $f^\circ(t) = 1/[\alpha^2(t) + \gamma^2(t)]$ as given in (3.13).

(See Lemmas 4, 6, and 8 of [83] for (P1), (P2), and (P3), respectively.) The proofs of (P1) and (P2) are rather standard. We give a proof of (P3) at the end of this section.

Proof (of Theorem 4.1.1) PROOF OF (i). By (3.42), which is valid since $1/|\gamma(t)|$ is summable over $[0, 1]$ because of (P2),

$$a(x) - a(-x) = \frac{2}{\pi} \int_0^\pi \frac{-\gamma(t)f^\circ(t)\sin xt}{t}\,dt.$$

By (P2)

$$f^\circ(t) \sim 1/A^2(1/t) \quad \text{and} \quad -\gamma(t)f^\circ(t) \sim 1/A(1/t) \tag{4.5}$$

---

[1] If $E|X| = \infty$, the integrals defining $\alpha(t)$ and $\gamma(t)$ are regarded as improper integrals, which are evidently convergent.

$(t \downarrow 0)$. Let $0 < \varepsilon < 1$. For the above integral restricted to $t \geq 1/\varepsilon x$ we have the bound $C\varepsilon/A(x)$ because of (P3), while that restricted to $t < \varepsilon/x$ is dominated in absolute value by a constant multiple of $x \int_0^{\varepsilon/x} dt/A(1/t) \sim \varepsilon/A(x)$. These reduce our task to showing that, for each ε,

$$\lim_{x \to \infty} A(x) \int_{\varepsilon/x}^{1/\varepsilon x} \frac{-\gamma(t)f^\circ(t) \sin xt}{t} \, dt = \int_\varepsilon^{1/\varepsilon} \frac{\sin t}{t} \, dt, \qquad (4.6)$$

for as $\varepsilon \downarrow 0$ the integral on the RHS converges to $\pi/2$. The change of the variable $t = w/x$ transforms the integral under the above limit to

$$\int_\varepsilon^{1/\varepsilon} g(w/x) \frac{\sin w}{w} \, dw \quad \text{where} \quad g(t) = -\gamma(t)f^\circ(t).$$

By the slow variation of A together with (4.5) it follows that $g(w/x)A(x) \to 1$ as $x \to \infty$ uniformly for $\varepsilon \leq w \leq 1/\varepsilon$. Thus we have (4.6).

PROOF OF (ii). Given a constant $0 < \varepsilon < 1$, we decompose the integral in the representation of $\bar{a}$ given in (3.10) as follows:

$$\pi \bar{a}(x) = I(x) - II(x) + III(x),$$

where

$$I(x) = \int_{1/\varepsilon x}^\pi \frac{f^\circ(t)\alpha(t)}{t} \, dt, \qquad II(x) = \int_{1/\varepsilon x}^\pi \frac{f^\circ(t)\alpha(t) \cos xt}{t} \, dt,$$

$$\text{and} \qquad III(x) = \int_0^{1/\varepsilon x} \frac{f^\circ(t)\alpha(t)(1 - \cos xt)}{t} \, dt.$$

By (2.20) it follows that

$$c(x) = \int_0^x y\mu(y) \, dy = o\,(xA(x))\,. \qquad (4.7)$$

Using $f^\circ(t) \sim 1/A^2(1/t)$ (see (4.5)) as well as $\alpha(t) \leq C_1 tc(1/t)$, we then deduce that

$$III(x) \leq C_1 x^2 \int_0^{1/\varepsilon x} \frac{c(1/t)t^2}{A^2(1/t)} \, dt = x^2 \int_{\varepsilon x}^\infty \frac{dy}{y^3 A(y)} \, dy \times o(1) = o\left(\frac{1}{A(x)}\right).$$

By (P3) we also have

$$|II(x)| \leq C\varepsilon/A(x).$$

Hence for the proof of (ii) it suffices to verify $I(x) \sim \frac{1}{2}\pi M(x)$. Since $\alpha(t)$ is positive, this follows if we can show

$$\tilde{I}(x) := \int_{1/\varepsilon x}^b \frac{\alpha(t)}{A^2(1/t)t} \, dt \sim \frac{\pi}{2} M(x) \qquad (4.8)$$

for some and, therefore, any positive constant $b \leq (1/x_0) \wedge \pi$. We decompose the integral that defines $\alpha(t)$ by splitting its range,

$$\alpha(t) = \left(\int_0^{\varepsilon/t} + \int_{\varepsilon/t}^{1/\varepsilon t} + \int_{1/\varepsilon t}^{\infty} \right) \mu(y) \sin ty \, dy$$

$$= \alpha_1(t) + \alpha_2(t) + \alpha_3(t) \qquad \text{(say)},$$

and accordingly

$$\tilde{I}(x) = \int_{1/\varepsilon x}^{b} [\alpha_1(t) + \alpha_2(t) + \alpha_3(t)] \, \frac{dt}{A^2(1/t)t}.$$

Obviously $\alpha_1(t) = \int_0^{\varepsilon/t} \mu(y) \sin ty \, dy \leq tc(\varepsilon/t)$, and integrating by parts leads to

$$\int_{1/\varepsilon x}^{b} \alpha_1(t) \frac{dt}{A^2(1/t)t} \leq \int_{1/b}^{\varepsilon x} \frac{c(\varepsilon y)}{A^2(y) y^2} \, dy$$

$$= -\frac{c(\varepsilon^2 x)\{1 + o(1)\}}{\varepsilon x A^2(\varepsilon x)} + O(1) + \int_{1/b}^{\varepsilon x} \frac{\varepsilon^2 \mu(\varepsilon y) \, dy}{A^2(y)} \{1 + o(1)\}$$

$$\leq \varepsilon M(x)\{1 + o(1)\}. \tag{4.9}$$

Noting that $\alpha_2(t) = \int_{\varepsilon/t}^{1/\varepsilon t} \mu(y) \sin ty \, dy = t^{-1} \int_{\varepsilon}^{1/\varepsilon} \mu(w/t) \sin w \, dw$, we infer

$$\int_{1/\varepsilon x}^{b} \alpha_2(t) \frac{dt}{A^2(1/t)t} = \int_{\varepsilon}^{1/\varepsilon} \sin w \, dw \int_{1/b}^{\varepsilon x} \frac{\mu(wy) \, dy}{A^2(y)}.$$

Changing the variable $y = z/w$ transforms the last repeated integral into

$$\int_{\varepsilon}^{1/\varepsilon} \frac{\sin w \, dw}{w} \int_{w/b}^{\varepsilon wx} \frac{\mu(z) \, dz}{A^2(z/w)}.$$

Because of the slow variation of A, the inner integral is asymptotically equivalent to $M(\varepsilon w x)$ as $x \to \infty$ uniformly for $\varepsilon \leq w \leq 1/\varepsilon$. Since $M(x)$ is s.v., we can conclude that for each ε, as $x \to \infty$,

$$\int_{1/\varepsilon x}^{b} \alpha_2(t) \frac{dt}{A^2(1/t)t} \sim \left[\int_{\varepsilon}^{1/\varepsilon} \frac{\sin w}{w} \, dw \right] M(x). \tag{4.10}$$

By definition of α_3, for any $0 < \delta \leq b$,

$$\int_{1/\varepsilon x}^{b} \alpha_3(t) \frac{dt}{A^2(1/t)t} = \int_{1/\varepsilon x}^{\delta} \frac{dt}{A^2(1/t)t} \int_{1/\varepsilon t}^{\infty} \mu(y) \sin ty \, dy + C_{b,\delta}.$$

Splitting the inner integral at $y = x$ and interchanging the order of integration transform the repeated integral on the RHS into

$$\int_{1/\varepsilon\delta}^{x} \mu(y)\,dy \int_{1/\varepsilon y}^{\delta} \frac{\sin ty\,dt}{A^2(1/t)t} + \int_{x}^{\infty} \mu(y)\,dy \int_{1/\varepsilon x}^{\delta} \frac{\sin ty\,dt}{A^2(1/t)t}.$$

For t small enough, $1/A^2(1/t)t$ is decreasing and we choose $\delta > 0$ so that this is true for $t \leq \delta$. Then the inner integrals of the first and second repeated integrals above are bounded by $\pi\varepsilon/A^2(\varepsilon y)$ and $\pi\varepsilon x/[yA^2(\varepsilon x)]$, respectively. Observe that

$$\int_{x}^{\infty} \frac{\mu(y)}{y}\,dy \leq \int_{x}^{\infty} \frac{A(y)}{y^2}\,dy \times o(1) = o\left(\frac{A(x)}{x}\right).$$

Now, we can easily see that

$$\int_{1/\varepsilon x}^{b} \alpha_3(t) \frac{dt}{A^2(1/t)t} \leq \varepsilon\,[CM(x) + o\,(1/A(x))] + C_{\delta,b}.$$

In conjunction with (4.9) and (4.10) this shows (4.8), finishing the proof of (ii).

PROOF OF (iii) AND (iv). (iii) follows by combining (i), (ii) and (4.2) since $M_-(x) \to \infty$. Let $EX = 0$. Then η_- is s.v. and $A(x) = \eta_-(x) - \eta_+(x)$. If $m_+/m_- \to 1$, then $m_\pm(x) \sim x\eta_-(x)$, so that $\eta_+(x) \sim \eta_-(x)$ by the monotone density theorem. Hence $A(x) = o(m(x)/x)$, and the result follows from (i).

The proof of Proposition 4.1.1 is complete.   □

Proof (of (P3)). Here we show that for some constant C and any $\varepsilon > 0$,

$$\left| \int_{1/\varepsilon x}^{\pi} \frac{f^{\circ}(t)\alpha(t)\cos xt}{t}\,dt \right| = \left| \int_{1/\varepsilon x}^{\pi} \frac{\cos xt}{1 - \psi(t)}\,dt \right| \leq \frac{C\varepsilon}{A(|x|)}; \qquad (4.11)$$

the bound of the integral with $\gamma(t)\sin xt$ in place of $\alpha(t)\cos xt$ is derived similarly.

We make the decomposition $1 - \psi(t) = \omega_x(t) - r_x(t)$, where

$$\omega_x(t) = 1 - E\left[e^{itX} : |X| < x\right], \qquad r_x(t) = E\left[e^{itX} : |X| \geq x\right].$$

Since $|r_x(t)| \leq \mu(x) = o(A(x)/x)$ $(x \to \infty)$, $1 - \psi(t) \sim -itA(1/|t|)$ $(t \to 0)$ by (P2), and $\Re\psi(t) < 1$ for $t \in (0,\pi]$, it follows that $r_x(t) = o(\omega_x(t))$ as $x \to \infty$ uniformly for $1/x < |t| < \pi$, and there exists a $\delta > 0$ such that

$$\omega_x(t) = -itA(1/|t|)\{1 + o(1)\} \qquad \text{uniformly for } 1/x < |t| < \delta,$$

where $o(1) \to 0$ as $t \wedge x^{-1} \to 0$. Using these bounds we find that as $x \to \infty$, $\int_{1/x}^{\pi} |r_x(t)/\omega_x^2(t)|\,dt \leq C\mu(x) \int_{x_0}^{x} [A(y)]^{-2}\,dy = o(1/A(x))$ (for some $x_0 > 1$), and hence

$$\int_{1/\varepsilon|x|<|t|<\pi} \frac{\cos xt}{1 - \psi(t)}\,dt = \int_{1/\varepsilon|x|<|t|<\pi} \frac{\cos xt}{\omega_x(t)}\,dt + o\left(\frac{1}{A(x)}\right). \qquad (4.12)$$

□

Lemma 4.1.3 *Under (2.20),* $|\omega'_x(t)| = A(x) \left\{ 1 + o\left(\sqrt{|t|x} \right) \right\}$ *as* $x \to \infty$ *uniformly for* $|t| > 1/x$.

Proof Performing differentiation we have

$$\omega'_x(t) = -i \int_{-x}^{x} y e^{ity} \, dF(y) = -i \int_{-x}^{x} y \, dF(y) + i \int_{-x}^{x} y(1 - e^{ity}) \, dF(y).$$

The first integral on the RHS is asymptotically equivalent to $A(x)$ under (2.20). As for the second one, on noting $|e^{iyt} - 1| \leq 2\sqrt{|t|y}$ $(y > 0)$, observe

$$\left| \int_{-x}^{x} y(1 - e^{ity}) \, dF(y) \right| \leq 2\sqrt{|t|} \int_{-x}^{x} |y|^{3/2} \, dF(y) \leq 3\sqrt{|t|} \int_{0}^{x} \sqrt{y} \, \mu(y) \, dy,$$

of which the final expression is $\sqrt{|t|x} \times o(A(x))$ since $\sqrt{y}\mu(y) = o\left(A(y)/\sqrt{y} \right)$. $\quad\square$

To the integral on the RHS of (4.12) we apply integration by parts. Noting that $x|\omega_x(1/\varepsilon x)| \sim A(x)/\varepsilon$ we infer

$$\int_{1/\varepsilon x}^{\pi} \frac{\cos xt}{\omega_x(t)} \, dt = O\left(\frac{\varepsilon}{A(x)} \right) + \frac{1}{x} \int_{1/\varepsilon x}^{\pi} \frac{\omega'_x(t) \sin xt}{\omega_x^2(t)} \, dt. \tag{4.13}$$

The contribution to the second term of the last integral restricted to $\delta < t \leq \pi$ is negligible, while by Lemma 4.1.3 the contribution of the other part of the integral is dominated in absolute value by a constant multiple of

$$\frac{A(x)}{x} \int_{1/\varepsilon x}^{\delta} \frac{1 + o\left(\sqrt{tx} \right)}{t^2 A^2(1/t)} \, dt = \frac{A(x)}{x} \int_{1/\delta}^{\varepsilon x} \frac{1 + o\left(\sqrt{x/t} \right)}{A^2(t)} \, dt \sim \frac{\varepsilon}{A(x)}.$$

This together with (4.13) verifies (4.11) since ε may be made arbitrarily small.

4.2 Distributions in Domains of Attraction

In this section we suppose that X belongs to the domain of attraction of a stable law with exponent $1 \leq \alpha \leq 2$ and let L, p and q be given as in the condition (2.16) of Section 2.3, namely

(a) $\int_{-x}^{x} y^2 \, dF(y) \sim L(x)$ if $\alpha = 2$,

(b) $\mu_-(x) \sim q x^{-\alpha} L(x)$ and $\mu_+(x) \sim p x^{-\alpha} L(x)$ if $1 \leq \alpha < 2$,   (4.14)

as $x \to \infty$ $(0 \leq p, q \leq 1, p + q = 1$ and L is s.v. at infinity). In the sequel we suppose, without causing any loss of generality, that L is positive and chosen so that L is absolutely continuous and $L'(x) = o(L(x)/x)$. It is assumed that $\int_{1}^{\infty} L(x) x^{-1} \, dx < \infty$ if $\alpha = 1$ and $\lim L(x) = \infty$ if $\alpha = 2$ so that $E|X| < \infty$ (hence $EX = 0$) and

$EX^2 = \infty$ unless the contrary is stated explicitly (in some cases we allow $E|X| = \infty$ as in Proposition 4.2.1(iv) and Remark 4.2.2(i)).

We use the notation $c(x)$, $\tilde{c}(x)$, $\bar{a}(x)$, $\alpha(t)$ and $\beta(t)$ introduced in Sections 3.1 and 2.2. Note that Spitzer's condition (b) of (2.16) is not assumed.

4.2.1 Asymptotics of $a(x)$ I

Put $L^*(x) = \int_x^\infty y^{-1} L(y)\, dy$. Then

$$\eta(x) = \begin{cases} o(c(x)/x) & \text{if } \alpha = 2, \\ (\alpha - 1)^{-1} x\mu(x)\{1 + o(1)\} & \text{if } 1 < \alpha < 2, \\ L^*(x)\{1 + o(1)\} & \text{if } \alpha = 1, \end{cases} \tag{4.15}$$

$$c(x) = \alpha^{-1} x^{2-\alpha} L(x)\{1 + o(1)\} \qquad (1 \le \alpha \le 2) \tag{4.16}$$

and

$$m(x) \sim \begin{cases} L(x)/2 & \text{if } \alpha = 2, \\ \dfrac{x^{2-\alpha} L(x)}{(\alpha - 1)(\alpha - 2)} & \text{if } 1 < \alpha < 2, \\ x L^*(x) & \text{if } \alpha = 1. \end{cases}$$

The derivation is straightforward. If $\alpha = 2$, condition (4.14a) is equivalent to $x^2\mu(x) = o(L(x))$ as well as to $c(x) \sim L(x)/2$ (Section A.1.1), which together show

$$\tilde{m}(x) = x\eta(x) + \tilde{c}(x) = o(m(x)) \quad \text{if } \alpha = 2. \tag{4.17}$$

The asymptotics of $\alpha(t)$ and $\beta(t)$ as $t \downarrow 0$ are given as follows:

$$\alpha(t) \sim \begin{cases} tL(1/t)/2 & \text{if } \alpha = 2, \\ \kappa_\alpha' t^{\alpha - 1} L(1/t) & \text{if } 1 < \alpha < 2, \\ \frac{1}{2}\pi L(1/t) & \text{if } \alpha = 1, \end{cases} \tag{4.18}$$

and

$$\beta(t) = \begin{cases} o(\alpha(t)) & \text{if } \alpha = 2, \\ \kappa_\alpha'' t^{\alpha - 1} L(1/t)\{1 + o(1)\} & \text{if } 1 < \alpha < 2, \\ L^*(1/t)\{1 + o(1)\} & \text{if } \alpha = 1, \end{cases} \tag{4.19}$$

where $\kappa_\alpha' = \Gamma(1 - \alpha)\cos\frac{1}{2}\pi\alpha$ and $\kappa_\alpha'' = -\Gamma(1 - \alpha)\sin\frac{1}{2}\pi\alpha$; in particular, for $1 < \alpha < 2$,

$$1 - Ee^{itX} = t\alpha(t) + it\gamma(t) \sim (\kappa_\alpha' + i(p - q)\kappa_\alpha'')t^\alpha L(1/t) \quad (t \downarrow 0). \tag{4.20}$$

For verification, see [58], [8, Theorems 4.3.1–2] if $1 \le \alpha < 2$. The estimate in the case $\alpha = 2$ is deduced from (4.15) and (4.16). Indeed, uniformly for $\varepsilon > 0$,

$$\alpha(t) = \int_0^{\varepsilon/t} \mu(x) \sin tx \, dx + O(\eta(\varepsilon/t)) = tc(\varepsilon/t) \left\{ 1 + O(\varepsilon^2) + o(1) \right\},$$

so that $\alpha(t) \sim tc(1/\varepsilon t) \sim tL(1/t)/2$; as for $\beta(t)$, use (3.12a) and (4.17).

In the case $\alpha = 1$, we shall need the following second-order estimate:

$$\beta(t) = \eta(1/t) + C^* L(1/t)\{1 + o(1)\}, \tag{4.21}$$

where C^* is Euler's constant, as is shown below. It holds that

$$\int_{1/t}^{\infty} \mu(y) \cos ty \, dy \sim C_1 L(1/t) \quad \text{and} \quad \int_0^{1/t} \mu(y)(1 - \cos ty) \, dy \sim C_2 L(1/t), \tag{4.22}$$

where $C_1 = \int_1^{\infty} \cos s \, ds/s$, $C_2 = \int_0^1 (1 - \cos s) \, ds/s$. Indeed, for each $M > 1$, by monotonicity of μ we have $\left| \int_{M/t}^{\infty} \mu(y) \cos ty \, dy \right| \leq 2\mu(M/t)/t \sim 2L(1/t)/M$, while

$$\int_{1/t}^{M/t} \mu(y) \cos ty \, dy = \int_1^M L(z/t) \cos z \, \frac{dz}{z} \sim C_1 L(1/t)$$

as $t \downarrow 0$ and $M \to \infty$ in this order, showing the first formula of (4.22). One can deduce the second one similarly. Since $C_2 - C_1 = C^*$, (4.21) follows.

The next proposition is also valid in the case $E|X| = \infty$, as long as S is *recurrent*. This remark is relevant only for its last assertion (iv), the r.w. being recurrent under $E|X| = \infty$ only if $\alpha = 2p = 1$ (see (4.33) for the recurrence criterion).

Proposition 4.2.1 *Suppose that (4.14) is satisfied. Then as $x \to \infty$*

(i) *except when $\alpha = 2p = 1$,*

$$\bar{a}(x) \sim \kappa_\alpha^{-1} x/m(x), \tag{4.23}$$

where $\kappa_\alpha = 2\left\{1 - 4pq \sin^2 \frac{1}{2}\pi\alpha\right\} \Gamma(\alpha)\Gamma(3 - \alpha)$ [note that $1 - 4pq \sin^2 t = \cos^2 t + [(p - q) \sin t]^2$ and $\kappa_\alpha = 0 \Leftrightarrow \alpha = 2p = 1$];
(ii) *if $1 \leq \alpha < 2$ and $p \neq 1/2$, then if $\alpha = 1$,*

$$\begin{cases} a(-x) \sim 2p\bar{a}(x), \\ a(x) \sim 2q\bar{a}(x), \end{cases} \tag{4.24}$$

where the sign '$\sim$' is interpreted in the obvious way if $pq = 0$; and
(iii) *if $\alpha = 2$ and $m_+/m \to p$ ($0 \leq p \leq 1$), then (4.24) holds;*
(iv) *if $\alpha = 2p = 1$ and there exists $\rho := \lim P_0[S_n > 0]$, then*

$$\bar{a}(x) \sim \begin{cases} \dfrac{2 \sin^2 \rho \pi}{\pi^2} \displaystyle\int_1^x \dfrac{dy}{y^2 \mu(y)} & \text{if} \quad 0 < \rho < 1, \\[4mm] \dfrac{1}{2} \displaystyle\int_{x_0}^x \dfrac{\mu(y)}{A^2(y)} \, dy & \text{if} \quad \rho = 0 \text{ or } 1; \end{cases} \tag{4.25}$$

$a(x) \sim a(-x);$ *and*

$$x/m(x) = o(\bar{a}(x)) \quad \text{if } E|X| < \infty.$$

[A(x) is given in (2.19) and x_0 is the same constant as in (4.1); see Remark 4.2.2(ii) when the existence of $\lim P_0[S_n > 0]$ is not assumed.]

If $\alpha = 1$, $p \neq q$ and $EX = 0$, then $\int_1^x \mu(y)\, dy/A^2(y) \sim x/(p-q)^2 m(x)$, so that the second expression on the RHS of (4.25) is a natural extension of that on the RHS of (4.23).

Proof (i) and (ii) are given in [3, Lemma 3.3] (cf. also [82, Lemma 3.1] for $1 < \alpha < 2$) except for the case $\alpha = 1$, and implied by Theorem 4.1.1 for $\alpha = 1$.

PROOF OF (iii). Recall that $a(x) = \bar{a}(x) + b_-(x) - b_+(x)$, where

$$b_\pm(x) = \frac{1}{\pi} \int_0^\pi \frac{\beta_\pm(t)}{[\alpha^2(t) + \gamma^2(t)]} \cdot \frac{\sin xt}{t}\, dt \qquad (4.26)$$

(see (3.44)). By Theorem 3.1.6 we can suppose $pq > 0$.

Let $\alpha = 2$. Since $\beta(t)/\alpha(t) \to 0$ $(t \to 0)$ and $\alpha(t) \sim \frac{1}{2}tL(1/t) \sim tm(1/t)$ (cf. Section A.1.1 for the latter equivalence), for each $\varepsilon > 0$, on using $|\sin xt| \leq 1$, the contribution from $t > \varepsilon/x$ to the above integral can be written as

$$\int_{\varepsilon/x}^\pi \frac{1}{t^2[L(1/t)]}\, dt \times o(1) = \int_{1/\pi}^{x/\varepsilon} \frac{1}{L(y)}\, dy \times o(1) = \frac{x/\varepsilon}{L(x/\varepsilon)} \times o(1) = o\left(\frac{x}{m(x)}\right).$$

(Note that the integral over $\delta < t \leq \pi$ tends to zero as $x \to 0$ for each δ.) Thus

$$b_+(x) = \frac{x}{\pi} \int_0^{\varepsilon/x} \frac{\beta_+(t)}{[tm(1/t)]^2}\, dt\{1 + O(\varepsilon^2)\} + o\left(\frac{x}{m(x)}\right), \qquad (4.27)$$

and similarly for $b_-(x)$.

To evaluate the above integral, we proceed analogously to the proof of Proposition 4.1.1(ii). We claim that if $m_+(x)/m(x) \to p$,

$$I(x) := \int_0^{\varepsilon/x} \frac{\beta_+(t)}{[tm(1/t)]^2}\, dt = \frac{\pi p}{2m(x)}\{1 + O(\varepsilon) + o(1)\} \qquad (4.28)$$

(as $x \to \infty$ and $\varepsilon \downarrow 0$ in this order). Define $\zeta_j(t)$, $j = 1, 2, 3$, by

$$\frac{\beta_+(t)}{t} = \left(\int_0^{\varepsilon/t} + \int_{\varepsilon/t}^{1/\varepsilon t} + \int_{1/\varepsilon t}^\infty\right)\eta_+(y)\sin ty\, dy$$
$$= \zeta_1(t) + \zeta_2(t) + \zeta_3(t),$$

so that

$$I(x) = \sum_{j=1}^3 I_j(x), \quad \text{where} \quad I_j(x) = \int_0^{\varepsilon/x} \frac{\zeta_j(t)}{tm^2(1/t)}\, dt.$$

Noting that the integrand of $I_1(x)$ is at most $\int_0^{\varepsilon/t} \eta_+(z)z\,dz/m^2(1/t)$ we see

$$
I_1(x) \le \int_{x/\varepsilon}^{\infty} \frac{\int_0^{\varepsilon y} \eta_+(z)z\,dz}{y^2 m^2(y)}\,dy
$$
$$
= \frac{\varepsilon \int_0^x \eta(z)z\,dz}{x m^2(x)}\{1 + o(1)\} + \varepsilon^2 \int_{x/\varepsilon}^{\infty} \frac{\eta(\varepsilon y)}{m^2(y)}\{1 + o(1)\}\,dy.
$$

The last integral being asymptotically equivalent to $\varepsilon \int_x^{\infty} \eta(z)\,dz/m^2(z) = \varepsilon/m(x)$, it follows that $I_1(x) \le \{\varepsilon + o(1)\}/m(x)$.

Changing the variable $ty = w$ we have $\zeta_2(t) = t^{-1}\int_\varepsilon^{1/\varepsilon} \eta_+(w/t)\sin w\,dw$, and hence

$$
I_2(x) = \int_\varepsilon^{1/\varepsilon} \sin w\,dw \int_{x/\varepsilon}^{\infty} \frac{\eta_+(wy)}{m^2(y)}\,dy = \int_\varepsilon^{1/\varepsilon} \frac{\sin w}{w}\,dw \int_{xw/\varepsilon}^{\infty} \frac{\eta_+(z)}{m^2(z/w)}\,dz.
$$

The inner integral of the last double integral is $\sim p/m(x)$ owing to the condition $m_+(x)/m(x) \to p$, for it implies that uniformly for $w \in [\varepsilon, 1/\varepsilon]$,

$$
\int_x^{\infty} \frac{\eta_+(y)}{m^2(y)} = -\frac{m_+(x)}{m^2(x)} + 2\int_x^{\infty} \frac{m_+(y)\eta(y)}{m^3(y)}\,dy = \frac{p}{m(x)}\{1 + o(1)\}.
$$

Since $\int_\varepsilon^{1/\varepsilon} \sin w\,dw/w = \frac{1}{2}\pi + O(\varepsilon)$, we can now conclude

$$
I_2(x) = \frac{p\pi/2}{m(x)}\{1 + O(\varepsilon) + o(1)\}.
$$

Since $\zeta_3(t) \le 2t^{-1}\eta_+(1/\varepsilon t)$,

$$
I_3(x) \le 2\int_0^{\varepsilon/x} \frac{\eta(1/\varepsilon t)}{t^2 m^2(1/t)}\,dt = 2\int_{x/\varepsilon}^{\infty} \frac{\eta(y/\varepsilon)}{m^2(y)}\,dy \sim 2\varepsilon/m(x).
$$

Combining the estimates of I_1 to I_3 obtained above yields (4.28). Thus $b_+(x) \sim \frac{1}{2}px/m(x)$. In the same way $b_-(x) \sim \frac{1}{2}qx/m(x)$ and by (i) $\bar{a}(x) \sim \frac{1}{2}x/m(x)$. Consequently $a(x) \sim qx/m(x) \sim 2q\bar{a}(x)$, and similarly $a(-x) \sim 2p\bar{a}(x)$. The proof of (iii) is complete.

PROOF OF (iv). Here $E|X|$ may be infinite. Note that $\alpha(t) \sim \frac{\pi}{2}L(1/t)$ remains valid under $E|X| = \infty$. If $\rho = 1$, then the assertion follows from Theorem 4.1.1 as a special case; the same is true when $\rho = 0$, by duality. Below, we let $0 < \rho < 1$.

Let c_n be determined by (2.17) with $c_\sharp = 1$, namely, $c_n/n \sim L(c_n)$ as $n \to \infty$. Then the laws of S_n/c_n constitute a tight family under P_0 if and only if $\rho_n := P_0[S_n > 0]$ is bounded away from 0 and 1. The convergence $\rho_n \to \rho \in (0, 1)$ is equivalent to the convergence

$$
b_n := nE[\sin\{X/c_n\}] \to b \quad \text{for some } -\infty < b < \infty. \tag{4.29}
$$

Observing $\gamma(t + s) - \gamma(t) = o(L(t))$ as $s/t \to 0$, from (4.29) we infer

$$- \gamma(t) = t^{-1}E[\sin Xt] \sim bL(1/|t|) \qquad (t \to 0). \qquad (4.30)$$

Recalling $\alpha(t) \sim \frac{1}{2}\pi L(1/t)$, we accordingly obtain

$$1 - \psi(t) \sim \left(\frac{1}{2}\pi|t| - ibt\right)L(1/|t|) \qquad (t \to 0).$$

By our choice of c_n, $C_\Phi = 1$ in (2.18) so that if $S_n/c_n - b_n \Rightarrow Y^\circ$, $E[e^{itY^\circ}] = \lim E[e^{it(S_n/c_n-b_n)}] = e^{-\frac{\pi}{2}|t|}$, showing that the values of ρ and b are related by

$$\rho = P[Y^\circ + b > 0]. \qquad (4.31)$$

The law of Y° has the density given by $\frac{1}{2}\left[x^2 + \frac{1}{4}\pi^2\right]^{-1}$, and solving (4.31) for b, one finds

$$b = \frac{1}{2}\pi \tan\left[\pi(\rho - \frac{1}{2})\right].$$

By (4.30)

$$\frac{\alpha(t)}{\alpha^2(t) + \gamma^2(t)} \sim \frac{2/\pi}{(1 + \tan^2[\pi(\rho - \frac{1}{2})])L(1/t)} = \frac{2\sin^2\rho\pi}{\pi L(1/t)},$$

and as before (see the argument used at $(*)$ in the next paragraph), we deduce

$$\bar{a}(x) \sim \frac{2\sin^2\rho\pi}{\pi^2}\int_1^x \frac{dy}{yL(y)}.$$

For the proof of $a(x) \sim a(-x)$, which is equivalent to $a(x) - a(-x) = o(\bar{a}(x))$, it suffices to show that

$$a(x) - a(-x) = \frac{2}{\pi}\int_0^\pi \frac{-\gamma(t)\sin xt}{[\alpha^2(t) + \gamma^2(t)]t}\,dt = o\left(\left[\frac{1}{L}\right]_*(x)\right), \qquad (4.32)$$

where $[1/L]_*(x) = \int_1^x [yL(y)]^{-1}\,dy$. Denote the above integral restricted to $[0, 1/x]$ and $[1/x, \pi]$ by $J_{<1/x}$ and $J_{>1/x}$, respectively. By (4.30) and $\alpha(t) \sim \frac{1}{2}\pi L(1/t)$ it follows that

$$J_{<1/x} \sim Cx\int_0^{1/x} \frac{dt}{L(1/t)} \sim \frac{C}{L(x)},$$

hence $J_{<1/x} = o\left([1/L]_*(x)\right)$. For the evaluation of $J_{>1/x}$, recall we have chosen L so that $L'(x)/L(x) = o(1/x)$. Let $\alpha_0(t) = \frac{1}{2}\pi L(1/t)$ and define $\tilde{L}$ and $\tilde{L}_0$ by

$$\tilde{L}(1/t) = \frac{-\gamma(t)}{\alpha^2(t) + \gamma^2(t)} \quad \text{and} \quad \tilde{L}_0(1/t) = \frac{bL(1/t)}{\alpha_0^2(t) + b^2L^2(1/t)}.$$

Then $\int_{1/x}^{\pi} \tilde{L}_0(1/t) \sin xt \, dt/t = O(\tilde{L}_0(x)) = O(1/L(x))$, whereas by (4.30)

$$\tilde{L}(1/t) - \tilde{L}_0(1/t) = o\left(1/L(1/t)\right),$$

showing

$$\left| \int_{1/x}^{\pi} \left[\tilde{L}(1/t) - \tilde{L}_0(1/t)\right] \sin xt \, dt/t \right| \le \int_1^x \left| \tilde{L}(y) - \tilde{L}_0(y) \right| dy/y = o\left([1/L]_*(x)\right).$$

It therefore follows that $J_{>1/x} = o\left([1/L]_*(x)\right)$. Thus (4.32) is verified.

In the case $E|X| < \infty$, by (4.25) and $m(x)/x \sim L^*(x)$ we can readily see that $x/m(x) = o(\bar{a}(x))$ (see Remark 4.2.2(ii) below). This finishes the proof of (iv). $\quad\square$

Remark 4.2.2 (i) Let $\alpha = 1$ in (4.14) and $E|X| = \infty$. Then F is recurrent if and only if

$$\int_1^{\infty} \frac{\mu(y)}{(L(y) \vee |A(y)|)^2} \, dy = \infty \qquad (4.33)$$

(cf. [83, Section 5.3]). If $p \ne 1/2$, then F is transient, $|A(y)| \sim |p-q| \int_0^y \mu(t) \, dt \to \infty$ and $\int^{\infty} \mu(x) \, dx / A^2(x) < \infty$.

(ii) Let $\alpha = 2p = 1$ in (4.14) and $EX = 0$. Observing that $\gamma(t) = -A(1/t) + o(L(1/t))$ (see (4.21)), and hence $\alpha^2(t) + \gamma^2(t) \sim (\frac{1}{2}\pi)^2 L^2(1/t) + A^2(1/t)$ $(t \downarrow 0)$, one infers that

$$\bar{a}(x) \sim \frac{1}{2} \int_1^x \frac{\mu(y) \, dy}{(\frac{1}{2}\pi)^2 L^2(y) + A^2(y)} \qquad (4.34)$$

without assuming the existence of $\lim P_0[S_n > 0]$. If $EX = 0$, then $L(y) \vee |A(y)| = o(L^*(y))$, which shows that $\bar{a}(x)/[x/m(x)] \to \infty$, for $x/m(x) \sim 1/L^*(x) \sim \int_1^x \mu(y) \, dy/[L^*(y)]^2$. This is also deduced from Theorem 3.1.2(ii).

4.2.2 Asymptotics of $a(x)$ II

For any subset B of $\mathbb{R}$ such that $B \cap \mathbb{Z}$ is non-empty, define

$$H_B^x(y) = P_x[S_{\sigma_B} = y] \qquad (y \in B), \qquad (4.35)$$

the hitting distribution of B for the r.w. S. We continue to assume (4.14) to hold.

Suppose $m_+/m \to 0$ throughout this subsection. In particular $E(-\hat{Z}) = \infty$, under which it holds that

$$a(-x) = \sum_{y=1}^{\infty} H_{[0,\infty)}^{-x}(y) a(y), \quad x > 0 \qquad (4.36)$$

(see (2.15)). Using this identity we will derive the asymptotic form of $a(-x)$ as $x \to \infty$ when $\mu_+(x)$ varies regularly at infinity. Recall that $V_d(x)$ $(U_a(x))$ denotes the renewal function for the weakly descending (strictly ascending) ladder height process and that, on writing $v°$ for $V_d(0)$,

$$\ell^*(x) = \int_0^x P[Z > t]\, dt \quad \text{and} \quad \hat{\ell}^*(x) = \frac{1}{v°} \int_0^x P[-\hat{Z} > t]\, dt.$$

We know that $\ell^*(x)$ (resp. $\hat{\ell}^*(x)$) is s.v. as $x \to \infty$ if $1 \le \alpha \le 2$ (resp. $\alpha = 2$) (under the present setting), that $P[-\hat{Z} > x]$ is s.v. if $\alpha = 1$ and that

$$U_a(x) \sim \frac{x}{\ell^*(x)} \quad \text{and} \quad V_d(x) \sim \begin{cases} x/\hat{\ell}^*(x) & \alpha = 2, \\ \kappa'_\alpha x^{\alpha-1} \ell^*(x)/L(x) & 1 < \alpha < 2, \\ v°/P[-\hat{Z} > x] & \alpha = 1, \end{cases} \quad (4.37)$$

where $\kappa'_\alpha = -\alpha[(2-\alpha)\pi]^{-1} \sin \alpha\pi$ – provided (4.14) holds (see[2] the table in Section 2.6.)

Proposition 4.2.3 *Suppose that (4.14) holds and $m_+/m \to 0$ and that $\mu_+(x)$ is regularly varying at infinity with index $-\beta$ and $\lim_{x\to\infty} a(-x) = \infty$. Then*

$$a(-x) \sim \begin{cases} U_a(x) \sum_{z=x}^{\infty} \dfrac{\mu_+(z)V_d(z)a(z)}{z} & \text{if} \quad \alpha = \beta = 2, \\[3mm] C_{\alpha,\beta} \dfrac{U_a(x)}{x} \sum_{z=1}^{x} \mu_+(z)V_d(z)a(z) & \text{otherwise} \end{cases}$$

as $x \to \infty$, where $C_{\alpha,\beta} = [\Gamma(\alpha)]^2 \Gamma(\beta - 2\alpha + 2)/\Gamma(\beta)$.

[More explicit expressions of the RHS are given in the proof: see (4.42) to (4.45). By virtue of Theorem 3.1.6 and Proposition 4.2.1(i) it follows, under $m_+/m \to 0$, that $a(x) \sim [\Gamma(\alpha)\Gamma(3 - \alpha)]^{-1} x/m(x)$.]

Proof Put $u_a(x) = U_a(x) - U_a(x - 1)$, $x > 0$, $v_d(z) = V_d(z) - V_d(z - 1)$, $u_a(0) = U_a(0) = 1$ and $v_d(0) = V_d(0) = v°$. Then $G(x_1, x_2) := u_a(x_2 - x_1)$ is the Green function of the strictly increasing ladder process killed on its exiting the half-line $(-\infty, 0]$ and by the last exit decomposition we obtain

$$H_{[0,\infty)}^{-x}(y) = \sum_{k=1}^{x} u_a(x - k)P[Z = y + k] \quad (x \ge 1, y \ge 0). \quad (4.38)$$

Suppose the conditions of the proposition to hold and let $\mu_+(x) \sim L_+(x)/x^\beta$ with $\beta \ge \alpha$ and L_+ slowly varying at infinity. Then noting $g_{[1,\infty)}(0, -z) = v_d(z)$ and summing by parts one deduces

---

[2] In Lemma 5.3.1(i)) we shall see that $V_d(x) \sim a(x)\ell^*(x)$ under $m_+/m \to 0$, so that the formula (4.37) for V_d with $1 < \alpha < 2$ also follows from Proposition 4.2.1.

$$P[Z > y] = \sum_{z=0}^{\infty} v_d(z)\mu_+(y+z) \sim \beta \sum_{z=0}^{\infty} V_d(z)\frac{\mu_+(y+z)}{y+z+1}$$

$$\sim C_0 V_d(y)\mu_+(y) \qquad (y \to \infty), \qquad (4.39)$$

and

$$C_0 = \beta \int_0^{\infty} t^{\alpha-1}(1+t)^{-\beta-1}\, dt = \frac{\Gamma(\alpha)\Gamma(\beta-\alpha+1)}{\Gamma(\beta)}.$$

If $\beta > 2\alpha - 1$, then $\sum \mu_+(x)[a(x)]^2 < \infty$, so that $a(-x)$ converges to a constant. Hence we may consider only the case $\alpha \le \beta \le 2\alpha - 1$.

Let $\alpha > 1$. Recall a varies regularly with index $\alpha - 1$ at infinity. Then returning to (4.36), performing summation by parts (w.r.t. y) and then substituting the above equivalence into (4.38) lead to

$$a(-x) \sim (\alpha-1)\sum_{y=1}^{\infty}\sum_{k=0}^{x} u_a(x-k)P[Z \ge y+k]\frac{a(y)}{y}$$

$$\sim (\alpha-1)C_0 \sum_{y=1}^{\infty}\sum_{k=0}^{x} u_a(x-k)V_d(y+k)\mu_+(y+k)\frac{a(y)}{y}. \qquad (4.40)$$

Owing to (2.5), namely $u_a(x) \sim 1/\ell^*(x)$, one can replace $u_a(x-k)$ by $1/\ell^*(x)$ in (4.40), the inner sum over $(1-\varepsilon)x < k \le x$ being negligible as $x \to \infty$ and $\varepsilon \to 0$. After changing the variables by $z = y+k$, the last double sum restricted to $y+k \le x$ accordingly becomes asymptotically equivalent to

$$\frac{1}{\ell^*(x)}\sum_{z=1}^{x} V_d(z)\mu_+(z)\sum_{k=0}^{z-1}\frac{a(z-k)}{z-k} \sim \frac{1}{(\alpha-1)\ell^*(x)}\sum_{z=1}^{x} V_d(z)\mu_+(z)a(z). \qquad (4.41)$$

If $\beta = 2\alpha - 1 > 1$ (entailing $\lim \ell^*(x) = EZ < \infty$), then $C_0 = C_{\alpha,\beta}$, the sum on the RHS of (4.41) varies slowly and the remaining part of the double sum in (4.40) is negligible, showing

$$a(-x) \sim \frac{C_0}{\ell^*(x)}\sum_{z=1}^{x} V_d(z)\mu_+(z)a(z) \sim \frac{C_{\alpha,\beta}}{\ell^*(x)}\sum_{z=1}^{x}\frac{L_+(z)}{\ell_*(z)z}, \qquad (4.42)$$

where

$$\ell_*(z) = \begin{cases} \hat{\ell}^*(z)L(z)/2 & \alpha = 2, \\ (\alpha^2\kappa_\alpha/[2(\alpha-1)(2-\alpha)^3\kappa_\alpha'])[L(z)]^2/\ell^*(z) & 1 < \alpha < 2. \end{cases}$$

Let $\alpha \le \beta < 2\alpha - 1$. If $|\alpha - 2| + |\beta - 2| \ne 0$, then the outer sum in (4.40) over $y > Mx$ as well as that over $y < x/M$ becomes negligibly small as M becomes large and one can easily infer that

$$a(-x) \sim \frac{(\alpha - 1)C_1 C_0 L_+(x) x^{2\alpha-1-\beta}}{\ell^*(x)\ell_*(x)(2\alpha - 1 - \beta)} \sim \frac{C_{\alpha,\beta}}{\ell^*(x)} \sum_{z=1}^{x} V_d(z)\mu_+(z)a(z), \qquad (4.43)$$

where ℓ_* is as above, and

$$\begin{cases} C_1 = (2\alpha - 1 - \beta) \int_0^1 ds \int_0^\infty (s+t)^{-\beta+\alpha-1} t^{\alpha-2}\, dt = \int_0^\infty (1+t)^{-\beta+\alpha-1} t^{\alpha-2}\, dt, \\ C_{\alpha,\beta} = (\alpha - 1)C_1 C_0 = C_0 \Gamma(\alpha)\Gamma(\beta - 2\alpha + 2)/\Gamma(\beta - \alpha + 1). \end{cases}$$

If $\alpha = \beta = 2$, then $a(x) \sim 2x/L(x)$ and uniformly for $0 \le k \le x$,

$$\sum_{y=x}^{\infty} V_d(y+k)\mu_+(y+k)\frac{a(y)}{y} = v^\circ \sum_{y=x}^{\infty} \frac{L_+(y+k)\{1+o(1)\}}{(y+k)v^\circ \hat{\ell}^*(y+k)L(y)/2} \sim \sum_{y=x}^{\infty} \frac{L_+(y)}{\ell_*(y)y},$$

while $\sum_{k=1}^{x} u_a(x-k) \sum_{y=1}^{x} V_d(y+k)\mu_+(y+k)a(y)/y \le CxL_+(x)/[\ell^*(x)\ell_*(x)]$ (split the outer sum at $k = x/2$ and use (4.41)). With these two bounds, from (4.40) we deduce that

$$a(-x) \sim \frac{C_0 x}{\ell^*(x)} \sum_{y=x}^{\infty} V_d(y)\mu_+(y)\frac{a(y)}{y} \sim \frac{x}{\ell^*(x)} \sum_{y=x}^{\infty} \frac{L_+(y)}{\ell_*(y)y} \qquad (4.44)$$

(since $(\alpha - 1)C_0 = C_0 = 1$).

The relations (4.42) to (4.44) together show those of Proposition 4.2.3 in the case $1 < \alpha \le 2$ since $U_a(x) \sim x/\ell^*(x)$.

It remains to deal with the case $\alpha = \beta = 1$ when in place of (4.40) we have

$$a(-x) \sim \sum_{y=1}^{\infty} \sum_{k=0}^{x} \frac{u_a(x-k)\mu_+(y+k)}{P[-\hat{Z} > y+k]/v^\circ} \cdot \frac{d}{dy}\frac{1}{L^*(y)}.$$

Note $P[-\hat{Z} > x]$ is s.v. and $\frac{d}{dy}[1/L^*(y)] = L(y)/y[L^*(y)]^2$. Then one sees that the above double sum restricted to $y + k \le x$ is asymptotically equivalent to

$$\frac{v^\circ}{\ell^*(x)} \sum_{z=1}^{x} \frac{L_+(z)}{zP[-\hat{Z} > z]L^*(z)}, \qquad (4.45)$$

hence s.v., while the outer sum over $y > x$ is negligible. It, therefore, follows that the above formula represents the asymptotic form of $a(-x)$ and may be written alternatively as $x^{-1}U_a(x) \sum_{y=1}^{x} V_d(y)\mu_+(y)a(y)$, as required. $\square$

Remark 4.2.4 Let $\alpha = \beta = 1$. Then $P[-\hat{Z} > x]/v^\circ \sim \int_x^\infty [\mu_-(t)/\ell^*(t)]\, dt$ according to Lemma 5.4.5. If $EZ < \infty$ and $\sum_{x>0} \mu_+(x)a^2(x) = \infty$ in addition (so that $\ell^*(x) \to EZ$ and $a(-x) \to \infty$), by evaluating the sum in (4.45) one finds

$$a(-x) \sim \int_1^x \mu_+(t)[1/L^*(t)]^2\, dt \sim \int_{x_0}^x \mu_+(t)[1/A(t)]^2\, dt.$$

Remark 4.2.5 Let the assumption of Proposition 4.2.3 be satisfied and M be an arbitrarily given number > 1. In Sections 5.1, 5.5 and 5.6 (see Lemma 5.1.3, Propositions 5.5.1 and 5.6.1) we shall be concerned with the condition

$$\frac{a(-R) - a(-R + x)}{a^{\dagger}(x)} \longrightarrow 0 \quad \text{as} \ R \to \infty \quad (x \leq R), \tag{4.46}$$

and, in particular, we will see that if $m_+/m \to 0$, then $P_x[\sigma_R < \sigma_0 \,|\, \sigma_{[R,\infty)} < \sigma_0] \to 1 \ (R \to \infty)$ uniformly for $x < R$ satisfying (4.46) (cf. Proposition 5.5.1).

Below we show the following.

(i) If $2\alpha - 1 \leq \beta$, then (4.46) holds uniformly for $-MR < x < R$, while if $\alpha \leq \beta < 2\alpha - 1$ (entailing $\alpha > 1$) (4.46) holds whenever $x/R \to 0$, but fails for $x < -R/M$.

(ii) Let $\alpha = 2$. Then $P_x[\sigma_R < \sigma_0 \,|\, \sigma_{[R,\infty)} < \sigma_0] \to 1$ as $R \to \infty$ uniformly for $-MR < x < R$, although (4.46) fails for $x < -R/M$ if $2 \leq \beta < 3$.

(iii) Let $\alpha \leq \beta < 2\alpha - 1 < 3$. Then $\delta < P_x[\sigma_R < \sigma_0 \,|\, \sigma_{[R,\infty)} < \sigma_0] < 1 - \delta$ for $-RM < x < -R/M$ for some $\delta = \delta_M > 0$.

[If $\alpha \leq \beta < 2\alpha - 1$, then for $\lim P_x[\sigma_R < \sigma_0 \,|\, \sigma_{[R,\infty)} < \sigma_0] = 1$ to hold (4.46) is necessary or not according as $\alpha < 2$ or $= 2$ as a consequence of (i) to (iii).]

For the proof we use results from Chapters 5 and 6. Write Z^* for $S_{\sigma[0,\infty)}$ and, first of all, note that Lemma 5.1.3 entails

$$\begin{cases} P_x[\sigma_{[R,\infty)} < \sigma_0] \sim E_x\,[a(Z^*); Z^* < R]\,/a(R) + P_x[Z^* \geq R], \\ P_x[\sigma_R < \sigma_0] \sim E_x\,[a(Z^*); Z^* < R]\,/a(R) + P_x[\sigma_R < \sigma_0, Z^* \geq R] \end{cases} \tag{4.47}$$

for $x < 0$ and also that for $-\frac{1}{2}R < x < 0$, $\sum_{z=1}^{R} g_{[0,\infty)}(x, -z) \asymp U_a(x)V_d(R)$ (by Lemma 6.2.1) and $\sum_R^{\infty} v_d(z)\mu_+(z) \sim [\alpha\hat{\rho}/(\beta - \alpha\hat{\rho})]V_d(R)\mu_+(R)$, yielding

$$P_x[Z^* > R] = \sum_{z=1}^{\infty} g_{[0,\infty)}(x, -z)\mu_+(z + R) \asymp U_a(x)V_d(R)\mu_+(R). \tag{4.48}$$

If $\beta < 2\alpha - 1$, then from Propositions 4.2.1 and 4.2.3 it follows that $a(-x)$ varies regularly with a positive index, which implies that (4.46) holds whenever $R/x \to -\infty$ because of the inequalities in (5.8), and fails for $x < -R/M$.

Suppose $\beta \geq 2\alpha - 1$. Then $a(-y)$, $y > 0$ is s.v., so that (4.46) holds uniformly for $-MR < x < -R/M$. If $\beta > 2\alpha - 1$, $a(-R)$ converges to a positive constant; hence the assertion is trivial. Let $\beta = 2\alpha - 1$ and $-\frac{1}{2}R < x < 0$, and put

$$L_1(y) = yV_d(y)\mu_+(y)a(y).$$

Then, noting that L_1 is s.v., we deduce from (4.48) and Proposition 4.2.3

$$\frac{E_x[a(Z^*); Z^* > R]}{a(x)} \asymp \frac{P_x[Z^* > R]a(R)}{a(x)} \asymp \frac{L_1(R)}{\int_1^{|x|} L_1(z)z^{-1}\,dz} \cdot \frac{|x|}{R}.$$

Since $\int_1^{|x|} L_1(z)z^{-1}\,dz \gg L_1(|x|)$, the rightmost member above approaches zero uniformly for $-\frac{1}{2}R < x < 0$. In view of (4.47) and the dual of (2.15), this shows $P_x[\sigma_R < \sigma_0]a(R) \sim a(x)$, an equivalent of (4.46). These verify (i).

For the proof of (ii), we show that if $\beta > 1$, then for any $\varepsilon > 0$ there exists a $n_\varepsilon \geq 1$ such that

$$P_x\left[Z^* > n_\varepsilon R \,\middle|\, Z^* \geq R\right] \leq \varepsilon \quad \text{for } -MR < x < -R/M. \tag{4.49}$$

To this end, use the inequality $g_{[0,\infty)}(x,z) \leq g_{\{0\}}(x,z) \leq g_{\{0\}}(x,x) = 2\bar{a}(x)$ to see

$$H^x_{[0,\infty)}(y) := P_x[Z^* = y] \leq 2\bar{a}(x)\mu_+(y),$$

showing the upper bound: for $n \geq 1$

$$P_x\left[Z^* > nR\right] \leq 2\bar{a}(x) \sum_{y \geq nR} \mu_+(y). \tag{4.50}$$

For the lower bound, noting that $g_B(x,-z) = g_{-B}(z,-x)$, we apply (5.24) to obtain

$$g_{[0,\infty)}(x,-z) \geq g_{\{0\}}(z,-x) - a(x)\{1+k\}$$
$$= a^\dagger(z) - a(z+x) - o(a(|x|)),$$

and hence that for $|x|/2 \leq z \leq |x|$, $g_{[0,\infty)}(x,-z) \geq a(|x|)\{1+o(1)\}$. This yields

$$H^x_{[0,\infty)}(y) \geq C_1 a(|x|)\mu_+(y) \quad \text{if } -x \asymp y, \tag{4.51}$$

provided μ_+ varies regularly with index $-\beta$. Thus for $-MR < x < -R/M$,

$$P_x[R < S_{\sigma[0,\infty)} \leq 2R] \geq C_1'\bar{a}(x) \sum_{y \geq R} \mu_+(y),$$

which together with (4.50) shows (4.49).

If $\alpha = 2$, then uniformly for $R \leq y \leq nR$ $(n = 1, 2, \ldots)$, $a(y) - a(-R+y) = a(R) + o(a(y))$ in view of Proposition 4.2.1(i), and hence $P_y[\sigma_R < \sigma_0] \to 1$, so that by (4.49) $P_x[\sigma_R < \sigma_0, Z^* \geq R] \sim P_x[Z^* \geq R]$ in (4.47). In conjunction with the result for $x \geq 0$ given in (5.53) (as well as with (4.47) and (4.49)) this shows (ii).

The proof of (iii) is similar. Note that $1 < \alpha < 2$ under the assumption of (iii). Observe that $P_y[\sigma_0 < \sigma_R] \wedge P_y[\sigma_0 > \sigma_R]$ is bounded away from zero for $2R \leq y \leq 3R$, which together with (4.51) shows that for some $c > 0$, for $-MR \leq x \leq -R/M$,

$$P_x[Z^* \geq R, \sigma_R < \sigma_0] \wedge P_x[Z^* \geq R, \sigma_0 < \sigma_R] \geq ca(|x|)\mu_+(R)R.$$

It also holds that $P_x[Z^* > R] \leq Ca(|x|)\mu_+(R)R$ (by (4.50)) and $P_x[\sigma_R < \sigma_0] \asymp a(-R)/a(R)$ (since by (4.43) $a(x)$ varies regularly with index $2\alpha - 1 - \beta \in (0,1)$). By (4.43) one has $a(-R)/a(R) \asymp a(R)\mu_+(R)R$. In view of (4.47), these together show $P_x[\sigma_{[R,\infty)} < \sigma_0] \asymp P_x[\sigma_R < \sigma_0] \asymp a(R)\mu_+(R)R$. Now one can conclude that both $P_x[\sigma_0 < \sigma_R \mid \sigma_{[R,\infty)} < \sigma_0]$ and $P_x[\sigma_0 > \sigma_R \mid \sigma_{[R,\infty)} < \sigma_0]$ are bounded away from zero. Thus (iii) is verified.

4.3 Asymptotics of $a(x + 1) - a(x)$

It is sometimes useful to know the bounds of the increments of $a(\pm x)$. From (3.10), whether S is recurrent or not,[3] it follows that if $EX = 0$,

$$\bar{a}(x+1) - \bar{a}(x) = \frac{1}{\pi} \int_0^\pi \Re \frac{\cos xt - \cos(x+1)t}{1 - \psi(t)} \, dt \tag{4.52}$$

$$= \frac{1}{\pi} \int_0^\pi \alpha(t) \frac{\sin t \sin xt + (1 - \cos t) \cos xt}{[\alpha^2(t) + \gamma^2(t)]t} \, dt.$$

Proposition 4.3.1 *Let $EX = 0$. If condition (H) (introduced in Section 3.1) holds, then for some constant C, as $|x| \to \infty$*

$$|a(x+1) - a(x)| \le \frac{C}{|x|} \int_1^{|x|} \frac{dy}{m(y)}$$

$$= O\left(|x|^{-\delta}\bar{a}(x)\right) \quad \text{(for each } \delta < 1\text{)};$$

in particular $|a(x + 1) - a(x)| = O\left(|x|^{-1}\bar{a}(x)\right)$ if $\liminf c(x)/m(x) > 0$.
[Cf. Proposition 4.3.5 below for the case $c/m \to 0$.]

Proof We split the range of the second integral in (4.52) at π/x. Denote by $I_*(x)$ and by $I^*(x)$ the contributions to it from $t \le \pi/x$ and from $t > \pi/x$, respectively. Under (H) we have $f^\circ(t) = 1/[\alpha^2(t)+\gamma^2(t)] \le C_1/[t^2 m^2(1/t)]$ for t small enough (Lemma 3.2.6). Using this together with $\alpha(t) \le 5tc(1/t)$ (Lemma 3.2.1), one deduces

$$|I_*(x)| \le x \int_0^{\pi/x} \frac{2t\alpha(t)}{\alpha^2(t) + \gamma^2(t)} \, dt$$

$$\le Cx \int_x^\infty \frac{c(y)\, dy}{m^2(y)y^2}$$

$$\le \frac{C}{m(x)},$$

where for the last inequality the monotonicity of m is used.

---

[3] Even if $E|X| = \infty$, γ may be well defined, although $\beta_\pm$ is not.

For the evaluation of $I^*(x)$ we apply Lemma 3.2.11, which entails

$$|\alpha(t) - \alpha(t + \pi/x)| \le Cm(1/t)/x \quad \text{and} \quad |f^\circ(t) - f^\circ(t + \pi/x)| \le Cf^\circ(t)/tx$$

for $1/x < t < \pi$. By a decomposition of the integral over $[\pi/x, \pi]$ similar to that made in (3.56) these bounds yield

$$|I^*(x)| \le \frac{C}{x} \int_{\pi/x}^{\pi} [m(1/t) + c(1/t)] \, f^\circ(t) \, dt$$

$$\le \frac{C'}{x} \int_{1/\pi}^{x/\pi} \frac{dy}{m(y)}.$$

Thus we have

$$|\bar{a}(x+1) - \bar{a}(x)| \le C_0 x^{-1} \int_1^x dy/m(y).$$

As for the increments of $a(\pm x)$, it suffices to obtain a similar bound for the increment of $\pi[b_+(x) - b_-(x)]$ because of (3.44). Split the range of the integral representing it at π/x and let $J_*(x) + J^*(x)$ be the corresponding decomposition. In the same way as above we get $|J^*(x)| \le Cx^{-1} \int_1^x dy/m(y)$. Putting

$$K_\pm(x) = \int_0^{\pi/x} \frac{\beta_\pm(t)}{[\alpha^2(t) + \gamma^2(t)]t} (\sin(x+1)t - \sin xt) \, dt, \qquad (4.53)$$

by elementary computation (see Lemma 4.3.2 below) we deduce that

$$|J_*(x)| = |K_-(x) - K_+(x)|$$
$$\le \pi|b_-(x) - b_+(x)|/x + O(1/m(x)).$$

Noting that $|b_-(x) - b_+(x)| = |a(x) - a(-x)|/2 \le C_2 x/m(x)$ under (H) and that $\int_0^x dy/m(y)$ is $o\left(x^{2-\delta}/m(x)\right)$ in general and bounded above by $x/m(x)$ if $m(x) \le O(c(x))$, we can now conclude the assertion of the proposition. $\qquad \square$

Lemma 4.3.2 *Let (H) hold and $K_\pm(x)$ be as given in (4.53). Then uniformly for $x > 1$,*

$$K_\pm(x) = \frac{1}{x} \int_0^{\pi/x} \frac{\beta_\pm(t) \sin xt}{[\alpha^2(t) + \gamma^2(t)]t} \, dt + O\left(\frac{1}{m(x)}\right). \qquad (4.54)$$

Proof The result follows from the equality

$$\sin(x+1)t - \sin xt = \sin t \cos xt + O(xt^3) = x^{-1} \sin xt + O(x^2 t^3),$$

valid uniformly for $x > 1, t > 0$ satisfying $0 < xt \le \pi$. $\qquad \square$

If the tails of F possess an appropriate regularity, we can obtain the exact asymptotics of the increments of $a(\pm x)$.

Proposition 4.3.3 *Suppose that (4.14) is satisfied with* $1 < \alpha \le 2$. *Then*

$$a(x+1) - a(x) \sim (\alpha - 1)a(x)/x \qquad if \; \liminf_{x\to\infty} a(x)/\bar{a}(x) > 0,$$

$$a(-x-1) - a(-x) \sim (\alpha - 1)a(-x)/x \quad if \; \liminf_{x\to\infty} a(-x)/\bar{a}(x) > 0,$$

as $x \to \infty$; *and*

$$\lim_{x\to\pm\infty} \frac{a(x+1) - a(x)}{\bar{a}(x)/|x|} = 0 \quad if \; \lim_{x\to\pm\infty} \frac{a(x)}{\bar{a}(x)} = 0,$$

where both upper or both lower signs should be chosen in the double signs.

Let $1 < \alpha \le 2$. If $1 < \alpha < 2$, the assertion of the above proposition is verified in [82, Lemma 3.1] by a relatively simple argument, whereas for $\alpha = 2$, the proof is somewhat involved. In preparation for this, we introduce some notation. Put

$$\phi_c(\theta) := \Re[1 - \psi(\theta)], \quad \phi_s(\theta) := \Im[1 - \psi(\theta)],$$

$$W_+(x) := \int_{x+0}^{\infty} y \, dF(y), \quad W_-(x) := -\int_{-\infty}^{-x-0} y \, dF(y),$$

$$\text{and} \quad W(x) := W_+(x) + W_-(x).$$

It follows that $1 - \psi(\theta) = \phi_c(\theta) + i\phi_s(\theta)$, $\phi_c(\theta) = \theta\alpha(\theta)$, $\phi_s(\theta) = \theta\gamma(\theta)$,

$$W_\pm(x) = x\mu_\pm(x) + \eta_\pm(x) \quad \text{and} \quad \int_0^x W_\pm(y) \, dy = c_\pm(x) + m_\pm(x).$$

Integrating by parts leads to

$$\phi_c'(\theta) = \theta \int_0^{\infty} W(x) \cos\theta x \, dx, \quad \phi_s'(\theta) = \theta \int_0^{\infty} [W_+(x) - W_-(x)] \sin\theta x \, dx,$$

for $\theta \ne 0$, while $\phi_c'(0+) = \phi_c'(0) = 0$, $\phi_s'(0+) = \phi_s'(0) = 0$. Here we have used the assumption $EX = W_+(0) - W_-(0) = 0$ in deriving the expression of $\phi_s'(\theta)$.

Lemma 4.3.4 *Let* $\alpha = 2$ *in (4.14).*

(i) $\phi_c'(\theta) \sim 2\theta m(1/\theta)\{1 + o(1)\}$, $\phi_s'(\theta) = o(\theta m(1/\theta))$ *as* $\theta \downarrow 0$.
(ii) *As* $x \to \infty$

$$\int_{M/x}^{\pi} \left| \left[\Re\frac{1}{1-\psi} \right]'(t) \right| t \, dt = O\left(\frac{x}{Mm(x/M)} \right) \quad uniformly \; for \; M > 1,$$

$$\int_{\varepsilon/x}^{\pi} \left| \left[\Im\frac{1}{1-\psi} \right]'(t) \right| t \, dt = o\left(\frac{x}{m(x)} \right) \quad for \; each \; \varepsilon > 0.$$

[The symbol $'$ *denotes differentiation.]*

Proof Split the range of the integral representing $\phi'_c(\theta)$ and $\phi'_s(\theta)$ at $1/\theta$, and note that $\int_{1/\theta}^{\infty} W(x) \cos\theta x\, dx \le 2\theta^{-1}W(1/\theta)$. Then, the formulae of (i) readily follow from $m(x) \sim c(x) \sim L(x)/2$ and $xW(x) = o(m(x))$ (cf. Section A.1.1).

Since $|\phi_s(t)| \ll \phi_c(t) \sim t\alpha(t) \sim t^2 m(1/t)$, by (i) one obtains

$$\left[\Re\frac{1}{1-\psi}\right]'(t) = \frac{[(\phi_s^2 - \phi_c^2)\phi'_c - 2\phi_c\phi_s\phi'_s\phi_c](t)}{[\phi_c^2 + \phi_s^2]^2(t)} \sim \frac{-\phi'_c(t)}{\phi_c^2(t)} \sim \frac{-2}{t^3 m(1/t)},$$

hence the first bound of (ii). The proof of the second one is similar. □

Proof (of Proposition 4.3.3 in the case $\alpha = 2$) Recall the increment $\bar{a}(x+1) - \bar{a}(x)$ is given by (4.52). Putting $\varepsilon = [(n+\frac{1}{2})\pi]^{-1}$ with a (large) integer n, we infer that its last integral restricted to $t \le 1/\varepsilon x$ is written as

$$\int_0^{1/\varepsilon x} \frac{1+o(1)}{m(1/t)} \left[\frac{\sin xt}{t} + O(1)\right] dt = \frac{\pi/2}{m(x)}\{1 + O(\varepsilon) + o(1)\}$$

(see the derivation of (4.28); also note that $\int_{\varepsilon/x}^{1/\varepsilon x} t^{-1} \sin xt\, dt = \frac{1}{2}\pi + O(\varepsilon)$), while by the first formula of Lemma 4.3.4(ii)

$$\int_{1/\varepsilon x}^{\pi} \Re\frac{t\sin xt}{1-\psi(t)} dt = \frac{1}{x}\int_{1/\varepsilon x}^{\pi} \left[\Re\frac{t}{1-\psi(t)}\right]' \cos xt\, dt + O(1/x)$$

$$= O\left(\frac{\varepsilon}{m(\varepsilon x)}\right).$$

From these estimates we can easily conclude that $\bar{a}(x+1) - \bar{a}(x) \sim \bar{a}(x)/x$.

We still need to identify the asymptotic form of increments of $b_\pm(x)$. From the second formula of Lemma 4.3.4(ii) we deduce, as above, that the integral representing $b_\pm(x+1) - b_\pm(x)$ over $\varepsilon/x \le t \le \pi$ is $o(1/m(x))$. Since $\tilde{m}/m \to 0$, we can replace $O(1/m(x))$ by $o(1/m(x))$ on the RHS of (4.54). Hence, recalling (4.26) (the integral representation of $b_+(x)$) we obtain

$$b_+(x+1) - b_+(x) = \frac{b_+(x)}{x} + o\left(\frac{1}{m(x)}\right),$$

and similarly for $b_-(x)$. Now the result follows immediately from $a(\pm x) = \bar{a}(x) \pm [b_-(x) - b_+(x)]$. □

If (4.14) is valid with $\alpha = 1$, one may expect to have $|a(x+1) - a(x)| = O(\bar{a}(x)/|x|)$ (or rather $= o(\bar{a}(x)/x)$) as $|x| \to \infty$, in Propositions 4.3.1 and 4.3.3. However, without some nice regularity condition of μ, it seems hard to show such a result. The difficulty arises mainly from that of getting a proper evaluation of $\int_{1/t}^{\infty} y\mu_\pm(y)e^{ity}\, dy$. Still, we can get one under a mild additional condition on $\mu_\pm$, as given below.

Put

$$L_\pm(x) = x\mu_\pm(x) \text{ and } L(x) = x\mu(x)$$

and bring in the condition: as $x \to \infty$

(∗) $\int_0^x L(y)\,dy = O(xL(x))$ and $\sum_{n>x} |L_\pm(n) - L_\pm(n-1)| = O(L(x))$.[4]

Proposition 4.3.5 *Suppose* (∗) *holds. Then* $a(x+1) - a(x) = O\left(\bar{a}(x)/|x|\right)$ *as* $|x| \to \infty$, *under each of the conditions*

(a) $EX = 0$ *and (H) holds;*
(b) F *is r.s. (F may be transient);*
(c) $x\mu(x)$ *is s.v. and* $|A(|x|)| = O(x\mu(x))$ $(x \to \infty)$.

[In case (c), F is recurrent under (∗) *(by the well-known recurrence criterion) and* $P[S_n > 0]$ *is bounded away from 0 and 1 if* $\mu_-/\mu_+ \to 1$ *(cf. [83, Section 4.3]).]*

Proof Under (∗) we have

$$\left| \int_x^\infty L_\pm(y)e^{ity}\,dy \right| \le \frac{C}{t}L(x) \quad \text{and} \quad \left| \int_0^x L_\pm(y)e^{ity}\,dy \right| \le \frac{C}{t}L(1/t) \quad (4.55)$$

for $1/x < t < \pi$, $x > 1$.[5] Before proving (4.55), we verify the assertion of the proposition by taking it for granted. In case (a), one has only to follow the proof of Proposition 4.3.1. For by the first bound of (4.55) the inner integral on the RHS of (3.29) is bounded above by a constant multiple of $\tau^{-1}L(1/s) \le c(1/s)$ $(\tau > s)$ so that the second bound of Lemma 3.2.11 can be replaced by const.$(t - s)c(1/s)$. For cases (b) and (c) we bring in the functions

$$\alpha_x(t) = \int_0^x \mu(y)\sin ty\,dy, \qquad \gamma_x(t) = \int_0^x [\mu_-(y) - \mu_+(y)]\cos ty\,dy.$$

Clearly $|\alpha(t) - \alpha_x(t)| \vee |\gamma(t) - \gamma_x(t)| \le 2\mu(x)/t$. By the second bound of (4.55)

$$|\alpha_x'(t)|, |\gamma_x'(t)| \le O\left(L(1/t)/t\right). \quad (4.56)$$

Let (b) hold. Then, $f^\circ(t) \sim |A(1/t)|^{-2}$ (see (P2) of Section 4.1) and, noting that by (∗) $L(x) \le O(L(1/t))$ for $t > 1/x$, we see

$$\int_{1/x}^\pi [\alpha(t) - \alpha_x(t)]\, f^\circ(t)\,dt \le \frac{C}{x} \int_{1/x}^{1/x_0} \frac{L(1/t)\,dt}{tA^2(1/t)} \sim \frac{2C\bar{a}(x)}{x}.$$

By the bound of α_x' in (4.56) it follows that for $t > 1/x$ and for a positive constant b both small enough

---

[4] The second condition holds if $n\mu_\pm(n)$ $(n = 1, 2, 3, \ldots)$ are ultimately non-increasing, while it implies that L is almost decreasing. To avoid any misunderstanding, we remark that none of $L_\pm$ or L are assumed to be r.s.

[5] Alternatively derive the corresponding bounds of $\sum_{n>x} L_\pm(n)e^{itn}$, $\sum_{0 \le n \le x} L_\pm(n)e^{itn}$ (much easier than (4.55)) and use the expression for $1 - \psi$ in the footnote on page 20.

$$\int_{1/x}^{b} \left[|\alpha'_x(t)| \vee |\gamma'_x(t)| \right] f^\circ(t)\, dt \le C' \int_{1/x}^{b} \frac{L(1/t)}{tA^2(1/t)}\, dt \sim 2C'\bar{a}(x),$$

showing that the contribution from $1/x < t \le \pi$ to the second integral in (4.52) is $O(\bar{a}(x)/x)$. The contribution from $0 \le t \le 1/x$ is at most

$$2 \int_0^{1/x} \alpha(t) f^\circ(t)\, dt = \int_0^{1/x} o\left(\frac{1}{|A(1/t)|}\right) dt = o\left(\frac{1}{xA(x)}\right) = o\left(\frac{\bar{a}(x)}{x}\right).$$

Thus $\bar{a}(x+1) - \bar{a}(x) = O(\bar{a}(x)/x)$. In the same way we see that the increment of $b_-(x) - b_+(x)$ is dominated, in absolute value, by a constant multiple of $\bar{a}(x)/x$.

If (c) holds, then $\alpha(t) \sim \frac{1}{2}\pi L(1/t)$, $\gamma(t) = O(L(1/t))$ and $\bar{a}(x) \asymp \int_1^x [yL(y)]^{-1}\, dy$ (see the derivation of (4.34)), and as above, one can easily obtain

$$\int_{1/x}^{\pi} [\alpha(t) - \alpha_x(t)] f^\circ(t)\, dt \le \frac{C_1}{x} \int_{1/x}^{\pi} \frac{dt}{tL(1/t)} \sim C_1 \frac{[1/L]_*(x)}{x},$$

$$\int_{1/x}^{b} \left[|\alpha'_x(t)| \vee |\gamma'_x(t)| \right] f^\circ(t)\, dt \le C_1 \int_{1/x}^{b} \frac{1}{tL(1/t)}\, dt \sim C_1 [1/L]_*(x),$$

and $\int_0^{1/x} \alpha f^\circ dt \le C \int_x^\infty [y^2 L(y)]^{-1}\, dy = o(\bar{a}(x)/x)$, to conclude the result. $\quad\square$

Proof (of (4.55)) We have only to prove the formulae for L_+. Note that $\mu(y) = \mu(n)$ for $n \le y < n+1$. Then, writing $ye^{ity} = (we^{itw} + ne^{itw})e^{itn}$, $w = y - n$, one infers that for $x > 1/t$,

$$\int_{\lfloor 1/t \rfloor}^{x} y\mu_+(y)e^{ity}\, dy = \lambda_1(t)I(t,x) + \lambda_2(t)II(t,x) + O(L(1/t)),$$

where $\lambda_1(t) = \int_0^1 we^{itw}\, dw = \frac{1}{2} + O(t)$, $\lambda_2(t) = \int_0^1 e^{itw}\, dw = 1 + O(t)$,

$$I(t,x) = \sum_{n=\lfloor 1/t \rfloor}^{\lfloor x \rfloor} \mu_+(n)e^{itn} \quad \text{and} \quad II(t,x) = \sum_{n=\lfloor 1/t \rfloor}^{\lfloor x \rfloor} L_+(n)e^{itn}.$$

Summing by parts and using $(*)$ one obtains $II(t,x) = O(L(1/t)/t)$, while $I(t,x) = O(\mu(1/t)/t)$ by monotonicity of μ_+. One can therefore conclude that under $(*)$

$$\left| \int_0^x L_+(y)e^{ity}\, dy \right| \le \int_0^{\lfloor 1/t \rfloor} L_+(y)\, dy + \left| \int_{\lfloor 1/t \rfloor}^{x} L_+(y)e^{ity}\, dy \right| \le \frac{C'}{t}L(1/t),$$

showing the second bound of (4.55). The first one is shown similarly. $\quad\square$

Chapter 5
Applications Under $m_+/m \to 0$

This chapter is devoted to applications of the results of Chapter 3. We apply Theorems 3.1.1, 3.1.2 and 3.1.6 to obtain some asymptotic estimates of the upwards overshoot distribution of the r.w. over a high level, R say. We also estimate the probability of the r.w. escaping the origin, i.e., going beyond the level $R > 1$ or either of the levels R or $-R$ without visiting zero. It seems hard to obtain sharp results for these things in a general setting. Under the condition $\lim_{x \to \infty} m_+(x)/m(x) = 0$ (abbreviated as $m_+/m \to 0$, as before), however, our theorems about $a(x)$ are effectively used to yield natural results. The overshoot distribution is related to the relative stability of the ladder height variables. We shall obtain a sufficient condition for the relative stability of the ascending ladder height (Proposition 5.2.1) and the asymptotic estimates of the escape probabilities as mentioned above (Proposition 5.5.1 (one-sided) and Propositions 5.6.1 and 5.6.6 (two-sided)). As a byproduct of these results, we deduce that if $m_+/m \to 0$, then both $a(x)$ and $a(-x)$ are asymptotically increasing as $x \to \infty$ (see Corollary 5.1.6) as well as the following result concerning the classical two-sided exit problem, which has not been satisfactorily answered in the case $\sigma^2 = \infty$. Denote by σ_B the first entrance time into a set $B \subset \mathbb{Z}$ of the r.w. and by $V_d(x)$ the renewal function for the weakly descending ladder height process of the r.w. It turns out that if $m_+/m \to 0$, then the ratio $V_d(x)/a(x)$ is s.v. at infinity and uniformly for $1 \le x \le R$, as $R \to \infty$,

$$P_x[\sigma_{[R,\infty)} < \sigma_{(-\infty,-1]}] \sim V_d(x)/V_d(R) \qquad (5.1)$$

(Lemma 5.3.1(i) and Proposition 5.3.2, respectively). For Lévy processes having no upward jumps, there is an identity for the corresponding probability (cf. [22, Section 9.4]), and (5.1) is an exact analogue of it for the r.w. In Chapter 6, the asymptotic equivalence (5.1) will be obtained under conditions other than $m_+/m \to 0$, and related matters will be addressed.

Recall that if the r.w. is *left-continuous* (i.e., $P[X \le -2] = 0$), then $a(x) = x/\sigma^2$ for $x > 0$; analogously $a(x) = -x/\sigma^2$ for $x < 0$ for *right-continuous* r.w.'s (i.e., r.w.'s satisfying $P[X > 2 = 0]$) and that there exists $\lim_{x \to \infty} a(x) \le \infty$ and $\inf_{x>0} a(x) > 0$ except for the left-continuous r.w.'s (cf. (2.12), (2.13), (2.15)). Most of the results

K. Uchiyama, *Potential Functions of Random Walks in $\mathbb{Z}$ with Infinite Variance*,
Lecture Notes in Mathematics 2338, https://doi.org/10.1007/978-3-031-41020-8_5

obtained in this chapter, except for those in the last two sections, are thought of as examining to what extent the results which are simple or well-established for the right-continuous r.w.'s can be generalised to those satisfying $m_+/m \to 0$. In later chapters, we shall discuss the subjects treated in this chapter in some other settings.

In [71], the Green function of the r.w. killed on hitting zero is defined by

$$g(x, y) = g_{\{0\}}(x, y) - \delta_{0,x},$$

so that $g(0, \cdot) = g(\cdot, 0) = 0$. Recall $a^\dagger(x) = a(x) + \delta_{0,x}$ and $\bar{a}(x) = \frac{1}{2}[a(x) + a(-x)]$. The function $a(x)$ bears relevance to $g_{\{0\}}$ through the identity (2.3), or, what is the same,

$$g_{\{0\}}(x, y) = a^\dagger(x) + a(-y) - a(x - y). \tag{5.2}$$

Since $g_{\{0\}}(x, y) = P_x[\sigma_y < \sigma_0]g(y, y)$ if $x \neq y \neq 0$, from (5.2) one infers that $P_x[\sigma_x > \sigma_0] = P_0[\sigma_x < \sigma_0] = 1/g(x, x)$ $(x \neq 0)$, and

$$P_x[\sigma_y < \sigma_0] = \frac{a^\dagger(x) + a(-y) - a^\dagger(x - y)}{2\bar{a}(y)} \qquad (y \neq 0). \tag{5.3}$$

The identities (5.2) and (5.3) will be fundamental in this chapter and will occasionally be used in later chapters.

Throughout this chapter, we shall tacitly assume $EX = 0$ (resp. S is recurrent) when the function $m(x)$ (resp. $a(x)$) is involved.

5.1 Some Asymptotic Estimates of $P_x[\sigma_R < \sigma_0]$

The potential function satisfies the functional equation

$$\sum_{y=-\infty}^{\infty} p(y - x)a(y) = a^\dagger(x) \tag{5.4}$$

(cf. [71, p.352]), which restricted to $x \neq 0$ states that a is harmonic there, so that the process $M_n := a(S_{\sigma_0 \wedge n})$ is a martingale (under P_x) for each $x \neq 0$, and by the optional sampling theorem and Fatou's lemma

$$a(x) \geq E_x[a(S_{\sigma_0 \wedge \sigma_y})] = a(y)P_x[\sigma_y < \sigma_0] \qquad (x \neq 0). \tag{5.5}$$

Lemma 5.1.1 *For all $x, y \in \mathbb{Z}$,*

$$-\frac{a(y)}{a(-y)}a(x) \leq a(x + y) - a(y) \leq a(x) \quad \text{if } a(-y) \neq 0. \tag{5.6}$$

This is Lemma 3.2 of [84]. The right-hand inequality of (5.6), stating the sub-additivity of $a(\cdot)$ that is given in [71], is the same as $g(x, -y) \geq 0$. The left-hand inequality, which seems much less familiar, will play a significant role in the sequel.

It follows readily from (5.3) and (5.5). Indeed, with variables suitably chosen, these relations together yield

$$\frac{a(x) + a(y) - a(x + y)}{a(y) + a(-y)} \le \frac{a(x)}{a(-y)} \quad (a(-y) \ne 0), \tag{5.7}$$

which, after rearrangement, becomes the left-hand inequality of (5.6).

Remark 5.1.2 (a) The left-hand inequality of (5.6) may yield useful upper as well as lower bounds of the middle term. Here we write down such bounds in a form to be used later:

$$-\frac{a(x - R)}{a(R - x)} a(-x) \le g_{\{0\}}(x, R) - a^\dagger(x)$$
$$= a(-R) - a(x - R) \le \frac{a(-R)a(x)}{a(R)}, \tag{5.8}$$

provided $a(R)a(R - x) \ne 0$. Both inequalities are deduced from the left-hand inequality of (5.6): the lower bound follows by replacing y and x with $x - R$ and $-x$, respectively, and the upper bound by replacing y with $-R$. For $x \ne R \ (\ge 1)$, since $P_x[\sigma_R < \sigma_0] = g_{\{0\}}(x, R)/2\bar{a}(R)$,

$$-\frac{a(x - R)}{a(R - x)} \cdot \frac{a(-x)}{2\bar{a}(R)} \le P_x[\sigma_R < \sigma_0] - \frac{a^\dagger(x)}{2\bar{a}(R)} \le \frac{a(-R)a(x)}{2a(R)\bar{a}(R)}. \tag{5.9}$$

(5.8) is quite efficient in the case $a(-R)/a(R) \to 0$ and will be used later.

(b) By (5.3) and the subadditivity of a,

$$P_x[\sigma_0 < \sigma_y] = \frac{a(x - y) + a(y) - a^\dagger(x)}{2\bar{a}(y)} \le \frac{\bar{a}(y - x)}{\bar{a}(y)} \quad (y \ne 0, x).$$

Lemma 5.1.3 *Suppose* $\lim_{z \to \infty} a(-z)/\bar{a}(z) = 0$. *Then*

(i) *uniformly for* $x \le -R$, *as* $R \to \infty$,

$$\frac{a(x) - a(x + R)}{a(R)} \longrightarrow 0 \quad and \quad P_x[\sigma_{-R} < \sigma_0] \longrightarrow 1;$$

(ii) *uniformly for* $-M < x < R$ *with any fixed* $M > 0$, *as* $R \to \infty$,

$$\frac{a(-R) - a(x - R)}{a^\dagger(x)} \longrightarrow 0 \quad and \quad P_x[\sigma_R < \sigma_0] = \frac{a^\dagger(x)}{a(R)}\{1 + o(1)\}.$$

Proof Suppose $\lim_{z \to \infty} a(-z)/\bar{a}(z) = 0$. This excludes the possibility of the left-continuity of the r.w. so that $a^\dagger(x) > 0$ for all x and (ii) follows immediately from (5.8) and (5.9). (i) is deduced from (5.6) as above (substitute $x + R$ and $-R$ for x and y, respectively, for the lower bound; use sub-additivity of a for the upper bound). □

By virtue of Proposition 3.1.5 and Theorem 3.1.6, Lemma 5.1.3(ii) entails the following

Corollary 5.1.4 *Suppose $m_+/m \to 0$. Then $P_x[\sigma_R < \sigma_0] \to 1$ as $x/R \uparrow 1$.*

By Remark 5.1.2(b) we know that $P_{R-y}[\sigma_R < \sigma_0] \geq 1 - \bar{a}(y)/\bar{a}(R)$, which is a better estimate than the one given above in most cases but does not generally imply the consequence in Corollary 5.1.4 since $\bar{a}(y)/\bar{a}(R)$ possibly approaches unity even if $y/R \downarrow 0$ (cf. Lemma 5.3.1(i)). By the same token $P_x[\sigma_R < \sigma_0]$ may approach zero when x/R increases to 1 in a suitable way (under $m_-/m \to 0$).

Lemma 5.1.5 *Suppose that $P[X \geq 2] > 0$ and*

$$\delta := \limsup_{y \to \infty} a(-y)/\bar{a}(y) < 1.$$

Then

$$\lim_{x \to \infty} \inf_{y \geq x} \frac{a(-y)}{a(-x)} \geq 1 - \delta.$$

In particular, if $\delta = 0$, then $a(-x)$ is asymptotically increasing with x in the sense that there exists an increasing function $f(x)$ such that $a(-x) = f(x)\{1 + o(1)\}$ $(x \to \infty)$.

Proof For any $\delta' \in (\delta, 1)$ choose $N > 0$ such that $a(-z)/a(z) < \delta'$ for all $z \geq N$ and let $N \leq x \leq y - N$ and $z = y - x$. If $a(-x) \leq a(-z)$, then on using (5.6)

$$
\begin{aligned}
a(-y) - a(-z) &= a(-x - z) - a(-z) \\
&\geq -\frac{a(-z)}{a(z)}a(-x) \geq -\frac{a(-z)}{a(z)}a(-z),
\end{aligned}
\tag{5.10}
$$

which by a simple rearrangement leads to

$$a(-y) \geq \left[1 - \frac{a(-z)}{a(z)}\right]a(-z) \geq (1 - \delta')a(-z) \geq (1 - \delta')a(-x).$$

If $a(-x) \geq a(-z)$, then interchanging the roles of x and z in (5.10), we have $a(-y) \geq (1 - \delta')a(-x)$ for $x > N$. Since $\lim_{x \to \infty} a(-y)/a(-x) = 1$ uniformly for $x \leq y < x + N$ and $\delta' - \delta$ can be made arbitrarily small, we conclude the first inequality of the lemma. $\square$

By the subadditivity of $a(-x)$ the inequality of Lemma 5.1.5 entails that for some positive constant C

$$C^{-1}a(-x) \leq a(-\lambda x) \leq (\lambda + C)a(-x) \quad \text{for } x > 0, \lambda \geq 1. \tag{5.11}$$

By Lemma 5.1.3 it follows that if $a(-x)/a(x) \to 0$, then $P_x[\sigma_{(-\infty, -R]} < \sigma_0] \sim P_x[\sigma_{-R} < \sigma_0]$, in particular

$$P_0[\sigma_{(-\infty, -R]} < \sigma_0] \sim 1/a(R). \tag{5.12}$$

This together with Lemma 5.1.5 yields

Corollary 5.1.6 *If $m_+/m \to 0$, then both $a(x)$ and $a(-x)$ are asymptotically increasing.*

If $a(-x), x > 0$, is asymptotically increasing, then by the subadditivity of $a(x)$ it follows that uniformly for $y \geq 0$, as $x \to \infty$,

$$o\left(E_x[a(S_{\sigma(-\infty,0]})]\right) + a(-y)) \leq E_x[a(S_{\sigma(-\infty,0]} - y)] - E_x[a(S_{\sigma(-\infty,0]})] \leq a(-y).$$

Hence, by (2.15), $E_x[a(S_{\sigma(-\infty,0]} - y)] - E_x[a(S_{\sigma(-\infty,0]})] = o(a(x)) + O(a(-y))$. Noting that $g_{(-\infty,0]}(x,y) = g(x,y) - E_x[g(S_{\sigma(-\infty,0]}, y)]$ and $a(\cdot)$ is subadditive, we have the second corollary of Lemma 5.1.5

Corollary 5.1.7 *If $a(-x)/\bar{a}(x) \to 0$ $(x \to \infty)$, then as $x \to \infty$, uniformly for $0 \leq y < Mx$ (for each $M > 1$), $g_{(-\infty,0]}(x,y) = a(x)\{1 + o(1)\} - a(x-y)$, in particular $g_{(-\infty,0]}(x,x) \sim a(x)$.*

Remark 5.1.8 If $\sigma^2 < \infty$ (with $EX = 0$), then uniformly for $x > 0$,

$$P_x[\sigma_R < \sigma_0] \sim \left\{(2R)^{-1}\left[\sigma^2 a^\dagger(x) + x\right]\right\} \wedge 1,$$

$$P_{-x}[\sigma_R < \sigma_0] \sim \begin{cases} (2R)^{-1}\left[\sigma^2 a^\dagger(-x) - x\right] & \text{if } E[X^3; X > 0] < \infty, \\ R^{-1}\sum_{y=0}^{R} P_{-x}[S_{\sigma[0,\infty)} > y] & \text{otherwise} \end{cases}$$

as $R \wedge x \to \infty$. The first equivalence (valid also for $x = 0$) is immediate from (2.10) and (5.3). For the second one, on noting $a(z) - a(z-R) \sim R/\sigma^2$ for $z > R$ $(R \to \infty)$, observe that for $x > 0$, $g(-x, R) = E_{-x}\left[g(S_{\sigma[0,\infty)}, R)\right]$ can be written as

$$\sum_{y=0}^{R} H_{[0,\infty)}^{-x}(y)\left[a(y) + \frac{y}{\sigma^2} + o(y)\right] + P_{-x}[S_{\sigma[0,\infty)} > R]\frac{2R + o(R)}{\sigma^2};$$

then use the fact that $\sum_{y=0}^{\infty} H_{[0,\infty)}^{-x}(y)[a(y) + y/\sigma^2] = a(-x) - x/\sigma^2$ $(x > 0)$ and $\sup_x E_{-x}|S_{\sigma[0,\infty)}| < \infty \Leftrightarrow E[X^3; X > 0] < \infty$ [76, Eq(2.9), Corollary 2.1] for the first case; perform summation by parts for the second case.

5.2 Relative Stability of Z and Overshoots

Here we are concerned with a sufficient condition for Z to be r.s. As stated in Section 2.4.2, this is related to the overshoot $Z(R) = S_{\sigma(R,\infty)} - R$. By (2.25) and what is stated at the end of Section 2.4.2 we have

> Z is r.s. if and only if $Z(R)/R \longrightarrow_P 0$; and for this
> to be the case it is sufficient that X is p.r.s. $\qquad(5.13)$

By combining Theorem 3.1.6 and a known criterion for the positive relative stability of X (see Section 2.4.1), we obtain a reasonably fine sufficient condition for Z to be r.s. For condition (C.I) in the following result, we neither assume $E|X| < \infty$ nor the recurrence of F. Recall $A(x) = \int_0^x \{\mu_+(y) - \mu_-(y)\}\,dy$ [as defined in (2.19)].

Proposition 5.2.1 *For Z to be r.s. each of the following conditions is sufficient.*

(C.I) $\mu(x) > 0$ *for all* $x \geq 0$ *and* $\lim_{x\to\infty} A(x)/[x\mu(x)] = \infty$,

(C.II) $EX = 0$ *and* $\lim_{x\to\infty} \dfrac{x\eta_+(x)}{m(x)} = 0.$

Proof As mentioned in Section 2.4, condition (C.I) is equivalent to the positive relative stability of X, and hence it is a sufficient condition for the relative stability of Z in view of (5.13). As for (C.II), expressing $P[Z(R) > \varepsilon R]$ as the infinite series

$$\sum_{w\geq 0} g_{(R,\infty)}(0, R - w)P[X > \varepsilon R + w],$$

one observes first that $g_{[R,\infty)}(0, R - w) < g_{\{0\}}(-R, -w) \leq g_{\{0\}}(-R, -R) = 2\bar{a}(R)$. Note that (C.II) entails (3.8) and accordingly (H) holds so that $\bar{a}(x) \asymp x/m(x)$ by Theorems 3.1.1 and 3.1.2. Hence for any $\varepsilon > 0$

$$P_0[Z(R) > \varepsilon R] \leq 2\bar{a}(R) \sum_{w\geq 0} P[X > \varepsilon R + w] \asymp \frac{R\eta_+(\varepsilon R)}{m(R)} \leq \frac{R\eta_+(\varepsilon R)}{m(\varepsilon R)}.$$

Thus (C.II) implies $Z(R)/R \to_P 0$, concluding the proof in view of (5.13) again. $\qquad\square$

Remark 5.2.2 (a) Condition (C.I) is stronger than (2.23), namely $\lim A(x)/x\mu_-(x) = \infty$, which is necessary and sufficient in order that $P[S_n > 0] \to 1$ according to [46], so that (C.I) is only of relevance in such a case. It is also pointed out that if S is recurrent and $\limsup a(-x)/a(x) < 1$, then the sufficiency of (C.I) for Z to be r.s. follows from the trivial bound $g_{-\Omega}(-R, x) \leq 2\bar{a}(R)$ (which entails $P_0[Z(R) > \varepsilon R] \leq 2R\bar{a}(R)\mu(\varepsilon R)$), since $a(x) - a(-x) \sim 1/A(x)$ and A is s.v. under (C.I) (cf. Theorem 4.1.1).

(b) Condition (C.II) is satisfied if $m_+/m \to 0$. The converse is of course not true – (C.II) may be fulfilled even if $m_+/m \to 1$ – and it seems hard to find any simpler substitute for (C.II). Under the restriction $m_+(x) \asymp m(x)$, however, (C.II) holds if and only if m_+ is s.v. (which is the case if $x^2\mu_+(x) \asymp L(x)$ with an s.v. L).

(c) If F belongs to the domain of attraction of a stable law and Spitzer's condition[1] holds, then that either (C.I) or (C.II) holds is also necessary for Z to be r.s., as will be discussed at the end of this section.

---

[1] The condition that there exists $\rho = \lim n^{-1} \sum_{k=1}^n P_0[S_k > 0]$ is called Spitzer's condition. It is equivalent to $\rho = \lim P[S_n > 0]$, according to [6].

(d) If $\mu_+(x)$ is positive for all $x > 0$ and of dominated variation, then

$$(*) \qquad Z \text{ is r.s.} \iff \lim U_a(R)E[V_d(X); X \geq R] = 0.$$

For the proof, splitting the sum $\sum_{y=0}^{\infty} g_{-\Omega}(-R, -y)\mu_+(y + \varepsilon R)$ at $y = 2R$ and using the dominated variation of μ_+ (with the help of (6.16)), one infers that for each $\varepsilon > 0$

$$P_0[Z(R) > \varepsilon R] \asymp U_a(R)V_d(R)\mu_+(R) + U_a(R) \sum_{y=R}^{\infty} v_d(y)\mu_+(y),$$

of which the RHS equals $U_a(R) \sum_{y=R}^{\infty} V_d(y)p(y)\{1 + o(1)\}$. Hence, $(*)$ follows.

(e) The situation is simplified if we are concerned with an (oscillatory) Lévy process $Y(t)$. According to [24] (cf. also [22]) $Z_R^Y/R \to 0$ in probability if and only if the law of Z_R^Y converges to a proper probability law, or, what amounts to the same, the ascending ladder height has a finite mean, which imposes $\int_1^{\infty} \mu_+^Y(x)x\,dx/m^Y(x) < \infty$ (see (d)), a more restrictive condition than that expected by considering the r.w. $(Y(n))_{n=0}^{\infty}$, a remarkable distinction from general r.w.'s. Here $m^Y(x)$ and $\mu_+^Y(x)$ stand for $\int_0^x dy \int_y^{\infty} P[|Y(1)| > t]\,dt < \infty$ and $P[Y(1) > x]$, respectively.

The following result, used in the next section, concerns an overshoot estimate for the r.w. S conditioned to avoid the origin. Put

$$\bar{a}^\dagger(x) := \frac{1}{2}\left[a^\dagger(x) + a(-x)\right].$$

Lemma 5.2.3 (i) *If $EX = 0$, then for $z \geq 0$ and $x \in \mathbb{Z}$,*

$$P_x\left[Z(R) > z, \sigma_{[R,\infty)} < \sigma_0\right] \leq 2\bar{a}^\dagger(x)\eta_+(z). \qquad (5.14)$$

(ii) *If $m_+/m \to 0$, then for each $\varepsilon > 0$, uniformly for $0 \leq x \leq R$,*

$$P_x\left[Z(R) \geq \varepsilon R \,\middle|\, \sigma_{[R,\infty)} < \sigma_0\right] \to 0 \quad (R \to \infty). \qquad (5.15)$$

[See Lemma 5.2.4, (2.28), Proposition 7.6.4 for (5.15) with σ_0 replaced by $T = \sigma_{(-\infty,0)}$ in the conditioning.]

By the trivial inequality $P_x[\sigma_{[R,\infty)} < \sigma_0] \geq P_x[\sigma_R < \sigma_0]$, (5.14) implies

$$P_x\left[Z(R) > z \,\middle|\, \sigma_{[R,\infty)} < \sigma_0\right] \leq 2\bar{a}^\dagger(x)\eta_+(z)/P_x[\sigma_R < \sigma_0]. \qquad (5.16)$$

Proof Put

$$r(z) = P_x\left[Z(R) > z, \sigma_{[R,\infty)} < \sigma_0\right].$$

Plainly $g_{\{0\}\cup[R,\infty)}(x, z) \leq g_{\{0\}}(x, z) \leq 2\bar{a}^\dagger(x)$. Hence

$$r(z) = \sum_{w>0} g_{\{0\}\cup[R,\infty)}(x, R - w)P[X > z + w] \leq 2\bar{a}^\dagger(x) \sum_{w>0} P[X > z + w].$$

Since $\sum_{w>0} P[X > z + w] = \eta_+(z)$ it therefore follows that $r(z) \le 2\bar{a}^\dagger(x)\eta_+(z)$.

Suppose $m_+/m \to 0$. Then $\bar{a}(x) \asymp x/m(x)$ and $a(-x)/a(x) \to 0 \ (x \to \infty)$ owing to Theorem 3.1.2 and Theorem 3.1.6, respectively. We can accordingly apply Lemma 5.1.3 to see that $P_x[\sigma_R < \sigma_0] = \bar{a}^\dagger(x)/\bar{a}(R)\{1 + o(1)\}$ uniformly for $0 \le x < R$. Hence, by (5.16), $P_x\left[Z(R) > z \,\middle|\, \sigma_{[R,\infty)} < \sigma_0\right] \le 2\bar{a}(R)\eta^+(z)\{1+o(1)\}$ on the one hand. On the other hand for $z > 0$, recalling $\eta_+(z) < m_+(z)/z$ we deduce that

$$\bar{a}(R)\eta_+(z) \le C\frac{m_+(z)R}{m(R)z}, \qquad (5.17)$$

of which the RHS with $z = \varepsilon R$ tends to zero. Thus (5.15) follows.   □

The following result – used only for the proof of Proposition 8.6.4(i) – is placed here, although, in its proof, we use some results from the next section and chapter.

Lemma 5.2.4 *Let* $\Lambda_R = \{\sigma_{[R+1,\infty)} < \sigma_\Omega\}$. *Each of the following*

(a) $m_+/m \to 0$;

(b) *S is p.r.s.;*

(c) *S is attracted to the standard normal law,*

implies that for each $\varepsilon > 0$,

$$P_x[Z(R) > \varepsilon R \,|\, \Lambda_R] \to 0 \qquad \text{uniformly for } 0 \le x \le R.$$

Proof If U_a is regularly varying, then by Lemma 6.2.1 there exists a positive constant c_0 such that for $0 \le x \le y$,

$$c_0 V_d(x/2)U_a(y) \le \sum_{z=0}^{y} g_\Omega(x, z) \le V_d(x)U_a(y). \qquad (5.18)$$

Under either of (a) to (c), $U_a(x) \sim x/\ell^*(x))$ in view of Proposition 5.2.1 and $P_x(\Lambda_R) \sim V_d(x)/V_d(R)$ according to Proposition 5.3.2 (for (a)) and Theorem 6.1.1 (for (b), (c)) (a p.r.s. S satisfies (C3) as noted in Remark 6.1.2); see also Remark 6.3.7. In particular (5.18) is applicable, and it easily follows that

$$P_x[Z(R) > \varepsilon R, \Lambda_R] \le \sum_{w=0}^{R} g_\Omega(x, R - w)\mu_+(w + \varepsilon R) \le V_d(x)U_a(R)\mu_+(\varepsilon R).$$

If (b) or (c) holds, then Lemmas 6.3.4 and 6.5.1(i) imply $V_d(R)U_a(R)\mu(\varepsilon R) \to 0$ and the result follows. By the left-hand inequality of (5.18) we see that

$$V_d(R/2)U_a(R) < 2R\bar{a}(R/2)/c_0.$$

If $m_+/m \to 0$, then $\bar{a}(x) \le C_1 x/m(x)$, and noting $x^2\mu_+(x) < 2c_+(x)$ we see that

$$P_x[Z(R) > \varepsilon R \,|\, \Lambda_R] \le C_2 R\bar{a}(R)\mu_+(\varepsilon R) \le C\varepsilon^{-2}c_+(\varepsilon R)/m(R) \to 0.$$

Thus the lemma is verified.   □

We shall need an estimate of overshoots as the r.w. exits from the half-line $(-\infty, -R]$ after having entering into it. Put

$$\tau(R) = \inf \left\{ n > \sigma_{(-\infty,-R]} : S_n \notin (-\infty, -R] \right\}, \qquad (5.19)$$

the first time when the r.w. exits from $(-\infty, -R]$ after entering it.

Lemma 5.2.5 *Suppose $m_+/m \to 0$. Then for each constant $\varepsilon > 0$, uniformly for $x > -R$ satisfying $P_x[\sigma_{(-\infty,-R]} < \sigma_0] \geq \varepsilon \bar{a}^\dagger(x)/\bar{a}(R)$, as $R \to \infty$,*

$$P_x \left[S_{\tau(R)} > -R + \varepsilon R \,\middle|\, \sigma_{(-\infty,-R]} < \sigma_0 \right] \to 0.$$

Proof Denoting by $\mathcal{E}_R$ the event $\{\sigma_{(-\infty,-R]} < \sigma_0\}$ we write down

$$P_x \left[S_{\tau(R)} > -R + \varepsilon R \,\middle|\, \mathcal{E}_R \right] \qquad (5.20)$$
$$= \sum_{w \leq -R} P_x \left[S_{\sigma(-\infty,-R]} = w \,\middle|\, \mathcal{E}_R \right] P_w \left[S_{\sigma(-R,\infty)} > -R + \varepsilon R \right].$$

If $P_x[\mathcal{E}_R] \geq \varepsilon \bar{a}^\dagger(x)/\bar{a}(R)$, by Lemma 5.2.3(i) (applied to $-S$.),

$$P_x \left[S_{\sigma(-\infty,-R]} < -R - z \,\middle|\, \mathcal{E}_R \right] \leq 2\varepsilon^{-1} m(z)\bar{a}(R)/z.$$

Given $\delta > 0$ (small enough) we define $\zeta = \zeta(\delta, R) \ (> R)$ by the equation

$$\frac{m(\zeta)R}{m(R)\zeta} = \delta$$

(uniquely determined since $x/m(x)$ is increasing), so that by $2\bar{a}(R) < CR/m(R)$

$$P_x \left[S_{\sigma(-\infty,-R]} < -R - \zeta \,\middle|\, \mathcal{E}_R \right] \leq 2\varepsilon^{-1} m(\zeta)\bar{a}(R)/\zeta \leq (C\varepsilon^{-1})\delta.$$

For $-R - \zeta \leq w \leq -R$,

$$P_w \left[S_{\sigma(-R,\infty)} > -R + \varepsilon R \right] = \sum_{y > -R + \varepsilon R} \sum_{z \leq -R} g_{(-R,\infty)}(w, z)p(y - z)$$
$$= \sum_{y > \varepsilon R} \sum_{z \leq 0} g_{[1,\infty)}(w + R, z)p(y - z)$$
$$\leq C_1 \bar{a}(\zeta)\eta_+(\varepsilon R)$$
$$\leq \delta^{-1} C' R\eta_+(\varepsilon R)/m(R),$$

where the first inequality follows from

$$g_{[1,\infty)}(w + R, z) \leq g_{\{1\}}(w + R, z) \leq \bar{a}(w + R - 1)$$

and the second from $\bar{a}(\zeta) \leq C_2 \zeta/m(\zeta)$ and the definition of ζ. Now, returning to

(5.20) we apply the bounds derived above to see that

$$P_x\left[S_{\tau(R)} > -R + \varepsilon R \,\middle|\, \sigma_{(-\infty,-R]} < \sigma_0\right] \le (C\varepsilon^{-1})\delta + \delta^{-1}C'R\eta_+(\varepsilon R)/m(R).$$

Since $R\eta_+(\varepsilon R)/m(R) \to 0$ and $\delta > 0$ may be arbitrarily small, this concludes the proof.   □

If $\mathcal{E}$ is an event depending only on $\{S_n, n \ge \tau(R)\}$, then $P_x[\mathcal{E}, \sigma_0 < \sigma_{(-\infty,-R]}] = P_0(\mathcal{E})P_x[\sigma_0 < \sigma_{(-\infty,-R]}]$. Putting $x = 0$ in this identity and subtracting both sides from $P_0[\mathcal{E}]$ one obtains $P_0[\mathcal{E} \,|\, \sigma_0 > \sigma_{(-\infty,-R]}] = P_0(\mathcal{E})$; in particular,

$$P_0\left[S_{\tau(R)} = x\right] = P_0\left[S_{\tau(R)} = x \,\middle|\, \sigma_{(-\infty,-R]} < \sigma_0\right]. \tag{5.21}$$

Since

$$P_0[\sigma(-\infty,-R] < \sigma_0] > P_0[\sigma_{-R} < \sigma_0] = \frac{1}{2}a^{\dagger}(0)/\bar{a}(R),$$

Lemma 5.2.5 yields the following

Corollary 5.2.6 *Suppose $m_+/m \to 0$. Then for any $\varepsilon > 0$, as $R \to \infty$,*

$$P_0[S_{\tau(R)} > -R + \varepsilon R] \to 0.$$

Overshoots for F in the domain of attraction

We suppose that (4.14) holds and examine the behaviour of the overshoot $Z(R)$. We neither assume the condition $E|X| < \infty$ nor the recurrence of the r.w., but assume that if $E|X| = \infty$, then $\sigma_{[1,\infty)} < \infty$ a.s.(P_0),[2] which, according to Erickson [29, Corollary 2]), is equivalent to

$$\int_0^{\infty} \frac{x\,\mathrm{d}F(x)}{1 + \int_0^x \mu_-(t)\,\mathrm{d}t} = \infty. \tag{5.22}$$

We shall observe that under (4.14), the sufficient condition of Proposition 5.2.1 is also necessary for Z to be r.s. Put $\rho = P_0^Y[Y > 0]$, which, if $\alpha \ne 1$, $P_0[S_n > 0]$ approaches as $n \to \infty$. Then we have the following:

(i) *If either $1 < \alpha < 2$ and $p = 0$ or $\alpha = 2$, then $x\eta_+(x) = o(m(x))$, so that $Z(R)/R \to_P 0$ as $R \to \infty$ according to Proposition 5.2.1.* (In this case, we have $\alpha\rho = 1$, and the same result also follows from Theorems A.2.3 and A.3.2.)

(ii) *Let $\alpha \in (0,1)\cup(1,2)$ and $p > 0$. Then $0 < \alpha\rho < 1$, which implies that $P[Z > x]$ is regularly varying of index $\alpha\rho$ and the distribution of $Z(R)/R$ converges weakly to the probability law determined by the density $C_{\alpha\rho}/x^{\alpha\rho}(1+x)$, $x > 0$ (see Theorems A.2.3 and A.3.2, [31, Theorem XIV.3]), so that Z is not r.s.*

---

[2] This condition, becoming relevant only in (iv) below, is automatic under our assumption of oscillation, which is not needed in (iii).

(iii) *Let $\alpha = 1$. If $p = 1/2$, we also suppose that Spitzer's condition is satisfied, or what amounts to the same thing, there exists*

$$\lim P_0[S_n > 0] = r. \tag{5.23}$$

For $p \neq 1/2$, (5.23) holds with $r \in \{0, 1\}$: $r = 1$ if either $p < 1/2$ and $EX = 0$ or $p > 1/2$ and $E|X| = \infty$; $r = 0$ in the other case of $p \neq 1/2$.[3] It follows that $r = 1$ if and only if (C.I) of Proposition 5.2.1 holds, and if this is the case, Z is r.s., so that $Z(R)/R \to_P 0$. If $p = 1/2$, r may take all values from $[0, 1]$ and if $0 < r < 1$, then the same convergence of $Z(R)/R$ as mentioned in (ii) holds. These together entail that Z is r.s. if and only if $r = 1$ (if $\alpha = 1$).

(iv) *Let $\alpha \leq 1$ and suppose that $r = 0$ if $\alpha = 1$ and $p = 0$ if $\alpha < 1$. Then, according to [81] (see the table in Section 2.6),[4] $P[Z > x]$ is s.v. at infinity, which is equivalent to $Z(R)/R \to_P \infty$.*

Thus, if $\alpha > 1$, condition (C.II) (which then becomes equivalent to $\alpha\rho = 1$) works as a criterion of whether Z is r.s., while if $\alpha = 1$, Z can be r.s. under $x\eta_+(x) \asymp m(x)$ (so that (C.II) does not hold) and condition (C.I) must be employed for the criterion. Since Z is not r.s. if either $r < 1 = \alpha$ or $\alpha\rho < 1 \neq \alpha$, the validity of (C.I) or (C.II) is necessary and sufficient for Z to be r.s.

5.3 The Two-Sided Exit Problem Under $m_+/m \to 0$

Put

$$k := \sup_{x \geq 1} \frac{a(-x)}{a(x)},$$

with the understanding that $k = \infty$ if the r.w. is left-continuous. The following bounds that play a crucial role in this section are taken from [84, Lemma 3.4]:

$$0 \leq g_{\{0\}}(x, y) - g_{(-\infty, 0]}(x, y) \leq (1 + k)a(-y) \qquad (x, y \in \mathbb{Z}). \tag{5.24}$$

For $0 \leq x \leq y$ we shall apply (5.24) in the slightly weaker form

$$-a(-x) \leq a^\dagger(x) - g_{(-\infty, 0]}(x, y) \leq (1 + k)a(-y) + ka(-x), \tag{5.25}$$

where the subadditivity $a(-y) - a(x - y) \leq a(-x)$ is used for the lower bound and the inequalities $a(x - y) - a(-y) \leq [a(x - y)/a(y - x)]a(-x) \leq ka(-x)$ for the upper bound.

---

[3] This is easily deduced by examining (C.I) (entailing $r = 1$; see the comment given after (2.21)).

[4] See also Remark 6.1.2 ($\alpha = 1$) and Lemma 6.3.1 ($\alpha < 1$) of Chapter 6 for another argument and related results.

Recall $u_a(0) = 1$, $u_a(x) = U_a(x) - U_a(x-1)$ $(x \geq 1)$, $v^\circ = v_d(0) = V_d(0)$ and

$$\ell^*(x) = \int_0^x P[Z > t]\, dt, \quad \hat{\ell}^*(x) = \frac{1}{v^\circ} \int_0^x P[-\hat{Z} > t]\, dt. \qquad (5.26)$$

In the rest of this section we assume

$$Z \text{ is r.s.} \quad \text{and} \quad \frac{a(-x)}{\bar{a}(x)} \to 0 \quad (x \to \infty). \qquad (5.27)$$

By Theorem 3.1.6 and Proposition 5.2.1 this condition holds if $m_+/m \to 0$. The first half of (5.27) is equivalent to the slow variation of ℓ^* [5] ([31, Theorem VII.7.3], Theorem A.2.3), and if it is the case, it follows (Section A.2.1) that

$$u_a(x) \sim 1/\ell^*(x). \qquad (5.28)$$

Lemma 5.3.1 *Suppose (5.27) holds. Then ℓ^* is s.v., and the following hold*

(i) $a(x) \sim V_d(x)/\ell^*(x)$ *as* $x \to \infty$;[6]
(ii) *uniformly for* $1 \leq x \leq y$, *as* $y \to \infty$,

$$g_{(-\infty,0]}(x,y) \sim V_d(x-1)/\ell^*(y); \qquad (5.29)$$

(iii) *for each $M > 1$, as $x \to \infty$,*[7]

$$g_{(-\infty,0]}(x,y)/a(x) \to 1 \quad \text{uniformly for} \ x \leq y \leq Mx. \qquad (5.30)$$

(iv) *For $x \leq y$, $\ell^*(x)/\ell^*(y) \to 1$ as $x \to \infty$ along with $a(-y)/a(x) \to 0$.*

Proof Under the second condition of (5.27), which allows us to apply the right-hand inequality of (5.11), (5.30) follows immediately from (5.25).

Let Z be r.s., so that ℓ^* varies slowly. We write Spitzer's formula as

$$g_{(-\infty,0]}(x,y) = \sum_{k=0}^{x-1} v_d(k)u_a(y-x+k) \quad (1 \leq x \leq y), \qquad (5.31)$$

$(v_d(x) = V_d(x) - V_d(x-1)$, $x \geq 1)$ and owing to (5.28) we see

$$g_{(-\infty,0]}(x,2x) \sim V_d(x-1)/\ell^*(x). \qquad (5.32)$$

---

[5] That ℓ^* is s.v. (equivalently, $xP[Z > x] = o(\ell^*(x))$) implies $f(x) := E[Z; Z \leq x]$ is s.v.; the converse is also true: since $\int_0^\infty e^{-\lambda x}\, df(x) = (E[e^{-\lambda Z}])'$, the slow variation of f implies $1 - E[e^{-\lambda Z}] = \int_0^\lambda s\, ds \int_0^\infty e^{-sx} f(x)\, dx \sim \lambda f(1/\lambda)$ $(\lambda \downarrow 0)$, so that $\ell^*(x) \sim f(x)$ by Karamata's Tauberian Theorem.

[6] See Remarks 6.4.6 and 7.1.2(b,d) for more about the behaviour of $a(x)\ell^*(x)/V_d(x)$.

[7] In fact, it suffices to assume $a(-Mx)/a(x) \to 0$ (even if $M \to \infty$) instead of $x \to \infty$.

If $a(-x)/a(x) \to 0$, by (5.30) we also have $g_{(-\infty,0]}(x, 2x) \sim a(x)$, whence the equivalence relation of (i) follows. (ii) follows from (5.31) and the slow variation of u_a for $1 \le x < y/2$, and from (5.30) in conjunction with (i) and the bound $g_{(-\infty,0]}(x, y) \le g_{(-\infty,0]}(y, y)$ for $y/2 \le x \le y$.

In view of (5.25), (iv) follows from (i) and (ii). □

Define a function f_r on $\mathbb{Z}$ by $f_r(x) = \sum_{y=1}^{\infty} V_d(y - 1)p(y - x)$ for $x \le 0$ and

$$f_r(x) = V_d(x - 1), \quad x \ge 1. \tag{5.33}$$

Then $f_r(x) = P[\hat{Z} < x]$ $(x \le 0)$ (cf. [84, Eq(2.3)]) and $f_r(x)$ is a harmonic function of the r.w. killed as it enters $(-\infty, 0]$, in the sense that $\sum_{y \ge 1} f_r(y)p(y - x) = f_r(x)$ $(x \ge 1)$. For each $x \in \mathbb{Z}$, $M_n := f_r(S_n)\mathbf{1}(n < \sigma_{(-\infty,0]})$ is a martingale under P_x, so that by optional stopping theorem

$$f_r(x) \ge \liminf_{n \to \infty} E_x\left[M_{n \wedge \sigma_{[R,\infty)}}\right] \ge f_r(R)P_x\left[\sigma_{[R,\infty)} < \sigma_{(-\infty,0]}\right].$$

Hence

$$f_r(x)/f_r(R) \ge P_x\left[\sigma_{[R,\infty)} < \sigma_{(-\infty,0]}\right] \ge P_x\left[\sigma_R < \sigma_{(-\infty,0]}\right]. \tag{5.34}$$

The last probability equals $g_{(-\infty,0]}(x, R)/g_{(-\infty,0]}(R, R)$ $(x \ne R)$, which is asymptotically equivalent to $f_r(x)/f_r(R)$ because of Lemma 5.3.1(ii) (note that (5.29) extends to $x \le 0$ if $f_r(x)$ replaces $V(x - 1)$). This leads to the following

Proposition 5.3.2 *If (5.27) holds, then uniformly for $x \le R$, as $R \to \infty$,*

$$P_x\left[\sigma_R < \sigma_{(-\infty,0]}\right] \sim P_x\left[\sigma_{[R,\infty)} < \sigma_{(-\infty,0]}\right] \sim f_r(x)/f_r(R), \tag{5.35}$$

and for $x = 0$, in particular, $P_0(\Lambda_R) \sim 1/V_d(R)$.

Remark 5.3.3 If (5.27) holds, then uniformly for $1 \le x \le R$, as $R \to \infty$,

$$\frac{f_r(x)/f_r(R)}{a(x)/a(R)} \longrightarrow \begin{array}{ll} 1 & \text{as } \ell^*(x)/\ell^*(R) \to 1, \\ 0 & \text{as } \ell^*(x)/\ell^*(R) \to 0, \end{array}$$

so that both probabilities in (5.35) are asymptotically equivalent to, or of a smaller order of magnitude than, $P_x[\sigma_R < \sigma_0]$ according as $\ell^*(x)/\ell^*(R)$ tends to 1 or 0, because of Lemma 5.1.3. If $m_+/m \to 0$ (entailing (5.27)), then by Proposition 3.1.5 it especially follows that as $x/R \to 1$, $V_d(x)/V_d(R) \to 1$ and $P_x[\sigma_R < \sigma_{(-\infty,0]}] \to 1$.

5.4 Spitzer's Condition and the Regular Variation of V_d

Combined with [77, Theorem 1.1], formula (5.1) will lead to the following result.

Theorem 5.4.1 *Suppose* $\sigma^2 = \infty$ *and* $m_+/m \to 0$. *Then*

(i) $\ell^*(x) := \int_0^x P[Z > t] \, dt$ *is s.v. and* $a(x) \sim V_d(x)/\ell^*(x)$ $(x \to \infty)$;
(ii) *for a constant* $\alpha \geq 1$ *the following (a) to (e) are equivalent*

  (a) $P_0[S_n > 0] \to 1/\alpha$,
  (b) $m(x)$ *is regularly varying with index* $2 - \alpha$,
  (c) $P_x[\sigma_{[R,\infty)} < \sigma_{(-\infty,0]}] \to \xi^{\alpha-1}$ *as* $x/R \to \xi$ *for each* $0 < \xi < 1$,
  (d) $V_d(x)$ *is regularly varying with index* $\alpha - 1$,
  (e) $a(x)$ *is regularly varying with index* $\alpha - 1$,

  and each of (a) to (e) implies that

$$a(x) \sim C_\alpha x/m(x), \quad \text{where} \quad C_\alpha = 1/\Gamma(3 - \alpha)\Gamma(\alpha). \tag{5.36}$$

[If $\sigma^2 < \infty$, *then all the statements (a) to (e) above are valid with* $\alpha = 2$, *and (i) and (5.36) are also valid but with* $a(x)$ *replaced by* $2a(x)$ *since* $\lim m(x) = \frac{1}{2}\sigma^2$.*]*

It is unclear how conditions (a) to (e) are related when the condition $m_+/m \to 0$ fails. For $\alpha > 1$, however, this condition seems to be a reasonable restriction – if no additional one on the tails of F is supposed – for describing the relation of the quantities of interest. The situation is different for $\alpha = 1$, where (a) and (c) are equivalent to each other quite generally, and under the condition $\liminf \mu_-(x)/\mu(x) > 0$, (c) implies that F is r.s. – hence the other conditions – owing to what is stated at (2.23) (cf. Theorems 4.1.1 and A.4.1).

Corollary 5.4.2 (i) *Suppose* $m_+/m \to 0$. *Then* F *is attracted to the normal law if and only if each of conditions (a) to (e) of Theorem 5.4.1 holds with* $\alpha = 2$.
(ii) *Suppose* $\mu_+/\mu \to 0$ *and* $1 < \alpha < 2$. *Then* F *belongs to the domain of attraction of a stable law of exponent* α *if and only if each of (a) to (e) holds.*

This corollary follows immediately from Theorem 5.4.1 because of the well-known characterisation theorem for the domain of attraction (for (i) see Lemma A.1.2 in Section A.1.1).

The condition $\lim P_0[S_n > 0] = \rho$ – apparently stronger than Spitzer's condition $n^{-1} \sum_{k=1}^n P_0[S_k > 0] \to \rho$ but equivalent to it (cf. [22], [6]) – plays an essential role in the study of fluctuations of r.w.'s. The condition holds if F belongs to the domain of attraction of a stable law of exponent $\alpha \neq 1$ (for $\alpha = 1$, see the comments after (AS) given in Section 2.3). It is of great interest to find a condition for the reverse implication to be true. Doney [20] and Emery [27] observed that for $\alpha > 1$, Spitzer's condition implies that F is attracted to a stable law if one tail outweighs the other overwhelmingly – conditions much stronger than $\mu_+/\mu \to 0$. (ii) of Corollary 5.4.2 sharpens their results. For $\alpha = 1$, however, the assertion corresponding to (ii) fails since the slow variation of η_- does not imply the regular variation of μ_-.

Remark 5.4.3 In Theorem 5.4.1 the assertion that (a) is equivalent to (b) under $m_+/m \to 0$ is essentially Theorem 1.1 of [77] (see (5.37) below), so that Corollary 5.4.2 restricted to conditions (a) and (b) (with "each of (a) to (e)" replaced by "each of (a) and (b)" in the statements) should be considered as its corollary, though not stated in [77] as such. The equivalence of (c) and (d) and that of (d) and (e) follow from (5.1) and Lemma 5.3.1 (i), respectively (recall Z is r.s. under $m_+/m \to 0$ by Proposition 5.2.1). The equivalence of (a) and (c) is valid if $\alpha = 1$, as mentioned above. Therefore the essential content supplemented by Theorem 5.4.1 is that (a) and (b) are equivalent to the conditions (c) to (e) in the case $\alpha > 1$.

According to [77, Theorem1.1] it follows under $m_+/m \to 0$ that for (a) in (ii) of Theorem to hold, i.e., $\lim P_0[S_n > 0] \to 1/\alpha$, it is necessary and sufficient that $1 \le \alpha \le 2$ and for some s.v. function $L(x)$ at infinity,

$$
\begin{array}{ll}
\int_{-x}^{0} t^2 \, dF(t) \sim 2L(x) & \text{if} \quad \alpha = 2, \\
\mu_-(x) \sim (\alpha - 1)(2 - \alpha)x^{-\alpha}L(x) & \text{if} \quad 1 < \alpha < 2, \\
\int_{-\infty}^{-x}(-t) \, dF(t) \sim L(x) & \text{if} \quad \alpha = 1.
\end{array}
\tag{5.37}
$$

Here we have chosen the coefficients – differently from [77] – so that the above condition is expressed for all $1 \le \alpha \le 2$ by the single formula

$$
m_-(x) \sim x^{2-\alpha}L(x).^8
\tag{5.38}
$$

This, in particular, shows that (a) and (b) are equivalent. Here we provide its proof adapted from [77].

Proof (of (a) ⇔ (b)) This equivalence is well-established for right-continuous r.w.'s (cf., e.g, [18], [8, Proposition 8.9.16] (in the case $\alpha > 1$); see also the comment made right before Corollary 5.4.2 for $\alpha = 1$). Let p_* be the probability defined in the proof of Lemma 3.5.1 and denote the corresponding objects by putting the suffix $*$ on them as before. Then $\sum xp_*(x) = 0$ and $p_*(x) = 0$, $x \ge 2$, and $m_*(x) \sim m(x)$, and by virtue of a Tauberian theorem (and the Bertoin–Doney result on Spitzer's condition) it suffices to show that $(1 - s)\Delta(s) \to 0$ as $s \uparrow 1$, where

$$
\Delta(s) := \sum_{n=0}^{\infty} s^n \sum_{y=-\infty}^{0} (p^n(y) - p_*^n(y)) = \frac{1}{2\pi} \int_{-\pi}^{\pi} \left[\frac{1}{1 - s\psi(t)} - \frac{1}{1 - s\psi_*(t)} \right] \frac{dt}{1 - e^{it}}.
$$

By (b) of Lemma 3.2.2 (applied with $\mu_\pm$ separately) we see that $\alpha_+(t) + \beta_+(t) \ll \alpha_-(t) + \beta_-(t) \asymp tm(1/t)$ ($t \to 0$), and the same relations for $\alpha_{*\pm}(t) + \beta_{*\pm}(t)$. Hence from (3.50) it follows that

---

[8] Standard arguments verify the equivalence between (5.38) and (5.37). Indeed, if $1 < \alpha < 2$ this is immediate by the monotone density theorem; for $\alpha = 1$ observe that the condition in (5.37) implies that $x\mu_-(x)/\int_{-\infty}^{-x}(-t) \, dF(t)$ approaches zero (see e.g., Theorem VIII.9.2 of [31]), and then that $\eta_-(x) \sim (2 - \alpha)L(x)$ for $\alpha < 2$ – the converse implication is easy; for $\alpha = 2$ see Lemma A.1.1 of Section A.1.1.

$$|\psi_*(t) - \psi(t)| \asymp O(t^2) + |t\beta_+(t)| \ll t^2 m(1/t) \asymp |1 - \psi(t)| \sim |1 - \psi_*(t)|$$

as $t \to 0$. Using $|1 - s\psi|^2 \geq (1 - s)^2 + s^2|1 - \psi|^2$, valid for any complex number ψ with $\mathfrak{R}\psi \leq 1$, for any $\varepsilon > 0$ one can accordingly choose a positive constant δ so that

$$(1 - s)|\Delta(s)| \leq \varepsilon \int_0^\delta \frac{(1 - s)m(1/t)t}{(1 - s)^2 + [m(1/t)t^2]^2}\, dt + o(1) \quad (s \uparrow 1).$$

Writing $h(t)$ for $m(1/t)t^2$, one sees that $h'(t) = [c(1/t) + m(1/t)]t > h(t)/t\ (t > 0)$, which entails that the above integral is less than $\pi/2$ (see Section A.5.2). Thus $(1 - s)|\Delta(s)| \to 0$ as $s \uparrow 1$, as desired.   □

Because of what has been mentioned in Remark 5.4.3, for the **proof of Theorem 5.4.1** now we have only to show the equivalence of (b) and (d), which is involved in the following

Proposition 5.4.4 *Let $m_+/m \to 0$. Then as $x \to \infty$,*

(i) $V_d(x)U_a(x) \sim xa(x) \asymp x^2/m(x)$ *and* $V_d(x)U_a(x)\mu_+(x) \to 0$; *and*
(ii) *in order for m to vary regularly with index $2 - \alpha$ ($\in [0, 1]$), it is necessary and sufficient that V_d varies regularly with index $\alpha - 1$; in this case it holds that*

$$V_d(x)U_a(x) \sim C_\alpha x^2/m(x), \tag{5.39}$$

where $C_\alpha = 1/\Gamma(\alpha)\Gamma(3 - \alpha)$.

When F is in the domain of attraction of a stable law of exponent $\alpha \in (0, 2]$ and there exists $\rho := \lim P_0[S_n > 0]$, one can readily derive from Eq(15) and Lemma 12 of Vatutin & Wachtel [88] the asymptotic form (explicit in terms of F) of $V_d(x)U_a(x)$ if $\rho(1 - \rho) \neq 0$ – in particular (5.39) follows as a special case of it except for the expression of C_α; our proof, quite different from theirs, rests on Lemma 5.3.1. In Lemma 6.5.1 the result will be extended to the case $\rho(1 - \rho) = 0$ unless $\alpha = 2p = 1$.

The first half of (i) of Proposition 5.4.4 follows immediately from Lemma 5.3.1(i) (since $U_a(x) \sim x/\ell^*(x)$). The second one follows from the first since $x^2\mu_+(x) \leq 2c_+(x)$. The proof of (ii) will be based on the identity

$$P[\hat{Z} \leq -x] = v_d(0) \sum_{y=0}^{\infty} u_a(y)F(-y - x) \quad (x > 0), \tag{5.40}$$

which one can derive by using (5.31) (or by the duality lemma [31, Section XII.1], which says $u_a(y) = g_{(-\infty,-1]}(0, y),\ y \geq 0$). Recall (5.28), i.e., $u_a(x) \sim 1/\ell^*(x)$ and that ℓ^* is s.v. (at least under $m_+/m \to 0$). Here we also note that for $0 \leq \gamma \leq 1$, if either $V_d(x)$ or $1/P[\hat{Z} \leq -x]$ varies regularly as $x \to \infty$ with index γ, then

$$P[\hat{Z} \leq -x]V_d(x)/v^\circ \longrightarrow 1/\Gamma(1 - \gamma)\Gamma(\gamma + 1), \tag{5.41}$$

where $v^\circ = v_d(0)$ (cf. Theorem A.2.3).

Proof (of the necessity part of Proposition 5.4.4(ii)) By virtue of (5.35) V_d is s.v. if and only if $P_x[\sigma_{[R,\infty)} < \sigma_{(-\infty,0]}] \to 1$ as $R \to \infty$ for $x \geq R/2$, which is equivalent to $\lim P_0[S_n > 0] = 1$ due to Kesten & Maller [46], as noted in (2.22). For $\alpha = 1$, by what is mentioned right after (5.28), this verifies the first half of (ii). [Here, the sufficiency part is also proved, which we shall prove without resorting to [46] (cf. Lemma 5.4.6).]

For $1 < \alpha < 2$, by (5.40) the condition (5.37) implies that

$$\frac{P[\hat{Z} < -x]}{v^\circ} \sim (\alpha - 1)(2 - \alpha) \sum_{y=0}^{\infty} \frac{L(x + y)}{\ell^*(y)(x + y)^\alpha} \sim \frac{(2 - \alpha)L(x)}{\ell^*(x)x^{\alpha-1}}, \tag{5.42}$$

hence V_d varies regularly with index $\alpha - 1$ because of (5.41).

Let $\alpha = 2$. Then the slow variation of m_- entails that F is in the domain of attraction of a normal law (cf. Section A.1.1)) and we have seen (Proposition 4.2.1(i)) that this entails $2\bar{a}(x) \sim x/m_{(}x)$, hence $V_d(x) \sim a(x)\ell^*(x) \sim x\ell^*(x)/m_-(x)$ owing to Lemma 5.3.1, showing $V_d(x)/x$ is s.v., as required.   □

In the following proof of the sufficiency part of Proposition 5.4.4(ii), we shall be concerned with the condition

$$V_d(x) \sim x^{\alpha-1}/\hat{\ell}(x) \quad \text{for some } \hat{\ell} \text{ that is s.v. at infinity.} \tag{5.43}$$

The next lemma deals with the case $1 \leq \alpha < 2$, when (5.41) is valid with $\gamma = \alpha - 1$ (under (5.43)). By (5.40) and (5.41) the above equivalence relation may be rewritten as

$$\sum_{y=0}^{\infty} u_a(y)F(-y - x) = C'_\alpha x^{1-\alpha}\hat{\ell}(x)\{1 + o(1)\}, \tag{5.44}$$

where $C'_\alpha = (2 - \alpha)C_\alpha = 1/\Gamma(\alpha)\Gamma(2 - \alpha)$. Put for $1 \leq \alpha < 2$,

$$\ell_\sharp(x) = x^{\alpha-1} \int_x^{\infty} \frac{\mu_-(t)}{\ell^*(t)} \, dt. \tag{5.45}$$

Lemma 5.4.5 *If Z is r.s. and (5.43) holds with $1 \leq \alpha < 2$, then*

$$\hat{\ell}(x) \sim \Gamma(\alpha)\Gamma(2 - \alpha)\ell_\sharp(x) \quad \text{and} \quad P[\hat{Z} \leq -x]/v^\circ \sim x^{1-\alpha}\ell_\sharp(x).$$

Proof For this proof, we need neither the condition $E|X| < \infty$ nor the recurrence of the r.w., but we do need the oscillation of the r.w. so that both Z and $\hat{Z}$ are well-defined proper random variables and Spitzer's representation (5.31) of $g_{(-\infty,0]}(x, y)$ is applicable. Let Z be r.s., so that ℓ^* is s.v. and $u_a(x) \sim 1/\ell^*(x)$. Put

$$\Sigma(x) = \int_1^{\infty} \frac{F(-t - x)}{\ell^*(t)} \, dt \quad \text{and} \quad \Delta(x) = \frac{\ell_\sharp(x)}{x^{\alpha-1}} = \int_x^{\infty} \frac{\mu_-(t)}{\ell^*(t)} \, dt.$$

In view of (5.44) it suffices to show the implication

$$\Sigma(x) \text{ varies regularly with index } 1 - \alpha \implies \Sigma(x) \sim \Delta(x). \qquad (5.46)$$

Clearly $\Sigma(x) \geq F(-2x) \int_1^x dt/\ell^*(t) \sim xF(-2x)/\ell^*(x)$. Replacing x by $x/2$ in this inequality and noting that $\Sigma(x/2) \sim 2^{\alpha-1}\Sigma(x)$ because of the premise of (5.46) we obtain

$$x\mu_-(x)/\ell^*(x) \leq 2^\alpha \Sigma(x)\{1 + o(1)\}. \qquad (5.47)$$

From the defining expressions of Σ and Δ one observes first that for each $\varepsilon > 0$, as $x \to \infty$,

$$\Sigma(x) \leq \frac{\varepsilon x}{\ell^*(x)}\mu_-(x)\{1 + o(1)\} + \Delta\left((1 + \varepsilon)x\right),$$

then, by substituting (5.47) into the RHS, that

$$[1 - 2^\alpha \varepsilon] \Sigma(x)\{1 + o(1)\} \leq \Delta\left((1 + \varepsilon)x\right) \leq \Delta(x) \leq \Sigma(x)\{1 + o(1)\}.$$

Since ε may be arbitrarily small, we can conclude $\Delta(x)/\Sigma(x) \to 1$, as desired. □

Lemma 5.4.6 *If V_d is s.v. and $x\eta_+(x)/m_-(x) \to 0$, then Z is r.s., $V_d(x) \sim 1/\ell_\sharp(x)$, η_- is s.v. and*

$$\eta_-(t) \sim A(t) \sim \ell^*(t)\ell_\sharp(t).$$

Proof The relative stability of Z follows from Proposition 5.2.1 and, together with the slow variation of V_d, implies $V_d(x) \sim 1/\ell_\sharp(x)$ (with $\alpha = 1$ in (5.45)) according to the preceding lemma. The slow variation of $\ell_\sharp$ implies $x\mu_-(x)/\ell^*(x) \ll \ell_\sharp(x)$, which in turn shows η_- is s.v. since $\ell_\sharp(x) \leq \eta_-(x)/\ell^*(x)$. It accordingly follows that $\eta_-(x) \sim m_-(x)/x \gg \eta_+(x)$, hence $A(x) \sim \eta_-(x)$. We postpone the proof of $\ell^*(x)\ell_\sharp(x) \sim A(x)$ to the next chapter (Lemma 6.4.3), where the result is extended to the case when V_d is s.v. and Z is r.s.   □

Completion of the proof of Proposition 5.4.4. We show that if $V_d(x)$ varies regularly, then (5.37) and (5.39) hold. Suppose that (5.43), that is $V_d(x) \sim x^{\alpha-1}/\hat{\ell}(x)$, holds for $1 \leq \alpha \leq 2$.

Case $1 \leq \alpha < 2$. Let $C'_\alpha = (2-\alpha)C_\alpha = 1/[\Gamma(\alpha)\Gamma(2-\alpha)]$ as before. By Lemma 5.4.5

$$\frac{C'_\alpha \hat{\ell}(x)}{x^{\alpha-1}} \sim \frac{\ell_\sharp(x)}{x^{\alpha-1}} = \int_x^\infty \frac{\mu_-(t)}{\ell^*(t)} dt. \qquad (5.48)$$

Let $1 < \alpha < 2$. Then by the monotone density theorem – applicable since $\mu_-(t)/\ell^*(t)$ is decreasing – the above relation leads to $\mu_-(x) \sim Cx^{-\alpha}\hat{\ell}(x)\ell^*(x)$ (with $C = [(2 - \alpha)(\alpha - 1)C_\alpha])$, which is equivalent to (5.37) with $L(x) = C_\alpha \hat{\ell}(x)\ell^*(x)$ and shows (5.39) again (which has been virtually seen already in (5.42)).

If $\alpha = 1$, then $u_a(x) \sim 1/\ell^*(x)$ and, by (5.48) and Lemma 5.4.6, $1/V_d(x) \sim \eta_-(x)/\ell^*(x)$, which implies (5.37) (with $L(x) = \hat{\ell}(x)\ell^*(x)$) and (5.39), as one can readily check.

CASE $\alpha = 2$. We have

$$\hat{\ell}^*(x) = \frac{1}{v^\circ} \int_0^x P[-\hat{Z} > t]\,dt = \int_0^x dt \sum_{y=0}^\infty u_a(y)F(-y-t) \qquad (5.49)$$

and, instead of (5.41), $\hat{\ell}^*(x)V_d(x) \sim x$ (Theorem A.2.3), so that

$$\hat{\ell}^*(x) \sim x/V_d(x) \sim \hat{\ell}(x). \qquad (5.50)$$

To complete the proof it suffices to show that m is s.v., or equivalently, $x\eta_-(x)/m(x) \to 0$, because the slow variation of m entails that of $\int_{-x}^x y^2\,dF(y)$ and hence (5.39) (see the proof of the necessity part). We shall verify the latter condition, and to this end, we shall apply the inequalities

$$1/C \le \hat{\ell}^*(x)\ell^*(x)/m(x) \le C \qquad (x > 1) \qquad (5.51)$$

(for some positive constant C), which follow from (5.50) and Lemma 5.3.1(i), the latter entailing $V_d(x)/\ell^*(x) \asymp x/m(x)$ in view of Theorems 3.1.1 and 3.1.2.

Using (5.49) and $u_a(x) \sim 1/\ell^*(x)$ as well as the monotonicity of ℓ^* and F, one observes that for each $M > 1$, as $x \to \infty$,

$$\hat{\ell}^*(x) - \hat{\ell}^*(x/2) \ge \int_{x/2}^x dt \int_0^{(M-1)x} \frac{F(-y-t)}{\ell^*(y) \vee 1}\{1+o(1)\}\,dy$$

$$\ge \int_{x/2}^x dt \int_0^{(M-1)x} \frac{F(-y-t)}{\ell^*(y+x)}\,dy\{1+o(1)\}$$

$$\ge \frac{x/2}{\ell^*(x)}\left[\eta_-(x) - \eta_-(Mx)\right]\{1+o(1)\},$$

which implies

$$x\eta_-(x) \le x\eta_-(Mx) + \hat{\ell}^*(x)\ell^*(x) \times o(1),$$

for by (5.50) $\hat{\ell}^*$ is s.v. Combining this with (5.51) one infers that

$$x\eta_-(x) \le \frac{m(Mx)}{M}\{1+o(1)\}$$

$$\le \frac{C[\hat{\ell}^*\ell^*](x)}{M}\{1+o(1)\}$$

$$\le \frac{C^2 m(x)}{M}\{1+o(1)\}$$

and concludes that $x\eta_-(x)/m(x) \to 0$, hence the desired slow variation of m. $\quad\square$

5.5 Comparison Between σ_R and $\sigma_{[R,\infty)}$ and One-Sided Escape From Zero

The following proposition gives the asymptotic form in terms of the potential kernel a of the probability of one-sided escape of the r.w. that is killed as it hits 0.

Proposition 5.5.1 (i) *If* $\lim_{x\to\infty} a(-x)/\bar{a}(x) = 0$, *then as* $R \to \infty$,

$$P_x[\sigma_{(-\infty,-R]} < \sigma_0] \sim P_x[\sigma_{-R} < \sigma_0] \quad \text{uniformly for } x \in \mathbb{Z}. \tag{5.52}$$

(ii) *Let* $m_+/m \to 0$ *and* $\sigma^2 = \infty$. *Then* $\lim_{x\to\infty} a(-x)/\bar{a}(x) = 0$ *and as* $R \to \infty$,

$$P_x[\sigma_{[R,\infty)} < \sigma_0] \sim P_x[\sigma_R < \sigma_0] \sim a^\dagger(x)/a(R) \tag{5.53}$$

uniformly for $x \le R$, *subject to the condition*

$$\frac{a(-R) - a(-R+x)}{a^\dagger(x)} \longrightarrow 0 \quad \text{and} \quad a^\dagger(x) > 0. \tag{5.54}$$

Under $m_+/m \to 0$, condition (5.54) holds at least for $0 \le x < R$ by Lemma 5.1.3(ii) and for $-R \ll x < 0$ subject to $[a(-x)/a(x)] \cdot [a(-R)/a(R)] \to 0$, $a(x) > 0$ by (5.8). [As to the necessity of (5.54) for the first equivalence of (5.53), see the exceptional cases addressed in Remark 4.2.5.] In Section 5.7 we shall state a corollary – Corollary 5.7.1 – of Proposition 5.5.1 that concerns the asymptotic distribution of $\#\{n < \sigma_{[R,\infty)} : S_n \in I\}$, the number of visits of a finite set $I \subset \mathbb{Z}$ by the r.w. before entering $[R, \infty)$.

Proof Let $\mathcal{E}_R$ stand for the event $\{\sigma_{[R,\infty)} < \sigma_0\}$. Note that for each x,

$$P_x[\sigma_{[R,\infty)} < \sigma_0] \sim P_x[\sigma_R < \sigma_0] \iff \lim_{R\to\infty} P_x[\sigma_R < \sigma_0 | \mathcal{E}_R] = 1. \tag{5.55}$$

By an analogous equivalence, (i) is immediate from Lemma 5.1.3(i).

For the proof of (ii), let $m_+/m \to 0$. By (5.3) condition (5.54) is then equivalent to the second equivalence of (5.53) under the assumption of (ii). For the proof of the first, take a small constant $\varepsilon > 0$. Then for any R large enough we can choose $u > 0$ so that

$$u/m(u) = \varepsilon R/m(R);$$

put $z = z(R, \varepsilon) = \lfloor u \rfloor$. Since $\bar{a}(x) \asymp x/m(x)$, this entails

$$\varepsilon C' \le \bar{a}(z)/\bar{a}(R) < \varepsilon C''$$

for some positive constants C', C''. Now define $h_{R,\varepsilon}$ via

$$P_x[\sigma_R < \sigma_0 | \mathcal{E}_R] = h_{R,\varepsilon} + \sum_{R \le y \le R+z} P_x\left[S_{\sigma_{[R,\infty)}} = y \,\big|\, \mathcal{E}_R\right] P_y[\tilde{\sigma}_R < \sigma_0]$$

with $\tilde{\sigma}_R = 0$ if $S_0 = R$ and $\tilde{\sigma}_R = \sigma_R$ otherwise. Then by using $\sigma_{[R,\infty)} \le \sigma_R$ together with (5.54) we infer from Lemma 5.2.3(i) (see (5.17) and its derivation)

$$h_{R,\varepsilon} \le P_x \left[S_{\sigma[R,\infty)} > R + z \,\middle|\, \mathcal{E}_R \right] \le C \frac{m_+(z)}{m(z)} \cdot \frac{\bar{a}(R)}{\bar{a}(z)} \le [C/\varepsilon C'] \frac{m_+(z)}{m(z)},$$

whereas $1 - P_y[\tilde{\sigma}_R < \sigma_0] \le \bar{a}(y - R)/\bar{a}(R) < \varepsilon C_1$ for $R \le y \le R + z$ (see Remark 5.1.2(b) for the first inequality). Since $m_+(z)/m(z) \to 0$ so that $h_{R,\varepsilon} \to 0$, and since $P_y[\tilde{\sigma}_R < \sigma_0] \sim a(y)/a(R)$, we conclude

$$\liminf_{R \to \infty} \inf_{0 \le x < R} P_x \left[\sigma_R < \sigma_0 \,\middle|\, \sigma_{[R,\infty)} < \sigma_0 \right] > 1 - \varepsilon C_1,$$

hence $P_x[\sigma_R < \sigma_0 \mid \sigma_{[R,\infty)} < \sigma_0] \to 1$, since ε can be made arbitrarily small. This verifies that the first equivalence in (5.53) holds uniformly for $0 \le x \le R$.

Letting $x = 0$ in (5.53) gives that as $R \to \infty$

$$P_0[\sigma_{[R,\infty)} < \sigma_0] \sim 1/a(R),$$

which shows the asymptotic monotonicity of $a(x)$, again, the probability of the LHS being a monotone function of R. With the help of Lemma 5.5.2 below and the trivial inequality $P_x[\sigma_{[R,\infty)} < \sigma_0] \ge P_x[\sigma_R < \sigma_0]$ as well as the second equivalence in (5.53) this shows that the first equivalence of (5.53) holds for $x < R$, subject to (5.54). $\qquad\square$

Lemma 5.5.2 *If $a(x)$ is asymptotically increasing as $x \to \infty$, then as $R \to \infty$,*

$$P_x[\sigma_{[R,\infty)} < \sigma_0] \le \frac{a^\dagger(x)}{a(R)}\{1 + o(1)\} \quad \text{uniformly for } x \in \mathbb{Z}.$$

Proof Since $a(S_{n \wedge \sigma_0})$ is a martingale under P_x (if $x \ne 0$), the optional stopping theorem with the help of Fatou's lemma shows

$$E_x \left[a(S_{\sigma[R,\infty)}); \sigma_{[R,\infty)} < \sigma_0 \right] \le a^\dagger(x).$$

The expectation on the LHS is bounded below by $a(R)P_x[\sigma_{[R,\infty)} < \sigma_0]\{1 + o(1)\}$ by the assumed monotonicity of $a(x)$, hence the asserted inequality follows. $\qquad\square$

5.6 Escape Into $(-\infty, -Q] \cup [R, \infty)$

Let Q, as well as R, be a positive integer. Here we consider the event

$$\sigma_{(-\infty, -Q] \cup [R, \infty)} < \sigma_0,$$

the escape into

$$\mathbb{Z} \setminus (-Q, R) = [(-\infty, -Q] \cup [R, \infty)] \cap \mathbb{Z}$$

from the killing at 0. Under $m^+/m \to 0$, we shall give asymptotic forms as $Q \wedge R \to \infty$ of its probability and the probability of the escape being on the upper/lower side. The next result is essentially a corollary of Proposition 5.5.1. For simplicity we write $\sigma_{y,z}$ for the first hitting time of the two-point set $\{y, z\}$, so that

$$\sigma_{y,z} = \sigma_y \wedge \sigma_z.$$

Proposition 5.6.1 *If $m_+/m \to 0$, then uniformly for $x \in (-Q, R)$ subject to (5.54), as $Q \wedge R \to \infty$,*

$$P_x[\sigma_{\mathbb{Z}\backslash(-Q,R)} < \sigma_0] = P_x[\sigma_{-Q,R} < \sigma_0]\{1 + o(1)\}. \tag{5.56}$$

Proof Put $\tau_- = \sigma_{(-\infty,-Q]}$ and $\tau_+ = \sigma_{[R,\infty)}$. It suffices to show that

$$\frac{P_x[\tau_- \wedge \tau_+ < \sigma_0 < \sigma_{\{-Q,R\}}]}{P_x[\tau_- \wedge \tau_+ < \sigma_0]} \to 0.$$

The numerator of the ratio above is less than

$$P_x[\tau_- < \sigma_0 < \sigma_{-Q}] + P_x[\tau_+ < \sigma_0 < \sigma_R]$$
$$= P_x[\tau_- < \sigma_0] \times o(1) + P_x[\tau_+ < \sigma_0] \times o(1)$$

under (5.54), where Proposition 5.5.1 – both (i) and (ii) – is applied. Hence it is of smaller order of magnitude than the denominator. □

Remark 5.6.2 Suppose (4.14) holds, so that F is in the domain of attraction of a stable law with exponent $1 \leq \alpha \leq 2$. If $\alpha = 2$ and the assumption of (iii) of Proposition 4.2.1 is satisfied, then equality (5.56) holds uniformly for $x \in (-Q, R)$ (see (i) in Remark 4.2.5). Also, (5.56) holds uniformly for $x \in \mathbb{Z}$, if $\alpha = q = 1$ and Q/R is bounded, since then $P_z[\sigma_{-Q} < \sigma_0] \to 1$ uniformly for $z \notin (-Q, R)$ by virtue of Lemma 5.1.3. Although (5.56) may fail for $-\delta Q < x < \delta R$ ($0 < \delta < 1$) if $1 \leq \alpha < 2$ and $pq \neq 0$, it generally holds that if F is recurrent and if Q/R is bounded away from zero and infinity,[9] then for all $1 \leq \alpha \leq 2$

$$P_x[\sigma_{\mathbb{Z}\backslash(-Q,R)} < \sigma_0] \asymp P_x[\sigma_{-Q,R} < \sigma_0] \qquad (z \in \mathbb{Z}).$$

We can compute the exact asymptotic form of the probability on the RHS for $x = 0$, provided Spitzer's condition is fulfilled and S is recurrent. Here we carry out the computation in the case $Q = R$ for simplicity, which leads to the following

$$\lambda_+ \sim \frac{1 - c_\alpha p}{1 - c_\alpha^2 pq} \cdot \frac{1}{2\bar{a}(R)}, \quad \lambda_- \sim \frac{1 - c_\alpha q}{1 - c_\alpha^2 pq} \cdot \frac{1}{2\bar{a}(R)}, \tag{5.57}$$

where $\lambda_\pm = \lambda_\pm(R) = P_0[\sigma_{\pm R} < \sigma_{\mp R,0}]$, and $c_\alpha = 2 - 2^{\alpha-1}$; in particular

---

[9] For $\alpha > 1$ this condition is superfluous. Note that $\inf_{z \notin (-Q,R)} P_z[\sigma_{-Q} < \sigma_0] > 0$ if $p = 0$ and Q/R is bounded above.

$$P_0[\sigma_{-R,R} < \sigma_0] \sim \frac{(2^{\alpha-1})/(1 - c_\alpha^2 pq)}{2\bar{a}(R)}.$$

The computation rests on Proposition 4.2.1. Combining (4.23) and (4.24) one observes that

$$P_R[\sigma_{-R} < \sigma_0] \to c_\alpha q, \quad P_{-R}[\sigma_R < \sigma_0] \to c_\alpha p$$

[in the case $\alpha = 2$, use the fact that both Z and $-\hat{Z}$ are r.s.], so that

$$P_0[\sigma_{-R} < \sigma_0] = P_0[\sigma_R < \sigma_0] \sim \lambda_- + c_\alpha q \lambda_+ \sim \lambda_+ + c_\alpha p \lambda_-,$$

showing $(1 - c_\alpha p)\lambda_- \sim (1 - c_\alpha q)\lambda_+$. This together with $P_0[\sigma_{-R} < \sigma_0] \sim 1/2\bar{a}(R)$, yields (5.57).

Recall that $H_B^x(\cdot)$ denotes the hitting distribution of a set B for the r.w. S under P_x:

$$H_B^x(y) = P_x[S_{\sigma_B} = y] \quad (y \in B)$$

(see (4.35)). Put

$$B(Q, R) = \{-Q, 0, R\}.$$

Then (5.56) is rephrased as

$$P_x[\sigma_{\mathbb{Z}\backslash(-Q,R)} < \sigma_0] \sim 1 - H_{B(Q,R)}^x(0). \tag{5.58}$$

By using Theorem 30.2 of Spitzer [71] one can compute an explicit expression of $H_{B(Q,R)}^x(0)$ in terms of $a(\cdot)$, which though useful, is pretty complicated. We derive the following result without using it.

Lemma 5.6.3 *For $x \in \mathbb{Z}$*

$$P_x[\sigma_{-Q} < \sigma_0] \geq 1 - H_{B(Q,R)}^x(0) - P_x[\sigma_R < \sigma_0]P_R[\sigma_0 < \sigma_{-Q}]$$
$$\geq P_x[\sigma_{-Q} < \sigma_0]P_{-Q}[\sigma_0 < \sigma_R].$$

Proof Write B for $B(Q, R)$. Plainly we have

$$1 - H_B^x(0) = P_x[\sigma_{-Q,R} < \sigma_0] \tag{5.59}$$
$$= P_x[\sigma_{-Q} < \sigma_0] + P_x[\sigma_R < \sigma_0] - P_x[\sigma_{-Q} \vee \sigma_R < \sigma_0].$$

Let $\mathcal{E}$ denote $\{\sigma_R < \sigma_0\} \cap \{S_{\sigma_R+\cdot} \text{ hits } -Q \text{ before } 0\}$, the event that $S.$ visits $-Q$ after the first hitting of R without visiting 0. Then

$$\mathcal{E} \subset \{\sigma_R \vee \sigma_{-Q} < \sigma_0\} \quad \text{and} \quad \{\sigma_{-Q} \vee \sigma_R < \sigma_0\} \backslash \mathcal{E} \subset \{\sigma_{-Q} < \sigma_R < \sigma_0\}.$$

By $P_x[\sigma_{-Q} < \sigma_R < \sigma_0] \leq P_x[\sigma_{-Q} < \sigma_0]P_{-Q}[\sigma_R < \sigma_0]$ it therefore follows that

$$0 \leq P_x[\sigma_{-Q} \vee \sigma_R < \sigma_0] - P_x(\mathcal{E}) \leq P_x[\sigma_{-Q} < \sigma_0]P_{-Q}[\sigma_R < \sigma_0]. \tag{5.60}$$

The left-hand inequality together with $P_x(\mathcal{E}) = P_x[\sigma_R < \sigma_0]P_R[\sigma_{-Q} < \sigma_0]$ leads to

$$P_x[\sigma_R < \sigma_0] - P_x[\sigma_{-Q} \vee \sigma_R < \sigma_0] \le P_x[\sigma_R < \sigma_0] - P_x(\mathcal{E})$$
$$= P_x[\sigma_R < \sigma_0]P_R[\sigma_0 < \sigma_{-Q}],$$

which, substituted into (5.59), yields the first inequality of the lemma after the trite transposition of a term. In a similar way, the second one is deduced by substituting the right-hand inequality of (5.60) into (5.59).   □

Lemma 5.6.4 *If $a(-x)/\bar{a}(x) \to 0$ $(x \to \infty)$, then as $Q \vee R \to \infty$,*

$$1 - H^0_{B(Q,R)}(0) \sim \frac{a(Q+R)}{a(Q)a(R)}. \tag{5.61}$$

Proof On taking $x = 0$ in Lemma 5.6.3, its second inequality reduces to

$$1 - H^0_{B(Q,R)}(0) \ge \frac{a(Q+R) + a(-Q) - a(R)}{4\bar{a}(R)\bar{a}(Q)} + \frac{a(-Q-R) + a(R) - a(-Q)}{4\bar{a}(Q)\bar{a}(R)}$$
$$= \frac{\bar{a}(Q+R)}{2\bar{a}(Q)\bar{a}(R)}$$
$$= \frac{a(Q+R)}{a(Q)a(R)}\{1 + o(1)\},$$

and its first inequality gives the upper bound

$$1 - H^0_{B(Q,R)}(0) \le \frac{a(Q+R) + a(-Q) + a(-R)}{4\bar{a}(Q)\bar{a}(R)} = \frac{a(Q+R)}{a(Q)a(R)}\{1 + o(1)\},$$

showing (5.61).   □

According to Lemma 3.10 of [84], for any finite set $B \subset \mathbb{Z}$ that contains 0 we have the identity

$$1 - H^x_B(0) \tag{5.62}$$
$$= [1 - H^0_B(0)]a^\dagger(x) + \sum_{z \in B \setminus \{0\}} [a(-z) - a(x-z)]H^z_B(0), \quad x \notin B \setminus \{0\}.$$

Recalling $P_x[\sigma_0 < \sigma_x] = 1/2\bar{a}(x) = P_0[\sigma_x < \sigma_0]$ $(x \ne 0)$ we see that

$$H^R_{B(Q,R)}(0) \vee H^{-Q}_{B(Q,R)}(0) \le P_0[\sigma_R < \sigma_0] \vee P_0[\sigma_{-Q} < \sigma_0] \le 1 - H^0_{B(Q,R)}(0).$$

For $B = B(Q, R)$, the second term on the RHS of (5.62) is negligible in comparison to the first under the following condition: as $Q \wedge R \to \infty$

$$|a(Q) - a(x + Q)| + |a(-R) - a(x - R)| = o(a^\dagger(x)) \quad \text{and} \quad a^\dagger(x) \neq 0.^{10} \quad (5.63)$$

This gives us the following result.

Lemma 5.6.5 *Uniformly for* $-Q < x < R$ *subject to (5.63), as* $Q \wedge R \to \infty$,

$$1 - H^x_{B(Q,R)}(0) \sim a^\dagger(x)\left[1 - H^0_{B(Q,R)}(0)\right]. \quad (5.64)$$

Condition (5.63) – always satisfied for each x (fixed) if $\sigma^2 = \infty$ and only for $x = 0$ otherwise – is necessary and sufficient for the following condition to hold:

$$P_x[\sigma_{-Q} < \sigma_0] \sim a^\dagger(x)/2\bar{a}(Q) \quad \text{and} \quad P_x[\sigma_R < \sigma_0] \sim a^\dagger(x)/2\bar{a}(R). \quad (5.65)$$

If $m_+/m \to 0$, then $|a(-R) - a(x - R)| = o(a(x) \vee 1)$ uniformly for $0 \leq x < R$ (as noted after Proposition 5.5.1) while by Proposition 4.3.1, $|a(Q) - a(x + Q)| = o(a(x) \vee 1)$ for $x \geq 0$ whenever $m(x)Q^{-1}\int_1^Q dy/m(y) \to 0$.

From Proposition 5.6.1 and Lemmas 5.6.4 and 5.6.5 we have the following

Proposition 5.6.6 *Let* $m_+/m \to 0$ *and* $\sigma^2 = \infty$. *Then, as* $Q \wedge R \to \infty$,

$$1 - H^0_{B(Q,R)}(0) \sim \frac{a(Q + R)}{a(Q)a(R)}$$

and uniformly for $-Q < x < R$ *subject to condition (5.63),*

$$P_x[\sigma_{\mathbb{Z}\setminus(-Q,R)} < \sigma_0] \sim a^\dagger(x)(1 - H^0_{B(Q,R)}(0)). \quad (5.66)$$

Proposition 5.6.7 *Suppose* $m_+/m \to 0$ *and* $\sigma^2 = \infty$. *Then uniformly for* $-Q < x < R$ *subject to (5.63), as* $Q \wedge R \to \infty$,

(i) $H^x_{B(Q,R)}(R) \sim a^\dagger(x)/a(R)$ [11] *and*

$$P_x\left[\sigma_R < \sigma_{-Q} \,\middle|\, \sigma_{-Q,R} < \sigma_0\right] \sim a(Q)/a(Q + R); \quad (5.67)$$

(ii) *if* $Q/R < M$ *for some* $M > 1$ *in addition,*

$$P_x\left[\sigma_{[R,\infty)} < \sigma_{(-\infty,-Q]} \,\middle|\, \sigma_{\mathbb{Z}\setminus(-Q,R)} < \sigma_0\right] \sim \frac{a(Q)}{a(Q + R)}.$$

Proof Let $m_+/m \to 0$ and (5.63) be valid. By decomposing

$$\{\sigma_R < \sigma_0\} = \{\sigma_{-Q} < \sigma_R < \sigma_0\} + \{\sigma_R < \sigma_{-Q} \wedge \sigma_0\} \quad (5.68)$$

('+' designates the disjoint union) it follows that

---

[10] The case $a^\dagger(x) = 0$ (i.e., S is right-continuous and $x < 0$), where $P_x[\sigma_{\mathbb{Z}\setminus(-Q,R)} < \sigma_0] = 1 - H^x_{B(Q,R)}(0) = [a(Q) - a(Q + x)]/a(Q)$, is not interesting.

[11] This is true for $-Q \ll x \leq R$ subject to $a(-Q)/a(Q) \ll a^\dagger(x)/a^\dagger(|x|)$ regardless of (5.63). [Use $P_x[\sigma_{-Q} < \sigma_0] \leq \bar{a}^\dagger(x)/\bar{a}(Q)$ instead of the bound used for (5.70).]

$$H^x_{B(Q,R)}(R) = P_x[\sigma_R < \sigma_0] - P_x[\sigma_{-Q} < \sigma_R < \sigma_0]. \qquad (5.69)$$

The second probability on the RHS is equal to $P_x[\sigma_{-Q} < \sigma_{0,R}]P_{-Q}[\sigma_R < \sigma_0]$, hence less than $P_x[\sigma_{-Q} < \sigma_0]P_{-Q}[\sigma_R < \sigma_0]$. By Lemma 5.1.1 (see (5.7)) $P_{-Q}[\sigma_R < \sigma_0] \leq a(-Q)/a(R)$, while owing to (5.63) we have $P_x[\sigma_{-Q} < \sigma_0] \sim a^\dagger(x)/a(Q)$. Hence

$$P_x[\sigma_{-Q} < \sigma_R < \sigma_0] \leq \frac{a^\dagger(x)a(-Q)}{a(R)a(Q)}\{1 + o(1)\}. \qquad (5.70)$$

The RHS being $o\left(a^\dagger(x)/a(R)\right)$, we obtain

$$H^x_{B(Q,R)}(R) \sim P_x[\sigma_R < \sigma_0] \sim a^\dagger(x)/a(R).$$

Now noting

$$P_x[\sigma_R < \sigma_{-Q} \mid \sigma_{\{-Q,R\}} < \sigma_0] = H^x_{B(Q,R)}(R)/\left[1 - H^x_{B(Q,R)}(0)\right]$$

one deduces (5.67) immediately from (5.64).

For the proof of (ii) let $\tau(Q)$ be the first time S exits from $(-\infty, -Q]$ after its entering this half-line (see (5.19)) and $\mathcal{E}_Q$ denote the event $\{\sigma_{(-\infty,-Q]} < \sigma_0\}$. Then

$$P_x[\sigma_{(-\infty,-Q]} < \sigma_{[R,\infty)} < \sigma_0] \leq \sum_{-Q < y < -Q/2} P_x[S_{\tau(Q)} = y, \mathcal{E}_Q]P_y[\sigma_{[R,\infty)} < \sigma_0]$$
$$+ P_x[S_{\tau(Q)} \geq -Q/2, \mathcal{E}_Q].$$

The second term on the RHS is $o\left(P_x(\mathcal{E}_Q)\right)$ owing to Lemma 5.2.5 (see also (5.65)) applied with $-Q$ in place of $-R$. As for the first term, we claim that under the condition $Q/R < M$,

$$P_y[\sigma_{[R,\infty)} < \sigma_0] \to 0 \quad \text{uniformly for} \quad -Q < y < -Q/2,$$

which, combined with the bounds above, yields

$$P_x[\sigma_{(-\infty,-Q]} < \sigma_{[R,\infty)} < \sigma_0] = P_x(\mathcal{E}_Q) \times o(1). \qquad (5.71)$$

Since for any $\varepsilon > 0$, $P_y[S_{\sigma[0,\infty)} \geq \varepsilon Q] \to 0$ uniformly for $-Q < y < 0$ owing to Proposition 5.2.1, it follows that under $Q/R < M$,

$$P_y[\sigma_{[R,\infty)} < \sigma_0] \leq \sum_{0 < z < R} P_y[S_{\sigma[0,\infty)} = z]P_z[\sigma_{[R,\infty)} < \sigma_0] + o(1) \to 0.$$

By Proposition 5.5.1(ii) the RHS is not larger than $P_y[\sigma_R < \sigma_0] + o(1)$ and by (5.65) $P_y[\sigma_R < \sigma_0] \sim a^\dagger(y)/2\bar{a}(R) \to 0$. Thus the claim is verified.

For the rest, we can proceed as in the proof of (i) above with an obvious analogue of (5.68). Under (5.63), by Proposition 5.5.1(ii) $P_x[\sigma_{[R,\infty)} < \sigma_0] \sim a^\dagger(x)/a(R)$ and by Proposition 5.5.1(i) $P_x(\mathcal{E}_Q) \sim a^\dagger(x)/a(Q)$. Using (5.66) and (5.71) we now obtain the result asserted in (ii) as in the proof of (i). The proof of Proposition 5.6.7 is finished.   □

Remark 5.6.8 If $\sigma^2 < \infty$ (with $EX = 0$), then as $Q \wedge R \to \infty$,

$$P_x[\sigma_{(-\infty,-Q] \cup [R,\infty)} < \sigma_0] = (P_x[\sigma_R < \sigma_0] + P_x[\sigma_{-Q} < \sigma_0]) \{1 + o(1)\}$$

uniformly for $-Q < x < R$. This follows by $\sup_{x>0} P_x[\sigma_{-Q} < \sigma_0] \to 0 \, (Q \to \infty)$ and the dual of it. [See Remark 5.1.8 for $P_x[\sigma_z < \sigma_0]$.]

5.7 Sojourn Time of a Finite Set for S with Absorbing Barriers

For a nonempty finite set $I \subset \mathbb{Z}$, define its sojourn time for S killed upon entering $[R, \infty)$ or $(-\infty, -Q] \cup [R, \infty)$ as

$$N_R(I) = \#\{n : n < \sigma_{[R,\infty)} : S_n \in I\}; \text{ and}$$

$$N_{Q,R}(I) = \#\{n : n < \sigma_{\mathbb{Z}\backslash(-Q,R)} : S_n \in I\},$$

respectively, where $\#$ designates the cardinality of a set. Spitzer and Stone [72] gave the asymptotic distribution of $N_R(I)$ and $N_{Q,R}(I)$ when $\#I = 1$, X is symmetric (i.e. $\mu_+ = \mu_-$) and $\sigma^2 < \infty$, which Kesten [39] generalised to arbitrary I with $\#I < \infty$ in some cases of $\sigma^2 = \infty$. Put

$$e_R = P_0[\sigma_{[R,\infty)} < \sigma_0] \quad \text{and} \quad e_{Q,R} = P_0[\sigma_{\mathbb{Z}\backslash(-Q,R)} < \sigma_0].$$

Then it is shown in [39, Lemma 1] that for any finite subset $I \subset \mathbb{Z}$ and $t > 0$

$$\begin{aligned} &(a) \; P_0\left[N_R(I) \ge (\#I)t/e_R\right] \to e^{-t} \qquad (R \to \infty), \\ &(b) \; P_0\left[N_{Q,R}(I) \ge (\#I)t/e_{Q,R}\right] \to e^{-t} \quad (R \wedge Q \to \infty). \end{aligned}$$

[Both (a) and (b) are valid for any recurrent r.w. that is irreducible on $\mathbb{Z}$.] Thus the problem of finding the asymptotic distributions of $N_R(R)$ and $N_{Q,R}(I)$ reduces to that of finding the asymptotics of e_R and $e_{Q,R}$. Kesten [39] derived explicit asymptotic forms of e_R when F is in the domain of normal attraction of a stable law with exponent $1 \le \alpha \le 2$, and those of $e_{Q,R}$ for $\alpha = 2$ and, by further restricting to symmetric distributions, for $1 \le \alpha < 2$. The next proposition extends his result significantly. We bring in the following conditions:

$$(\mathrm{I}_+) \begin{cases} \text{either} & EX = 0, \; \sigma^2 = \infty \text{ and } m_+/m \to 0, \\ \text{or} & S \text{ is p.r.s.} \end{cases}$$

$$(\mathrm{I}_-) \begin{cases} \text{either} & EX = 0, \; \sigma^2 = \infty \text{ and } m_-/m \to 0, \\ \text{or} & S \text{ is n.r.s.} \end{cases}$$

Proposition 5.7.1 *If (I_+) holds, then*

$$e_R \sim 1/a(R) \qquad (R \to \infty); \; and \tag{5.72}$$

$$e_{Q,R} \sim \frac{a(Q+R)}{a(Q)a(R)} \qquad (Q \wedge R \to \infty). \tag{5.73}$$

The obvious analogue holds under (I_-): one may replace $a(\cdot)$ by $a(-\cdot)$.

Let B be a non-empty subset of $\mathbb{Z}$ such that $0 \notin B$ and put

$$e(B) = P_0[\sigma_B < \sigma_0],$$

so that $e_R = e([R, \infty))$ and $e_{Q,R} = e((-\infty, -Q] \cup [R, \infty))$. Then

$$e(B) = 1/g_B(0,0) \quad \text{and} \quad e(B) = e(-B). \tag{5.74}$$

Thus $e_R = P_0[\sigma_{(-\infty,-R]} < \sigma_0]$ and $e_{Q,R} = P_0[\sigma_{(-\infty,-R]\cup[Q,\infty)} < \sigma_0]$.

If $m_+/m \to 0$, (5.72) is a special case of Proposition 5.5.1, and (5.73) is that of Proposition 5.6.6l.

For the r.s. walks, we consider the case (I_-), the other one being disposed of by duality. Let S be n.r.s. Then $e_R \sim 1/a(-R)$ by Theorem 6.1.1. Since $a(-x)$ is s.v. as $x \to \infty$ (Corollary 4.1.2), it suffices to show that $e_{Q \wedge R} \sim e_{Q,R}$ as $R \wedge Q \to \infty$. This follows from Lemma 5.7.3 below (which will also be of later usage).

Lemma 5.7.2 *Suppose $\hat{Z}$ is r.s. Then, for each $\delta < 1$, as $z \to \infty$,*

$$\frac{P_z[\sigma_y < T]}{P_x[\sigma_y < T]} \le \frac{\hat{\ell}^*(x)}{\hat{\ell}^*(z)}\{1+o(1)\} \quad \text{uniformly for } 0 \le y \le x < \delta z.$$

Proof Let $\hat{Z}$ be r.s., so that $v_d(x) \sim 1/\hat{\ell}^*(x)$. Then, by Spitzer's formula, $g_\Omega(z,y) \sim U_a(y)/\hat{\ell}^*(z)$ for $0 < y < \delta z$, while $g_\Omega(x,y) \ge U_a(y)/\hat{\ell}^*(x)\{1+o(1)\}$ ($0 \le y \le x$) since $\hat{\ell}^*$ is increasing. This shows the inequality of the lemma.   □

Lemma 5.7.3 *If F is r.s., then for each $\delta < 1$, as $R \to \infty$,*

$$g_{B(R)}(x,y) \sim g_\Omega(x,y) \qquad \text{uniformly for } \begin{cases} 0 \le x \le y < \delta R & \text{if } S \text{ is p.r.s.,} \\ 0 \le y \le x < \delta R & \text{if } S \text{ is n.r.s.,} \end{cases}$$

where $B(R) := (-\infty, 0) \cup (R, \infty)$.

Proof Let S be n.r.s. and $0 \le y \le x < \delta R$. Then $\hat{Z}$ is r.s. and by the preceding lemma $g_\Omega(z,y) \le \{1+o(1)\}g_\Omega(x,y)$ for $z > R$. This together with $P_x(\Lambda_R) \to 0$ shows $g_{B(R)}(x,y) \sim g_\Omega(x,y)$ (cf. (7.54)). The case when S is p.r.s. is disposed of by duality.   □

Remark 5.7.4 (i) Using the relation noted in Remark 5.6.8 we see that if $\sigma^2 < \infty$,

$$e_R \sim 1/[2a(R)] \quad \text{and} \quad e_{Q,R} \sim a(Q+R)/[2a(Q)a(R)].$$

For asymptotically stable walks with exponent $1 \leq \alpha \leq 2$ having $\rho = \lim P_0[S_n > 0]$, we shall compute the asymptotics of $g_\Omega(x,y)$ and $g_{B(R)}(x,y)$ in Chapter 8 (Theorems 8.2.1 and 8.4.3), which give asymptotic forms of e_R and $e_{R,Q}$ in view of (5.74) (see (8.14) and Theorem 8.4.1) and in particular show that the equivalences in (5.72) and (5.73) hold for $\alpha = 1$ with $0 < \rho < 1$ (where $a(x) \sim a(-x)$). See also Proposition 9.5.1 and Theorem 9.5.5.

(ii) If $m_+/m \to 0$, then by Theorem 3.1.6 and (5.24), as $x \to \infty$

$$g_\Omega(x,x) \sim a(x).$$

(iii) If $a(-x)$, $x \geq 1$, is almost increasing, then by using (2.15) (see also (7.35) and (7.34)) one can easily deduce the lower bound $g_\Omega(x,x) \geq c[a(x) \vee a(-x)]$ (with $c > 0$), entailing $g_\Omega(x,x) \asymp g_{\{0\}}(x,x)$. This "$\asymp$" is not generally true, since $g_\Omega(x,x) = \sum_0^x v_d(k)u_a(k)$ is increasing, whereas it is possible that $\liminf \bar{a}(2x)/\bar{a}(x) = 0$ (see Section 3.6), so that $\liminf[g_\Omega(x,x)/g(x,x)] = 0$.

Chapter 6
The Two-Sided Exit Problem – General Case

Put

$$\Omega = (-\infty, -1] \quad \text{and} \quad T = \sigma_\Omega.$$

In this and the next chapter, we are concerned with the asymptotic form as $R \to \infty$ of the probability $P_x(\Lambda_R)$, where

$$\Lambda_R = \{\sigma_{[R+1,\infty)} < T\},$$

the event that the r.w. S exits from an interval $[0, R]$ on the upper side. The classical result given in [71, Theorem 22.1] says that if the variance $\sigma^2 := EX^2$ is finite, then $P_x(\Lambda_R) - x/R \to 0$ uniformly for $0 \le x \le R$. This can be refined to

$$P_x(\Lambda_R) \sim \frac{V_{\mathrm{d}}(x)}{V_{\mathrm{d}}(R)} \quad \text{uniformly for } 0 \le x \le R \text{ as } R \to \infty \tag{6.1}$$

as is given in [76, Proposition 2.2] (the proof is easy: see Section 6.3.1 of the present chapter). Here $V_{\mathrm{d}}(x)$ denotes as before the renewal function of the weakly descending ladder height process. Our primary purpose in this chapter is to find a sufficient condition for (6.1) to hold that applies to a wide class of r.w.'s in $\mathbb{Z}$ with $\sigma^2 = \infty$. When S is attracted to a stable process of index $0 < \alpha \le 2$ and there exists $\rho = \lim P_0[S_n > 0]$, we shall see that the sufficient condition obtained is also necessary for (6.1) and fulfilled if and only if $(\alpha \vee 1)\rho = 1$, and also provide some asymptotic estimates of $P_x(\Lambda_R)$ in the case $(\alpha \vee 1)\rho \ne 1$.

If the distribution of X is symmetric and belongs to the domain of normal attraction of a stable law with exponent $0 < \alpha \le 2$, the problem is treated by Kesten [40]: for $0 < \alpha \le 2$ he derived (among other things) the explicit analytic expression of the limit of the law of the suitably scaled overshoot (beyond the level -1 or $R + 1$) and thereby identified the limit of $P_x(\Lambda_R)$ as $x \wedge R \to \infty$ under $x/R \to \lambda \in (0, 1)$. Rogozin [64] studied the corresponding problem for strictly stable processes and obtained an analytic expression of the probability of the process exiting the unit interval on the upper side (based on the law of the overshoot distribution given in (2.27)) and, as an application of the result, obtained the scaled limit of $P_x(\Lambda_R)$ for asymptotically stable r.w.'s (see Remark 6.1.4), extending the result of [40] mentioned above. There are a few other results concerning the asymptotic behaviour of $P_x(\Lambda_R)$. Griffin and McConnell [35] and Kesten and Maller [46] gave a criterion for $P_x(\Lambda_{2x})$ to approach unity (see Section 2.4). When X is in the domain of attraction of a stable law, Doney [19] derived the asymptotic form of $P_0(\Lambda_R)$. Kesten and Maller [47] gave an analytic condition for $\liminf P_x(\Lambda_{2x}) \wedge [1 - P_x(\Lambda_{2x})] > 0$, i.e., for S started at the centre of a long interval to exit either side of it with positive probability (see Remark 6.1.6).

6.1 Statements of Results

In the case $E|X| < \infty$, we use the notations $\eta_\pm(x) = \int_x^\infty \mu_\pm(t)\,dt$, $\eta = \eta_- + \eta_+$, as well as $m_\pm(x) = \int_0^x \eta_\pm(t)\,dt$ and $m = m_- + m_+$. In the sequel we always assume $\sigma^2 = EX^2 = \infty$ for simplicity. Because of our basic assumption that S is oscillatory, $EX = 0$ if $E|X| < \infty$. The main result of this chapter is stated in the following theorem. We provide some results complementary to it in Proposition 6.1.3.

Theorem 6.1.1 *For (6.1) to be true each of the following is sufficient.*

(C1) $EX = 0$ *and* $m_+(x)/m(x) \to 0$ *as* $x \to \infty$.
(C2) $EX = 0$ *and* $m(x)$ *is s.v. as* $x \to \infty$.
(C3) Z *is r.s. and* V_d *is s.v.*
(C4) $\mu_-(x)/\mu_+(x) \to 0$ *as* $x \to \infty$ *and* $\mu_+(x)$ *is regularly varying at infinity with index* $-\alpha$, $0 < \alpha < 1$.

Recall that the ladder height Z is *relatively stable (r.s.)* if and only if $u_a(x)$ is s.v., in which case $u_a(x) \sim 1/\ell^*(x)$ (cf. (5.28)). On the other hand V_d is s.v. if and only if $P_0[\hat{Z}(R)/R > M] \to 1$ for every $M > 1$ (see Theorem A.2.3).

The sufficiency of (C1) follows from Proposition 5.3.2, and the primary task of this chapter is to verify the sufficiency of (C2) to (C4); the proof will be made independently of the result under (C1).

It is warned that the second condition in (C4) does not mean that $\mu_-(x)$ may get small arbitrarily fast as $x \to \infty$, the random walk S being assumed to oscillate, so that the dual of the growth condition (5.22) must be satisfied. In case (C3) (6.1) is especially of interest when $x/R \to 0$, for if V_d is s.v., (6.1) entails $P_x(\Lambda_R) \to 1$ whenever x/R is bounded away from zero.

Condition (C2) holds if and only if F belongs to the domain of attraction of the standard normal law (cf. Section A.1.1). If F is attracted to a stable law of exponent α, then (C3) is equivalent to $\alpha = 1 = \lim P_0[S_n > 0]$ [81] (see the table in Section 2.6). (C4) holds if and only if F is attracted to a stable law of exponent $\alpha \ (\in (0,1))$ and $\mu_+/\mu \to 1$, and if this is the case $\lim P_0[S_n > 0] = 1$.

Remark 6.1.2 The condition (C3) holds if S is p.r.s., or, what amounts to the same under $\sigma^2 = \infty$, $A(x)/[x\mu(x)] \longrightarrow \infty$ ($x \to \infty$) (see Section 2.4.1), because the positive relative stability entails that Z is r.s. and $P_x(\Lambda_{2x}) \to 1$ (see (2.21)), the latter implies V_d is s.v. by virtue of (6.12) below. Note that in view of Theorem 6.1.1, (C3) implies $P_x(\Lambda_{2x}) \to 1$. Combining this with the result given at (2.22) we shall see, in Theorem A.4.1, that (C.3) implies – hence is equivalent to – the positive relative stability of S (this fact will not be used throughout).

As before put $v° = v_d(0)(= V_d(0))$,

$$\ell^*(x) = \int_0^x P[Z > t]\, dt \quad \text{and} \quad \hat{\ell}^*(x) = \frac{1}{v°} \int_0^x P[-\hat{Z} > t]\, dt.$$

Under (C3) and (C4), we bring in the function

$$\ell_\sharp(x) = \alpha \int_x^\infty \frac{U_a(t)\mu_-(t)}{t}\, dt \quad (x > 0, \, \alpha < 1), \tag{6.2}$$

in which the summability of the integral is assured by $U_a(x) = \sum_{-\infty}^0 U_a(y)p(y + x)$ (see (6.10) given in the dual form and (6.19)). Then

$$V_d(x) \sim \begin{cases} a(x)\ell^*(x) & \text{in case} \quad (C1), \\ x/\hat{\ell}^*(x) & \text{in case} \quad (C2), \\ 1/\ell_\sharp(x) & \text{in cases} \quad (C3), (C4). \end{cases} \tag{6.3}$$

[See Lemma 5.3.1, Theorem A.3.2 and Lemma 6.3.1 for the first, second and third formulae of (6.3), respectively.] In each case of (6.3), ℓ^* or $\hat{\ell}^*$ or $\ell_\sharp$ is s.v. – hence in case (C4) (like (C3)) $P_x(\Lambda_{2x}) \to 1$. Under (C1) $a(x) \asymp x/m(x)$ (cf. Section 3.1). Since (C3) holds if S is p.r.s., combining the third equivalence of (6.3) with Theorem 4.1.1 leads to the following extension of the first one (see Remark 6.4.4):

if $p := \lim \mu_+(x)/\mu(x) < 1/2$ and F is recurrent and p.r.s., then

$$V_d(x) \sim \left(1 - q^{-1}p\right) a(x)\ell^*(x) \qquad (q = 1 - p).$$

In all cases (C1) to (C4) it follows that $U_a(x) \sim x^{\alpha \wedge 1}/\ell(x)$, where $\alpha \geq 1$ in cases (C1) to (C3) and ℓ is some s.v. function which is chosen to be a normalised one[1] and that either Z is r.s. or $P[S_n > 0] \to 1$ (see Theorem 5.4.1 (i) for the case (C1)).

In the rest of this section, we shall suppose the asymptotic stability condition (AS) together with (2.16) introduced in Section 2.3 to hold and state results as to upper and lower bounds of $P_x(\Lambda_R)$.

If (AS) holds, then it follows that $\alpha\rho \leq 1$; S is p.r.s. if and only if $\rho = 1$ (see (2.23)); Z is r.s. if and only if $\alpha\rho = 1$; $\rho = 1/2$ for $\alpha = 2$, and that

$$(C2) \Leftrightarrow \alpha = 2; \quad (C3) \Leftrightarrow \alpha = \rho = 1; \quad (C4) \Leftrightarrow \alpha < 1 = p \Rightarrow \rho = 1 \tag{6.4}$$

---

[1] An s.v. function of the form $e^{\int_1^x f(t)\, dt}$ with $f(t) = o(1/t)$ is called normalised (see [8]).

(cf. Section 2.6); in particular, either one of (C1) to (C4) holds if and only if $(\alpha \vee 1)\rho = 1$. We consider the case $(\alpha \vee 1)\rho < 1$ in the next proposition, which combined with Theorem 6.1.1 shows that (6.1) holds if and only if $(\alpha \vee 1)\rho = 1$.

Proposition 6.1.3 *Suppose (AS) holds with* $0 < \alpha < 2$. *Let* δ *be a constant arbitrarily chosen so that* $\frac{1}{2} < \delta < 1$.

(i) *If* $0 < (\alpha \vee 1)\rho < 1$ *(equivalently, $p > 0$ for $\alpha > 1$ and $0 < \rho < 1$ for $\alpha \leq 1$), then there exist constants $\theta_* > 0$ and $\theta^* < 1$ such that for all sufficiently large R,*

$$\theta_* \leq \frac{P_x(\Lambda_R)V_d(R)}{V_d(x)} \leq \theta^* \qquad \text{for} \quad 0 \leq x \leq \delta R. \qquad (6.5)$$

(ii) *Let* $\alpha \leq 1$. *In (6.5), one can choose θ_* and θ^* so that $\theta^* \to 0$ as $\rho \to 0$ and $\theta_* \to 1$ as $\rho \to 1$. [The statement "$\theta^* \to 0$ as $\rho \to 0$" means that for any $\varepsilon > 0$ there exists a $\delta > 0$ such that $\theta^* < \varepsilon$ for any 'admissible' F with $0 < \rho < \delta$, and similarly for "$\theta^* \to 1$ as $\rho \to 1$".] If $\rho = 0$, then $P_x(\Lambda_R)V_d(R)/V_d(x) \to 0$ uniformly for $0 \leq x < \delta R$, and $V_d(x) \sim x^\alpha / \hat{\ell}(x)$ for some s.v. function $\hat{\ell}$ (for $\alpha = 1$, one can take $\hat{\ell}^*$ for $\hat{\ell}$ (see (6.49).)*

Remark 6.1.4 Suppose that (AS) holds and that $0 < \rho < 1$ if $\alpha = 1$ (so that $P_0[S_n/c_n \in \cdot]$ converges to a non-degenerate stable law for appropriate constants c_n; the results below do not depend on the choice of c_n). Let $Y = (Y(t))_{t \geq 0}$ be a limit stable process with probabilities P_ξ^Y, $\xi \in \mathbb{R}$ (i.e., the limit law for $S_{\lfloor nt \rfloor}/c_n$ as $S_0/c_n \to \xi$). Denote by σ_Δ^Y the first hitting time of Y to a 'nice' set $\Delta \subset \mathbb{R}$ and by Λ_r^Y $(r > 0)$ the event corresponding to Λ_R for Y: $\Lambda_r^Y = \{\sigma_{(r,\infty)}^Y < \sigma_{(-\infty,0)}^Y\}$, and put $Q_r(\xi) = P_\xi^Y(\Lambda_r^Y)$ $(0 < \xi < 1)$. Rogozin [64] established the overshoot law (2.27) of Section 2.5 and from it derived an analytic expression of $Q_1(\xi)$ which, if $0 < \rho < 1$, must read

$$Q_1(\xi) = \frac{1}{B(\alpha\rho, \alpha\hat{\rho})} \int_0^\xi t^{\alpha\hat{\rho}-1}(1-t)^{\alpha\rho-1}\, dt \quad (0 < \xi < 1), \qquad (6.6)$$

where $\hat{\rho} = 1 - \rho$ and $B(s,t) = \Gamma(s)\Gamma(t)/\Gamma(s+t)$, the beta function. (When $\rho\hat{\rho} = 0$, $Q_1(\xi) = 1$ or 0 $(\xi > 0)$ according as $\rho = 1$ or 0.) By the functional limit theorem, one deduces that for each $0 < \xi < 1$, $P_{\xi c_n}(\Lambda_{c_n}) \to P_\xi^Y(\Lambda_1^Y)$, as is noted in [64]. This convergence is uniform since $P_x(\Lambda_R)$ is monotone in x and $Q(\xi)$ is continuous.

The verification of (6.6), given in [64], seems to be unfinished or wrongly given, one more non-trivial step to evaluate a certain integral being skipped that results in the incorrect formula – ρ and $\hat{\rho}$ are interchanged in (6.6). Here we provide a proof of (6.6), based on (2.27). Suggested by the proof of Lemma 1 of [40], we prove the following equivalent of (6.6):

$$Q_{1+c}(1) = \frac{1}{B(\alpha\rho, \alpha\hat{\rho})} \int_0^{1/(1+c)} t^{\alpha\hat{\rho}-1}(1-t)^{\alpha\rho-1}\, dt \quad (c > 0), \qquad (6.7)$$

an approach which makes the necessary computations plain to see. By (2.27) the conditional probability on the LHS is expressed as

$$\frac{\sin \pi \alpha \rho}{\pi} c^{\alpha \rho} \int_0^\infty \frac{t^{-\alpha \rho}(t+1+c)^{-\alpha \hat{\rho}}}{t+c} \, dt.$$

On changing the variable $t = cs$, this yields

$$P_0^Y(\Lambda_{1+c}^Y) = \frac{\sin \pi \alpha \rho}{\pi} \int_0^\infty \frac{s^{-\alpha \rho}(1+(1+s)c)^{-\alpha \hat{\rho}}}{1+s} \, ds.$$

Using the identity $\int_0^\infty s^{\gamma-1}(1+s)^{-\nu} \, ds = B(\gamma, \nu - \gamma)$ $(\nu > \gamma > 0)$ one can easily deduce that the above expression and the RHS of (6.7) are both equal to 1 for $c = 0$ and their derivatives with respect to c coincide, showing (6.7).

Remark 6.1.5 If $\alpha \rho = 1$, from Lemma 5.2.4 it follows that

$$\inf_{0 \leq x \leq R} P_0[Z(R) \leq \eta R | \Lambda_R] \to 1 \text{ for } \eta > 0.$$

If $\hat{\rho} = \alpha = 1$, $P_x[Z(R) \leq \eta R] \to 0$ for every $\eta > 0$ while, as we shall see in Proposition 7.6.4, $P_{\xi R}[Z(R)/R \in \cdot | \Lambda_R]$ converges to a proper probability law for every $\xi \in (0, 1)$ under some additional condition on μ_+ if F is transient.

Remark 6.1.6 For general random walks (not restricted to arithmetic ones), Kesten and Maller [47] gave an analytic equivalent for the r.w. S to exit from symmetric intervals on the upper side with positive P_0-probability. Theorem 1 of [47] entails that $\limsup P_x(\Lambda_{2x}) < 1$ if and only if $\limsup P_0[S_n \geq 0] < 1$, provided that $\mu_-(x) > 0$ for all $x > 0$, and each of these two is equivalent to $\limsup_{x \to \infty} A(x)[\mu_-(x)c(x)]^{-1/2} < \infty$ (recall $c(x) = \int_0^x t\mu(t) \, dt$). Under $\mu_+(x) > 0$ for all $x > 0$, this result may be paraphrased as follows:

$$\liminf_{x \to \infty} P_x(\Lambda_{2x}) > 0 \quad \text{if and only if} \quad \liminf_{n \to \infty} P_0[S_n \geq 0] > 0, \qquad (6.8)$$

and this is the case if and only if $\liminf_{x \to \infty} A(x)[\mu_+(x)c(x)]^{-1/2} > -\infty$.

As mentioned in Remark 6.1.4, the formula (6.6) yields the explicit asymptotic form of $P_x(\Lambda_R)$ if x/R is bounded away from zero unless $\alpha = 1$ and $\rho \hat{\rho} = 0$, but when $x/R \to 0$ it says only $P_x(\Lambda_R) \to 0$. One may especially be interested in the case $x = 0$ (cf. Remark 8.3.7). By extending the idea of Bolthausen [10], Doney [19, Corollary 3] proved that if (AS) holds with $0 < \rho < 1$, then

$$P_0\left[\max_{n<T} S_n \geq b_n\right] = P_0(\Lambda_{b_n}) \sim c/n, \qquad (6.9)$$

where c is a positive constant and $b_1, b_2, \ldots$ are norming constants characterised by the relation $V_d(b_n)/n \to v^\circ$,[2] so that (6.9) may be written as $P_0(\Lambda_R) \sim c/V_d(R)$.

---

[2] Here, using (6.13), we have rephrased the condition in [19] where it is expressed by means of $P[-\hat{Z} > t]$ instead of $V_d(x)$.

The corollary below gives the corresponding consequence of Proposition 6.1.3, which improves on the result of [19] mentioned right above by extending it to the case $\rho \in \{0, 1\}$ and providing information on how the limit depends on ρ.

Suppose (AS) holds. We distinguish the following three cases:

Case *I*: $(\alpha \vee 1)\rho = 1$.

Case *II*: $0 < (\alpha \vee 1)\rho < 1$.

Case *III*: $\rho = 0$.

If $\alpha = 2$, *I* is always the case. For $1 < \alpha < 2$, *I* or *II* is the case according as $p = 0$ or $p > 0$. If $\alpha = 1$, we have Cases *I* (*III*) if $p < 1/2$ ($p > 1/2$), and all the cases *I* to *III* are possible if $p = 1/2$. For $\alpha < 1$, *I* or *III* is the case according as $p = 1$ or $p = 0$, otherwise we have Case *II*.

The following result is obtained as a corollary of Theorem 6.1.1 and Proposition 6.1.3, except for the case $\alpha > 1$ of its last assertion.

Corollary 6.1.7 *Suppose (AS) holds with* $\alpha < 2$. *Then*

(i) *there exists* $c := \lim P_0(\Lambda_R)V_d(R)/v^\circ$;
(ii) $c = 1$ *in Case I*, $0 < c < 1$ *in Case II and* $c = 0$ *in Case III;*
(iii) $c \to 0$ *as* $\rho \to 0$ *(necessarily* $\alpha \leq 1$*), and* $c \to 1$ *as* $\rho \to 1/(\alpha \vee 1)$.

Proof As to the proof of (i) and (ii), Cases *I* and *III* are covered by Theorem 6.1.1 and by the second half of Proposition 6.1.3(ii), respectively, and Case *II* by the above mentioned result of [19][3] combined with Proposition 6.1.3. The assertion (iii) follows from Proposition 6.1.3(ii) if $\alpha \leq 1$. If $1 < \alpha < 2$, we need to use a result from Chapter 8. There we shall obtain the asymptotic form of $P_x[\sigma_y < T \,|\, \Lambda_R]$ as $R \to \infty$ for $x < y$ in Proposition 8.3.1(i) (deduced from the estimates of u_a and v_d (see Lemma 8.1.1) proved independently of the present arguments), which entails that as $R \to \infty$,

$$\frac{P_0(\Lambda_R)V_d(R)}{v^\circ} \geq \frac{P_0[\sigma_R < T]V_d(R)}{v^\circ} \longrightarrow \frac{\alpha - 1}{\alpha(1 - \rho)}.$$

Thus the second half of (iii) also follows for $\alpha > 1$.   □

Remark 6.1.8 For $\alpha \geq 1$ (with an additional condition if $\alpha = 2\rho = 1$), we shall obtain an explicit asymptotic form of $P_x(\Lambda_R)$ as $x/R \to 0$, which will in particular give $c = [\alpha\hat{\rho}B(\alpha\rho, \alpha\hat{\rho})]^{-1}$ (see (8.69) of Proposition 8.6.4 in the case $0 < \alpha\rho < 1$).

The rest of this chapter is organised as follows. Section 6.2 states several known facts and provides a lemma; these will be fundamental in the remainder of this treatise. A proof of Theorem 6.1.1 is given in Section 6.3. In Section 6.4, we suppose (C3) or (C4) and prove lemmas as to miscellaneous matters. In Section 6.5 we prove Proposition 6.1.3 by further showing several lemmas.

---

[3] In [19] $EX = 0$ is assumed if $\alpha = 1$. However, by examining the arguments therein, one sees that this assumption is superfluous, so that (6.9) is valid under (AS) with $0 < \rho < 1$.

6.2 Upper Bounds of $P_x(\Lambda_R)$ and Partial Sums of $g_\Omega(x, y)$

Let v_d denote the renewal sequence for the weakly descending ladder height process: $v_d(x) = V_d(x) - V_d(x-1)$ and $v_d(0) = v^\circ$. Recall $V_d(x)$ is set to be zero for $x \le -1$. Similarly, let $U_a(x)$ and $u_a(x)$ be the renewal function and sequence for the strictly ascending ladder height process.

The function V_d is harmonic for the r.w. S killed as it enters Ω in the sense that

$$\sum_{y=0}^{\infty} V_d(y)p(y - x) = V_d(x) \qquad (x \ge 0), \qquad (6.10)$$

where $p(x) = P[X = x]$ [71, Proposition 19.5], so that the process $M_n = V_d(S_{n \wedge T})$ is a martingale under P_x for $x \ge 0$. By the optional stopping theorem one deduces

$$E_x\left[V_d(S_{\sigma(R,\infty)}); \Lambda_R\right] = V_d(x). \qquad (6.11)$$

Indeed, on passing to the limit in $E_x[M_{n \wedge \sigma(R,\infty)}] = V_d(x)$, Fatou's lemma shows that the expectation on the LHS is less than or equal to $V_d(x)$, which, in turn, shows that the martingale $M_{n \wedge \sigma(R,\infty)}$ is uniformly integrable since it is bounded by the summable random variable $V_d(R) \vee M_{\sigma(R,\infty)}$. Obviously the expectation in (6.11) is not less than $V_d(R)P_x(\Lambda_R)$, so that

$$P_x(\Lambda_R) \le V_d(x)/V_d(R). \qquad (6.12)$$

If either V_d or $\int_0^x P[-\hat{Z} > t]\, dt$ is regularly varying with index γ, it follows that

$$\frac{V_d(x)}{v^\circ x} \int_0^x P[-\hat{Z} > t]\, dt \longrightarrow \frac{1}{\Gamma(1 + \gamma)\Gamma(2 - \gamma)} \qquad (6.13)$$

(see Theorem A.2.3).

For a set $B \subset \mathbb{R}$ with $B \cap \mathbb{Z} \ne \emptyset$ we have defined by (2.1) the Green function $g_B(x, y)$ of the r.w. killed as it hits B. We restate Spitzer's formula

$$\text{Eq(2.6)} \quad g_\Omega(x, y) = \sum_{k=0}^{x \wedge y} v_d(x - k)u_a(y - k) \quad \text{for } x, y \ge 0.$$

It follows that for $x \ge 1$,

$$P[\hat{Z} = -x] = \sum_{y=0}^{\infty} g_\Omega(0, y)p(-x - y) = v^\circ \sum_{y=0}^{\infty} u_a(y)p(-x - y) \qquad (6.14)$$

and, by the dual relation $g_{[1,\infty)}(0, -y) = \hat{g}_\Omega(0, y) = v_d(y)$,

$$P[Z = x] = \sum_{y=0}^{\infty} g_{[1,\infty)}(0, -y)p(x + y) = \sum_{y=0}^{\infty} v_d(y)p(x + y). \qquad (6.15)$$

Lemma 6.2.1 *For $x \geq 0$ and $R \geq 1$,*

$$\sum_{y=0}^{R} g_{\Omega}(x, y) < V_{\mathrm{d}}(x) U_{\mathrm{a}}(R). \tag{6.16}$$

Take a positive constant $\delta < 1$ arbitrarily. Then

$$\sum_{w=0}^{\delta x} g_{\Omega}(x - w, x) \leq \sum_{w=0}^{\delta x} g_{\Omega}(x, x + w) \leq V_{\mathrm{d}}(x) U_{\mathrm{a}}(\delta x). \tag{6.17}$$

If U_{a} is regularly varying, then there exists a constant $c > 0$ such that for $0 \leq x \leq \delta R$,

$$\sum_{y=0}^{R} g_{\Omega}(x, y) \geq c \, V_{\mathrm{d}}(x) U_{\mathrm{a}}(R),$$

and if U_{a} is s.v., then uniformly for $0 \leq x < \delta R$,

$$\sum_{y=0}^{R} g_{\Omega}(x, y) \sim V_{\mathrm{d}}(x) U_{\mathrm{a}}(R).$$

Proof One has $g_{\Omega}(x - w, x) \leq \sum_{k=0}^{x} v_{\mathrm{d}}(k) u_{\mathrm{a}}(w + k) = g_{\Omega}(x, x + w)$ for $w \geq 0$. By the subadditivity of $U_{\mathrm{a}}(\cdot - 1)$ that entails $U_{\mathrm{a}}(k + \delta x) - U_{\mathrm{a}}(k - 1) \leq U_{\mathrm{a}}(\delta x)$, summing over w gives (6.17).

Split the sum on the LHS of (6.16) at x. By (2.6) the sum over $y > x$ and that over $y \leq x$ are equal to, respectively,

$$\sum_{y=x+1}^{R} \sum_{k=0}^{x} v_{\mathrm{d}}(k) u_{\mathrm{a}}(y - x + k) = \sum_{k=0}^{x} v_{\mathrm{d}}(k) \left[U_{\mathrm{a}}(R - x + k) - U_{\mathrm{a}}(k) \right]$$

and

$$\sum_{k=0}^{x} \sum_{y=k}^{x} v_{\mathrm{d}}(x - y + k) u_{\mathrm{a}}(k) = \sum_{k=0}^{x} \left[V_{\mathrm{d}}(x) - V_{\mathrm{d}}(k - 1) \right] u_{\mathrm{a}}(k).$$

Summing by parts yields

$$\sum_{k=0}^{x} \left[V_{\mathrm{d}}(x) - V_{\mathrm{d}}(k - 1) \right] u_{\mathrm{a}}(k) = \sum_{k=0}^{x} v_{\mathrm{d}}(k) U_{\mathrm{a}}(k),$$

so that

$$\sum_{y=0}^{R} g_{\Omega}(x, y) = \sum_{k=0}^{x} v_{\mathrm{d}}(k) U_{\mathrm{a}}(R - x + k),$$

from which the assertions of the lemma other than (6.17) follow immediately. □

Lemma 6.2.2 *Suppose S is recurrent.*

(i) *For each $z \in \mathbb{Z}$, as $R \to \infty$, $P_{R+z}[\sigma_R < T] \to 1$, or, what is the same thing, $g_\Omega(R + z, R)/g_\Omega(R, R) \to 1$.*

(ii) *If u_a is asymptotically monotone, then $g_\Omega(x, R)$ is asymptotically monotone in $0 \leq x \leq R$ in the sense that $g_\Omega(x, R) \leq g_\Omega(x', R)\{1 + o(1)\}$ where $o(1) \to 0$ as $R \to \infty$ uniformly for $0 \leq x \leq x' \leq R$. Similarly if v_d is asymptotically monotone, then $g_\Omega(x, R)$ is asymptotically monotone in $x \geq R$ in the sense that $g_\Omega(x, R) \geq g_\Omega(x', R)\{1 + o(1)\}$ for $R \leq x < x'$.*

Proof One can easily verify (i) by comparing the hitting time distributions of σ_R and T under P_{R+z} as $R \to \infty$. Let S be recurrent and u_a asymptotically monotone. By (i) we may suppose $R - x' > M$ for an arbitrarily chosen constant M. For any $\varepsilon > 0$, choose M so that $u_a(z') < u_a(z)(1 + \varepsilon)$ if $z > z' > M$. Then for $M < x \leq R$,

$$g_\Omega(x, R) = \sum_{k=0}^{x} v_d(k)u_a(R - x + k) \leq \sum_{k=0}^{x} v_d(k)u_a(R - x' + k)(1 + \varepsilon).$$

The RHS being less than $g_\Omega(x', R)(1 + \varepsilon)$ shows the first half of (ii). The proof of the second half is similar.   □

6.3 Proof of Theorem 6.1.1

The proof of Theorem 6.1.1 is given in two subsections, the first for case (C2) and the second for cases (C3) and (C4).

6.3.1 Case (C2)

Suppose $EX = 0$ and recall $\eta_\pm(x) = \int_x^\infty P[\pm X > t]\, dt$ and $\eta = \eta_- + \eta_+$. It follows that if $\lim x\eta_+(x)/m(x) = 0$, then Z is r.s. (Proposition 5.2.1), so that $u_a(x) \sim 1/\ell^*(x)$ is s.v. (see (2.5)). Thus (C2), which is equivalent to $\lim x\eta(x)/m(x) = 0$, entails both v_d and u_a are s.v., so that $\sum_{k=0}^{y} v_d(k)u_a(k) \sim yv_d(y)u_a(y) \sim V_d(y)u_a(y)$ $(y \to \infty)$ and one can easily deduce that as $R \to \infty$,

$$g_\Omega(x, R) \sim \sum_{k=0}^{x} \frac{v_d(k)}{\ell^*(R - x + k)} \sim \frac{V_d(x)}{\ell^*(R)} \qquad \text{uniformly for } 0 \leq x \leq R.$$

This leads to the lower bound

$$P_x(\Lambda_{R-1}) \geq P_x[\sigma_{\{R\}} < T] = \frac{g_\Omega(x, R)}{g_\Omega(R, R)} = \frac{V_d(x)}{V_d(R)}\{1 + o(1)\}, \qquad (6.18)$$

which combined with (6.12) shows (6.1), as desired.

6.3.2 Cases (C3) and (C4)

Each of conditions (C3) and (C4) implies that $U_a(x)/x^\alpha$ (necessarily $0 < \alpha \le 1$) is s.v. Let ℓ be an s.v. function such that

$$U_a(x) \sim x^\alpha/\ell(x) \quad \text{if } \alpha < 1, \quad \text{and} \quad \ell = \ell^* \text{ if } \alpha = 1. \tag{6.19}$$

Here and in the sequel $\alpha = 1$ in case (C3). In cases (C1) and (C2) we have $P_x[\sigma_R < T] \sim V_d(x)/V_d(R)$ which, in conjunction with (6.12), entails (6.1), but in cases (C3) and (C4) this equivalence does not generally hold (see Proposition 8.3.1) and for the proof of (6.1) we shall make use of (6.11) not only for the upper bound but for the lower bound.

Lemma 6.3.1 *Suppose that either (C3) or (C4) holds. Then $\ell_\sharp(x)$ is s.v., and*

$$(a) \; P[-\hat{Z} \ge x] \sim v^\circ \ell_\sharp(x) \quad and \quad (b) \; V_d(x) \sim 1/\ell_\sharp(x).$$

Proof We have only to show (a), in addition to $\ell_\sharp$ being s.v., (a) entailing (b) in view of (6.13) applied with $\alpha\rho = 0$. If (C3) or (C4) holds, then V_d is s.v. (cf. [83] in case (C4)), hence by (5.41) $P[-\hat{Z} \ge x]$ is also s.v. Because of this fact, together with the expression $P[-\hat{Z} \ge x] = v^\circ \sum_{y=0}^\infty u_a(y)F(-x-y)$, the assertion follows from a general result involving regularly varying functions that we give in Section A.1.2 (Lemma A.1.3) so as not to break the flow of the discourse. [Under (C3), Lemma 6.3.1 is a special case of Lemma 5.4.5, where a different proof is employed.] □

Lemma 6.3.2 (i) *Let (C3) hold. Then for any $\delta > 0$,*

$$P[Z > x] \ge V_d(x)\mu_+(x + \delta x)\{1 + o(1)\}. \tag{6.20}$$

If in addition $\limsup \mu_+(\lambda x)/\mu_+(x) < 1$ for some $\lambda > 1$, then

$$P[Z > x] \le V_d(x)\mu_+(x)\{1 + o(1)\}. \tag{6.21}$$

(ii) *If (C4) holds, then $P[Z > x] \sim V_d(x)\mu_+(x)$.*

Proof First, we prove (ii). Suppose (C4) holds and let $\mu_+(x) \sim L_+(x)/x^\alpha$ with some s.v. function L_+. By (6.15)

$$P[Z > x] = \sum_{y=0}^\infty v_d(y)\mu_+(x + y). \tag{6.22}$$

The sum over $y \le x$ is asymptotically equivalent to

$$\sum_{y=0}^x \frac{v_d(y)L_+(x+y)}{(x+y)^\alpha} \sim L_+(x) \sum_{y=0}^x \frac{v_d(y)}{(x+y)^\alpha} = \frac{V_d(x)L_+(x)}{x^\alpha}\{1 + o(1)\},$$

since V_d is s.v. owing to Lemma 6.3.1 and hence $\sum_{y=\varepsilon x}^{x} v_d(y) = o(V_d(x))$ for any $\varepsilon > 0$. Choosing L_+ so that $L'_+(x) = o(L_+(x)/x)$ we also see that the sum over $y > x$ is $o(V_d(x)\mu_+(x))$. Thus the asserted equivalence follows.

The inequality (6.20) is obtained by restricting the range of summation in (6.20) to $y \le \delta x$. (6.21) follows from

$$P[Z > x] \le V_d(x)\mu_+(x) + \sum_{y=x}^{\infty} v_d(y)\mu_+(y). \tag{6.23}$$

Indeed, under the assumption for (6.21), the sum of the infinite series is of a smaller order of magnitude than $V_d(x)\mu_+(x)$, according to a general result, Lemma A.1.4 of the Appendix, concerning s.v. functions. □

Remark 6.3.3 Lemma 6.3.2 says that if the positive tail of F satisfies a mild regularity condition (such as (2.16)), then $P[Z > x] \sim \mu_+(x)/\ell_\sharp(x)$ under (C3). As for the integrals $\int_0^x [\mu_+(t)/\ell_\sharp(t)]\, dt$ and $\int_0^x P[Z > t]\ell_\sharp(t)\, dt$, we have the corresponding relations without assuming any additional condition (see Lemmas 6.4.1 and 6.4.2).

The slow variation of $\ell^*(x) = \int_0^x P[Z > t]\, dt$ implies $P[Z > x] = o(\ell^*(x)/x)$, hence $\lim U_a(x)P[Z > x] = 0$ if Z is r.s. (the converse is true (see (A.24)). This combined with $P[Z > x] = \sum_{y=0}^{\infty} v_d(y)\mu_+(y+x) \ge V_d(x)\mu_+(x)$ shows

$$Z \text{ is r.s.} \implies \lim V_d(x)U_a(x)\mu_+(x) = 0. \tag{6.24}$$

Lemma 6.3.4 (i) *Under (C3)*,

$$V_d(x)U_a(x)\mu(x) \to 0. \tag{6.25}$$

(ii) *Under (C4)*, $V_d(x)U_a(x)\mu(x) \to (\sin \pi\alpha)/\pi\alpha$.

One may compare the results above with the known result under (AS) with $0 < \rho < 1$ stated in (6.55) below; (i) will be refined in Lemma 6.5.1(iii) under (AS) and in Remark 7.1.2(d) for p.r.s. r.w.'s.

Proof Under (C4), Lemma 6.3.2(ii) says $P[Z > x] \sim V_d(x)\mu_+(x)$ so that by Lemma 6.3.1 $P[Z > x]$ is regularly varying with index $-\alpha$, which implies that $P[Z > x]U_a(x)$ approaches $(\sin \pi\alpha)/\pi\alpha$ (cf. Theorem A.2.3). Hence we have (ii).

As for (i), on recalling the definition of $\ell_\sharp$, the slow variation of $\ell_\sharp$ entails $x\mu_-(x)/\ell^*(x) = o(\ell_\sharp(x))$, so that

$$V_d(x)U_a(x)\mu_-(x) \sim V_d(x)x\mu_-(x)/\ell^*(x) \sim x\mu_-(x)/[\ell^*(x)\ell_\sharp(x)] \to 0. \tag{6.26}$$

Thus by (6.24) (i) follows. □

Remark 6.3.5 (6.25) holds also under (C2). Indeed, we know

$$V_d(x)U_a(x) \sim x^2 \bigg/ \int_0^x t\mu(t)\, dt$$

(see Lemma 6.5.1(ii)), and hence (6.25) follows since $x^2\mu(x) = o\left(\int_0^x t\mu(t)\, dt\right)$.

Recall $Z(R) = S_{\sigma[R+1,\infty)} - R$, the overshoot beyond the level R.

Lemma 6.3.6   (i) *Under (C3), for each $\varepsilon > 0$ as $R \to \infty$,*

$$E_x\left[V_d(S_{\sigma(R,\infty)}); Z(R) \geq \varepsilon R, \Lambda_R\right]/V_d(x) \to 0 \text{ uniformly for } 0 \leq x \leq R.$$
$$(6.27)$$

(ii) *Under (C4), as $M \wedge R \to \infty$,*

$$E_x\left[V_d(S_{\sigma(R,\infty)}); Z(R) \geq MR, \Lambda_R\right]/V_d(x) \to 0 \text{ uniformly for } 0 \leq x \leq R.$$
$$(6.28)$$

Proof We first prove (ii). The expectation on the LHS of (6.28) is less than

$$\sum_{w \geq R+MR} \sum_{z=0}^{R-1} g_\Omega(x,z)p(w-z)V_d(w) \qquad (6.29)$$

because of the trivial inequality $g_{\mathbb{Z}\setminus[0,R)}(x,z) < g_\Omega(x,z)$. By Lemma 6.3.1 $V_d(x) \sim 1/\ell_\sharp(x)$ and the derivative $(1/\ell_\sharp)'(x) = o(1/x\ell_\sharp(x))$. Hence for each $M > 0$ there exists a constant R_0 such that for all $R > R_0$ and $z < R$,

$$\sum_{w \geq R+MR} p(w-z)V_d(w) \leq \frac{\mu_+(MR)}{\ell_\sharp(R)\{1+o(1)\}} + \sum_{w \geq MR} \frac{\mu_+(w)}{w\ell_\sharp(w)} \times o(1) \qquad (6.30)$$

$$= \frac{\mu_+(R)}{M^\alpha \ell_\sharp(R)}\{1+o(1)\} \leq 2\frac{V_d(R)\mu_+(R)}{M^\alpha},$$

where we have employed summation by parts and the trivial bound $w < R+w$ for the inequality and (C4) for the equality. Owing to Lemma 6.3.4(ii) the last expression is at most $3/M^\alpha U_a(R)$ and hence on applying Lemma 6.2.1 the double sum in (6.29) is dominated by $3V_d(x)/M^\alpha$, showing (6.28).

In case (C3), by Lemma 6.3.4(i) $\mu_+(x)/\ell_\sharp(x) \ll 1/U_a(x) \sim \ell^*(x)/x$, and we infer that the second expression in (6.30) is $o(1/U_a(R))$. Thus the sum on the left of (6.30) is $o(1/U_a(R))$ for any $M > 0$ and on returning to (6.29) we conclude (6.27) by Lemma 6.2.1 again.   □

Proof (of Theorem 6.1.1 in cases (C3), (C4)) We have the upper bound (6.12) of $P_x(\Lambda_R)$. To obtain the lower bound, we observe that for $x \leq R/2$,

$$\frac{V_d(R)}{1+o(1)} \geq E_x\left[V_d(S_{\sigma(R,\infty)}); Z(R) < MR \big| \Lambda_R\right] \geq \frac{(1-\varepsilon)V_d(x)}{P_x(\Lambda_R)\{1+o(1)\}}.$$

The left-hand inequality follows immediately from the slow variation of V_d (see Lemma 6.3.1 in case (C4)), and the right-hand one from (6.11) and Lemma 6.3.6. The inequalities above yield the lower bound for $x \leq R/2$ since ε can be made arbitrarily small. Because of the slow variation of V_d the result for $\lfloor R/2 \rfloor \leq x < R$ follows from $P_{\lfloor R/2 \rfloor}(\Lambda_R) \to 1$. The proof is complete. $\square$

Remark 6.3.7 (a) Theorem 6.1.1 together with Lemma 6.3.6 shows that, uniformly for $0 \leq x \leq R$,

$$\text{if (C3) holds, } P_x[Z(R) > \varepsilon R \mid \Lambda_R] \to 0 \quad \text{as } R \to \infty \text{ for each } \varepsilon > 0;$$
$$\text{if (C4) holds, } P_x[Z(R) > MR \mid \Lambda_R] \to 0 \quad \text{as } R \wedge M \to \infty. \tag{6.31}$$

(b) Let (AS) hold with $\alpha\rho = 1$. Then (6.27) holds, since $V_d(x)U_a(x)\mu_+(x) \to 0$ (see Remark 6.3.5), entailing $V_d((1 + \varepsilon)R)\mu_+(\varepsilon(R)) = o(1/U_a(R))$, and the same proof given under (C3) applies. The first formula of (6.31) also holds.

6.4 Miscellaneous Lemmas Under (C3), (C4)

In this section, we suppose (C3) or (C4) to hold and provide several lemmas for later usage, not only in this chapter but also in Chapters 6 and 7.

Lemma 6.4.1 *Under (C3), $\ell^*(x) \sim \int_0^x V_d(t)\mu_+(t) \, dt$.*

Proof Summation by parts yields

$$P[Z > s] = \sum_{y=0}^{\infty} v_d(y)\mu_+(y + s) = \sum_{y=0}^{\infty} V_d(y)p(y + 1 + \lfloor s \rfloor).$$

Since $\int_0^x p(y + 1 + \lfloor s \rfloor) \, ds = F(y + 1 + x) - F(y) + O(p(\lfloor x \rfloor + y + 1))$, putting

$$\tilde{\ell}(x) := \int_0^{\infty} V_d(t) \left[F(t + x) - F(t) \right] dt$$

and noting $\sum_{y \geq 0} V_d(y)p(y) < \infty$ we see that

$$\ell^*(x) = \int_0^x P[Z > s] \, ds = \tilde{\ell}(x) + o(1).$$

After replacing $F(t + x) - F(t)$ by $\mu_+(t) - \mu_+(t + x)$, a rearrangement of terms and a change of the variable yield

$$\tilde{\ell}(x) = \int_0^x V_d(t)\mu_+(t) \, dt + \int_0^{\infty} (V_d(t + x) - V_d(t)) \mu_+(t + x) \, dt. \tag{6.32}$$

Similarly, taking the difference for the integral defining $\tilde{\ell}(x)$, we deduce

$$\tilde{\ell}(x) - \tilde{\ell}(\tfrac{1}{2}x)$$

$$= \int_0^\infty V_d(t) \left[\mu_+(t + \tfrac{1}{2}x) - \mu_+(t+x) \right] dt \tag{6.33}$$

$$= \int_0^{x/2} V_d(t)\mu_+(t + \tfrac{1}{2}x)\, dt + \int_0^\infty \left(V_d(t + \tfrac{1}{2}x) - V_d(t) \right) \mu_+(t+x)\, dt.$$

Since the integrals in the last expression of (6.33) are positive and $\tilde{\ell}$ is s.v., both must be of smaller order than $\tilde{\ell}(x)$, in particular the second integral restricted to $t \geq \tfrac{1}{2}x$ is $o(\tilde{\ell}(x))$, hence, by the monotonicity of $\mu_+(t)$, $\int_x^\infty (V_d(t+x) - V_d(t))\, \mu_+(t+x)\, dt = o(\tilde{\ell}(x))$. We also have $\int_0^x V_d(t+x)\mu_+(t+x)\, dt \leq xV_d(2x)\mu_+(x) \ll x/U_a(x) \sim \ell^*(x)$. Thus

$$\int_0^\infty (V_d(t+x) - V_d(t))\, \mu_+(t+x)\, dt = o(\ell^*(x)).$$

Returning to (6.32) we can now conclude $\int_0^x V_d(t)\mu_+(t)\, dt \sim \ell^*(x)$ as desired. $\square$

Lemma 6.4.2 *Suppose (C3) holds. Then*

$$\int_0^x \ell_\sharp(t) P[Z > t]\, dt \sim \int_0^x \mu_+(t)\, dt; \tag{6.34}$$

and if in addition $EX = 0$, then

$$\int_x^\infty \ell_\sharp(t) P[Z > t]\, dt = \eta_+(x)\{1 + o(1)\} + o\,(x\mu_+(x))\,. \tag{6.35}$$

[In case (C4), (6.34) holds, following immediately from Lemma 6.3.2(ii).]

Proof Denote by $J(x)$ the integral on the LHS of (6.34). We first derive the lower bound for it. By (6.20) and $\lim \ell_\sharp(x)V_d(x) \to 1$, for any $\lambda > 1$,

$$\frac{J(x)}{1 + o(1)} \geq \int_0^x \ell_\sharp(t)V_d(t)\mu_+(\lambda t)\, dt = \frac{1}{\lambda} \int_0^{\lambda x} \mu_+(t)\, dt\{1 + o(1)\} \quad (x \to \infty).$$

Since λ may be arbitrarily close to unity, we have $J(x) \geq \int_0^x \mu_+(t)\, dt\{1 + o(1)\}$.

According to Lemma A.1.5 given in the Appendix, by the slow variation of V_d and the monotonicity of μ_+ it follows that

$$\sum_{y=x}^\infty v_d(y)\mu_+(y) = o\left(V_d(x)\mu_+(x) + \int_x^\infty V_d(t)\mu_+(t)\frac{dt}{t} \right).$$

In view of (6.23), it therefore suffices for the upper bound to show that

$$\int_0^x \ell_\sharp(s)\, ds \int_s^\infty V_d(t)\mu_+(t)\frac{dt}{t} \leq C \int_0^x \mu_+(t)\, dt. \tag{6.36}$$

Integrating by parts we may write the LHS as

$$\left[x\ell_\#(x) \int_x^\infty V_d(t)\mu_+(t)\frac{dt}{t} + \int_0^x \mu_+(t)\,dt \right]\{1 + o(1)\}.$$

Integrating by parts again and using Lemma 6.4.1, one can easily deduce that the first integral in the large square brackets is less than $\int_x^\infty \ell^*(t)t^{-2}\{1 + o(1)\}\,dt \sim$ $\ell^*(x)/x \le V_d(x) \int_0^x \mu_+(t)\,dt \times O(1/x)$. Thus we obtain (6.36), hence (6.34).

Let $EX = 0$. One can analogously show (6.35). Indeed, observing that

$$\frac{1}{\lambda} \int_{\lambda x}^\infty \mu_+(t)\,dt \ge \frac{1}{\lambda}\left[\eta_+(x) - (\lambda - 1)x\mu_+(x) \right]$$

one can readily obtain the lower bound, while the proof of the upper bound is only simplified.   □

Lemma 6.4.3 *If (C3) holds, then as $t \to \infty$,*

$$\ell^*(t)\ell_\#(t) = - \int_0^t \mu_-(s)\,ds + \int_0^t P[Z > s]\ell_\#(s)\,ds \tag{6.37}$$

$$= A(t) + \int_0^t \mu_+(s)\,ds \times o(1)$$

and if $EX = 0$, both η_- and η are s.v. and

$$\ell^*(t)\ell_\#(t) = \int_t^\infty \left(\mu_-(s) - P[Z > s]\ell_\#(s) \right)\,ds = A(t)\{1 + o(1)\} + o(\eta_+(t)). \tag{6.38}$$

Proof Recall $\ell^*(t) = \int_0^t P[Z > s]\,ds$ and note that as $t \downarrow 0$, $\ell^*(t) = O(t)$ and $\ell_\#(t) \sim O(\log 1/t)$. Then integration by parts leads to

$$\int_0^t P[Z > s]\ell_\#(s)\,ds = \ell^*(t)\ell_\#(t) + \int_0^t \mu_-(s)\,ds,$$

hence the first equality of (6.37). The second one follows from (6.34).

Let $EX = 0$. Then, $\ell^*(t)\ell_\#(t) \to 0$ because of (6.37), and the first equality of (6.38) is deduced from the identity

$$[\ell^*(t)\ell_\#(t)]' = P[Z > t]\ell_\#(t) - \mu_-(t) \tag{6.39}$$

(valid for almost every $t > 1$). For the second one, noting that $x\mu(x) = o\left(\ell_\#(x)\ell^*(x)\right)$ by virtue of Lemma 6.3.4(i), we have only to apply (6.35). Lemma 6.3.4(i) together with (b) of Lemma 6.3.1 entails that under (C3)

$$x\mu(x) \ll \ell^*(x)\ell_\#(x) \le \eta_-(x).$$

Thus both η_- and η are s.v.   □

Remark 6.4.4 Suppose that S is p.r.s. and $\mu_+(x)/\mu(x) \to p$, and let $q = 1 - p$. Then combining Lemmas 6.3.1 and 6.4.3 shows that for $p \neq 1/2$,

$$A(x) \sim \ell_\sharp(x)\ell^*(x) \sim x/[U_a(x)V_d(x)].$$

(See Section 7.7 for more details.) If F is recurrent (necessarily $p \leq 1/2$), we have $a(x) \sim \int_{x_0}^x \mu_-(t)\,dt / A^2(t)$ and $a(-x) = \int_{x_0}^x \mu_+(t)\,dt / A^2(t) + o(a(x))$ as $x \to \infty$ according to Corollary 4.1.2. Hence

$$\frac{1}{A(x)} = \int_{x_0}^x \frac{\mu_-(t) - \mu_+(t)}{A^2(t)}\,dt + \frac{1}{A(x_0)} \sim (1 - q^{-1}p)a(x)$$

(with the usual interpretation if $p = q$), so that $U_a(x)V_d(x)/x \sim (1 - q^{-1}p)a(x)$ ($p < q$).

Lemma 6.4.5 *If F is p.r.s., $EX = 0$, $\liminf \mu_-(x)/\mu(x) > 0$, and either μ_+ is regularly varying at infinity with index -1 or $\limsup a(-x)/a(x) < 1$, then*

$$\int_1^x \frac{\mu_-(t)\,dt}{[\ell^*(t)\ell_\sharp(t)]^2} \sim g_\Omega(x,x).$$

Proof Suppose $x\mu_+(x)$ is s.v. [See Remark 6.4.6(a) below for the other case.] Then $[\ell^*(t)]' = P[Z > t] \sim \mu_+(t)/\ell_\sharp(t)$ (Remark 6.3.3) and we observe that

$$g_\Omega(x,x) = \sum_{k=0}^x v_d(k)u_a(k) = \sum_{k=0}^x \frac{v_d(k)}{\ell^*(k)}\{1 + o_k(1)\}$$

$$= \frac{V_d(x)}{\ell^*(x+1)}\{1 + o(1)\} + \sum_{k=0}^x V_d(k)\left(\frac{1}{\ell^*(k)} - \frac{1}{\ell^*(k+1)}\right)\{1 + o_k(1)\}$$

$$= \frac{1}{\tilde{A}(x)}\{1 + o(1)\} + \sum_{k=1}^x \frac{\mu_+(k)}{[\tilde{A}(k)]^2}\{1 + o_k(1)\},$$

where $\tilde{A}(t) = \ell^*(t)\ell_\sharp(t)$ and $o_t(1) \to 0$ as $t \to \infty$. By $P[Z > t]\ell_\sharp(t) \sim \mu_+(t)$ again and (6.39) it follows that

$$\frac{1}{\tilde{A}(x)} = \int_1^x \frac{\mu_-(t) - \mu_+(t)}{[\tilde{A}(t)]^2}\,dt + \int_1^x \frac{\mu_+(t) \times o_t(1)}{[\tilde{A}(t)]^2}\,dt + \frac{1}{\tilde{A}(1)} \qquad (6.40)$$

and hence, on absorbing $1/\tilde{A}(1)$ into the second integral on the RHS,

$$g_\Omega(x,x) = \int_1^x \frac{\mu_-(t) - \mu_+(t)}{[\tilde{A}(t)]^2}\,dt\{1 + o(1)\} + \int_1^x \frac{\mu_+(t)}{[\tilde{A}(t)]^2}\{1 + o_t(1)\}\,dt, \quad (6.41)$$

which leads to the equivalence of the lemma.   □

Remark 6.4.6 (a) Let F be recurrent and p.r.s. We shall see that $g_\Omega(x,x) \sim a(x)$ ($x \to \infty$) in Chapter 7 (see Theorem 7.1.1), so that the equivalence in Lemma 6.4.5, which we do not use in this treatise, gives

$$\int_1^x \frac{\mu_-(t)\, dt}{[\ell^*(t)\ell_\sharp(t)]^2} \sim a(x) \sim \int_{x_0}^x \frac{\mu_-(t)\, dt}{A^2(t)}.$$

These asymptotic relations are of interest in the critical case $\lim a(-x)/a(x) = 1$, when we have little information about the behaviour of $A(x)/\ell^*(x)\ell_\sharp(x)$.

(b) Under (C3) we have $V_{\rm d}(x) \sim 1/\ell_\sharp(x)$ and $[1/\ell_\sharp]'(t) = \mu_-(t)/[\ell_\sharp^2(t)\ell^*(t)]$, so that $V_{\rm d}(x) \sim \int_1^x \ell^*(t)\mu_-(t)[\ell_\sharp(t)\ell^*(t)]^{-2}\, dt$. Combined with the first equivalence in (a) this shows (see also (8.11) if $\rho > 0$) that under the assumption of Lemma 6.4.5,

$$a(x)/V_{\rm d}(x) \longrightarrow 1/EZ \quad \text{and} \quad V_{\rm d}(x) \le \ell^*(x)a(x)\{1 + o(1)\}.$$

Lemma 6.4.7 *If either (C3) or (C4) holds, then for any $\delta < 1$, as $R \to \infty$*

$$1 - P_{R-x}(\Lambda_R) = o\left(\frac{U_{\rm a}(x)}{U_{\rm a}(R)}\right) \quad \textit{uniformly for } 0 \le x < \delta R.$$

Proof We prove the dual assertion. Suppose $U_{\rm a}$ and $v_{\rm d}$ are s.v. We show

$$P_x(\Lambda_R)V_{\rm d}(R)/V_{\rm d}(x) \to 0 \quad \text{uniformly for } 0 \le x < \delta R. \tag{6.42}$$

Since for $\frac13 R \le x < \delta R$, $V_{\rm d}(R)/V_{\rm d}(x)$ is bounded and $P_x(\Lambda_R) \to 0$ as $R \to \infty$, we have only to consider the case $x < \frac13 R$. Putting $R' = \lfloor R/3 \rfloor$ and $R'' = 2R'$, we have

$$P_x(\Lambda_R) = J_1(R) + J_2(R),$$

where $J_1(R) = P_x\left[S_{\sigma(R',\infty)} > R'', \Lambda_R\right]$ and $J_2(R) = P_x\left[S_{\sigma(R',\infty)} \le R'', \Lambda_R\right]$. Obviously

$$J_1(R) \le \sum_{z=0}^{R'} g_{Z\setminus[0,R']}(x,z)P[X > R'' - z] \le \mu_+(R') \sum_{z=0}^{R'} g_\Omega(x,z).$$

The last sum is at most $V_{\rm d}(x)U_{\rm a}(R)$ by Lemma 6.2.1. Using Lemma 6.3.4(i) we accordingly infer that

$$J_1(R)\frac{V_{\rm d}(R)}{V_{\rm d}(x)} \le V_{\rm d}(R)U_{\rm a}(R)\mu_+(R') \longrightarrow 0.$$

As for J_2, we decompose $J_2(R) = P_x(\Lambda_{R'}) \sum_{z=R'}^{R''} P_x\left[S_{[R',\infty)} = z \,\big|\, \Lambda_{R'}\right] P_z(\Lambda_R)$. On applying the upper bound (6.12) to $P_x(\Lambda_{R'})$ it then follows that as $R \to \infty$

$$J_2(R)\frac{V_{\rm d}(R)}{V_{\rm d}(x)} \le \frac{V_{\rm d}(R)}{V_{\rm d}(R')} \sum_{z=R'}^{R''} P_x\left[S_{[R',\infty)} = z \,\big|\, \Lambda_{R'}\right] P_z(\Lambda_R) \longrightarrow 0, \tag{6.43}$$

for $P_z(\Lambda_R) \to 0$ uniformly for $R' \le z \le R''$. Thus (6.42) is verified.

In case (C4), $V_d(x) \sim x^\alpha/\hat{\ell}(x)$ and $U_a(x) \sim 1/\ell_\sharp(x)$ (see (6.48) and (6.49)) and we can proceed as above.   □

The following lemma presents some of the results obtained above in a neat form under condition (AS).

Lemma 6.4.8 *Suppose (AS) holds with $\alpha = 1$.*

(i) *The following are equivalent*

    (a) $P[-\hat{Z} \geq x]$ *is s.v.;*
    (b) Z *is r.s. ;*
    (c) $\rho = 1$,

and each of (a) to (c) above implies $P[-\hat{Z} \geq x] \sim v^\circ/V_d(x) \sim v^\circ\ell_\sharp(x)$; in particular these two asymptotic equivalences hold if either $EX = 0, p < 1/2$ or $E|X| = \infty, p > 1/2$.

(ii) *If either of (a) to (c) of (i) holds, then*

$$\sup_{x>(1+\varepsilon)R} P_x[\sigma_{\{R\}} < T] \to 0 \qquad \text{for each } \varepsilon > 0; \text{ and}$$

$$\begin{cases} P[Z > x] \sim V_d(x)\mu_+(x) & \text{if } p > 0, \\ P[Z > x] = o\,(V_d(x)\mu_-(x)) & \text{if } p = 0. \end{cases} \qquad (6.44)$$

Proof Each of (a) and (b) follows from (c) as noted in Remark 6.1.2 (since S is p.r.s. under (AS) with $\rho = 1$; cf. (2.23)). The converses follow from Theorem A.3.2. The other assertion of (i) is contained in Lemma 6.3.1. From the slow variation of V_d it follows that

$$\lim_{y\to\infty} P_y[S_{\sigma(-\infty,0]} < -My] = 1$$

for each $M > 1$ (the dual of the third case in (A.16)), which entails the first relation of (ii). The second one follows immediately from Lemma 6.3.2(i) if $p > 0$, since one can take $\delta = 0$ in (6.20) under (AS). In the case $p = 0$, examine the proof of (6.21) by noting that $\sum_y v_d(y)\mu_+(y) \ll \sum_y v_d(y)\mu_-(y)$.   □

Under (C3) one can derive the exact asymptotic form of v_d if the negative tail of F satisfies the following regularity conditions:

    (a) $\exists \lambda > 1$, $\displaystyle\limsup_{x\to\infty} \frac{\mu_-(\lambda x)}{\mu_-(x)} < 1$ and   (b) $\displaystyle\limsup_{x\to\infty} \frac{p(-x)x}{\mu_-(x)} < \infty.$[4]

The next result, not used in this chapter, plays a crucial role in Section 7.5 (where S is transient). Its proof is based on an extension – given in Section A.2.1.2 – of Theorem 1.1 of Nagaev [54].

---

[4] (a) implies that there exist positive constants C, θ and x_1 such that

$$\mu_-(y)/\mu_-(x) \leq C(y/x)^{-\theta} \qquad \text{whenever} \quad y > x > x_1.$$

(Obviously the converse is true.) If $\limsup_{x\to\infty} \frac{\mu(\lambda x)}{\mu(x)} < \delta < 1$, one may take $\theta = -(\log \delta)/\log \lambda$ and $C = 1/\delta$.

Lemma 6.4.9 *If (a) and (b) above hold in addition to (C3), then*

(i) $P[-\hat{Z} = x]/v° \sim \mu_-(x)/\ell^*(x)$; *and*
(ii) $v_d(x) \sim [V_d(x)]^2 P[-\hat{Z} = x]/v° \sim [V_d(x)]^2 u_a(x)\mu_-(x)$.

[If $EX = 0$ and the positive and negative tails of F are not balanced, condition (b) can be replaced by much weaker one in order for (ii) to be valid (see Theorem 7.1.4).]

Proof After changing the variable of summation, write (6.14) in the form

$$P[-\hat{Z} = x]/v° = \sum_{z=x}^{\infty} u_a(z-x)p(-z) = \sum_{z=x}^{(1+\varepsilon)x} + \sum_{z>(1+\varepsilon)x}$$
$$= I_*(x) + I^*(x) \quad \text{(say)}.$$

Since $u_a(x) \sim 1/\ell^*(x)$ under (C3), for each $\varepsilon \in (0, \tfrac{1}{2})$, we have

$$I^*(x) = \sum_{z=x+\varepsilon x}^{x/\varepsilon} \frac{p(-z)\{1+o(1)\}}{\ell^*(z-x)} + O\left(\frac{\mu_-(x/\varepsilon)}{\ell^*(x)}\right) = \frac{\mu_-(x+\varepsilon x) + O(\mu_-(x/\varepsilon))}{\ell^*(x)\{1+o(1)\}},$$

while (b) entails that $I_*(x) \le CU_a(\varepsilon x)\mu_-(x)/x \le C'\varepsilon\mu_-(x)/\ell^*(x)$ and $\mu_-(x+\varepsilon x) = \mu_-(x)\{1+O(\varepsilon)\}$. Hence by (a) we can conclude (i).

Using the 'continuity' of $\mu_-(x)$ noted right above, one deduces from (i) that

$$\limsup_{x\to\infty} \sup_{\delta x < y < x} \left|P[-\hat{Z} = y] - P[-\hat{Z} = x]\right| / P[-\hat{Z} = x] \to 0 \quad \text{as } \delta \uparrow 1. \quad (6.45)$$

In Section A.2 (Lemma A.2.2) we shall show that this condition together with the slow variation of $P[\hat{Z} < -x]/v° \sim 1/V_a(x)$ implies (ii). □

Lemma 6.4.10 *Suppose that (a), (b) above and (C4) hold, and that $u_a(x) \sim \alpha x^{\alpha-1}/\ell(x)$ as $x \to \infty$. Then it holds that $v_d(x) \sim [V_d(x)]^2 P[-\hat{Z} = x]/v°$ and $P[-\hat{Z} = x] = O(u_a(x)\mu_-(x))$. If $\mu_-(x)$ is further assumed to be regularly varying with index $-\beta$ ($\alpha \le \beta \le 1$), then*

$$P[-\hat{Z} = x]/v° \sim C_{\alpha,\beta}u_a(x)\mu_-(x), \quad \text{where } C_{\alpha,\beta} = \alpha\beta B(\alpha, \alpha+1-\beta).$$

[We shall give in Lemma 8.1.3 (ii) a sufficient condition for $u_a(x) \sim \alpha x^{\alpha-1}/\ell(x)$.]

Proof Put $J^*(x, y) = \sum_{z>(1+\varepsilon)x} u_a(z-y)p(-z)$ and let I_* and I^* be as in proof of Lemma 6.4.9. Using the assumption on u_a, one deduces for $(1-\varepsilon^2)x < y \le x$,

$$|J^*(x,y) - I^*(x)| = \left| \sum_{z=x+\varepsilon x}^{\infty} [u_a(z-y) - u_a(z-x)]\, p(-z) \right| = o\,(I^*(x))$$

as $x \to \infty$ and $\varepsilon \downarrow 0$ in this order. By (b), it accordingly follows that

$$|I^*(y) - I^*(x)| \le |I^*(y) - J^*(x, y)| + |I^*(x) - J^*(x, y)| \le C\varepsilon^2 u_a(\varepsilon x)\mu_-(x) + o(I^*(x)).$$

By (a), one can choose $\lambda > 1$ so that

$$I^*(x) \ge u_a(\lambda x)[\mu_-(x + \varepsilon x) - \mu_-(x + \lambda x)] \ge 2^{-1}\lambda^{\alpha-1}(u_a\mu_-)(x + \varepsilon x),$$

hence $I_*(x) + I_*(y) \le C'\varepsilon^\alpha x^{\alpha-1}\mu_-(x) \le C\varepsilon^\alpha I^*(x)$. Now we obtain (6.45), and applying Lemma A.2.2 gives the first half of the lemma.

Let μ_- vary regularly with index $-\alpha$. Then as $x \to \infty$, the measure $\zeta_x(dt) := -d_t\mu_-(x(1+t))/\mu_-(x)$ weakly converges to $\alpha(1+t)^{-\alpha-1} dt$ on the interval $[\varepsilon, 1/\varepsilon]$ for each $\varepsilon > 0$. Hence

$$\sum_{y=\varepsilon x}^{x/\varepsilon} u_a(y)p(-x-y) \sim \int_\varepsilon^{1/\varepsilon} u_a(xt)\zeta_x(dt) \sim u_a(x)\mu_-(x)\int_\varepsilon^{1/\varepsilon} \frac{\alpha t^{-1+\alpha}\, dt}{(1+t)^{1+\alpha}}.$$

As $\varepsilon \downarrow 0$, the last integral tends to $C_{\alpha,\beta}$, and the sums over $y < \varepsilon x$ and $y > x/\varepsilon$ become negligible because of (b). Thus the second half of the lemma follows.   □

6.5 Some Properties of the Renewal Functions U_a and V_d

Here we make a slight digression from the main topic of the present discussion to collect some facts regarding U_a and V_d that are used in the next section and Chapter 8 as well. Throughout this and the next section, we suppose (AS) to hold. Let p, q and L be as in (2.16). Recall that $\alpha\rho$ ranges exactly over the interval $[(\alpha - 1) \vee 0, \alpha \wedge 1]$.

It is known that $P[Z > x]$ varies regularly at infinity with index $-\alpha\rho$ if $\alpha\rho < 1$ (which, for $1 < \alpha < 2$, is equivalent to $p > 0$), and Z is r.s. if $\alpha\rho = 1$ (see [63, Theorem 9] for $\rho > 0$ and Lemma 6.3.1 for $\rho = 0$, see also Theorems A.2.3 and A.3.2); and in either case

$$\frac{U_a(x)\int_0^x P[Z > t]\, dt}{x} \longrightarrow \frac{1}{\Gamma(1 + \alpha\rho)\Gamma(2 - \alpha\rho)} \tag{6.46}$$

(the dual of (6.13)). We choose an s.v. function $\ell(x)$ so that

$$\begin{aligned}
\ell(x) &= \ell^*(x) &&\text{if } \alpha\rho = 1, \\
P[Z > x] &\sim \frac{\sin \pi\alpha\rho}{\pi\alpha\rho}x^{-\alpha\rho}\ell(x) &&\text{if } \alpha\rho < 1.
\end{aligned} \tag{6.47}$$

Here $t^{-1}\sin t$ is understood to equal unity for $t = 0$. [The constant factor in the second formula above is chosen so as to have the choice conform to that in (6.19) – see (6.49) below.] As the dual relations we have an s.v. function $\hat{\ell}$ such that

$$\hat{\ell} = \hat{\ell}^* \quad (\alpha\hat{\rho} = 1); \quad P[-\hat{Z} > x]/v^\circ \sim \frac{\sin \pi\alpha\hat{\rho}}{\pi\alpha\hat{\rho}}x^{-\alpha\hat{\rho}}\hat{\ell}(x) \quad (\alpha\hat{\rho} < 1).$$

Note that according to Lemma 6.3.1,

$$\hat{\ell}(x) \sim \ell_\sharp(x) \quad \text{if } \rho = 1;$$

similarly $\ell(x) \sim \hat{\ell}_\sharp(x)$ if $\rho = 0$ (necessarily $\alpha \le 1$), where

$$\hat{\ell}_\sharp(x) = \alpha \int_x^\infty \frac{\mu_+(t)}{t^{1-\alpha}\hat{\ell}(t)} \, dt \qquad (x > 0). \tag{6.48}$$

It then follows that for all $0 \le \rho \le 1$,

$$U_a(x) \sim x^{\alpha\rho}/\ell(x) \quad \text{and} \quad V_d(x) \sim x^{\alpha\hat{\rho}}/\hat{\ell}(x). \tag{6.49}$$

The two s.v. functions ℓ and $\hat{\ell}$ are linked, as shown below in Lemma 6.5.1. We bring in the constant

$$\begin{aligned}
\kappa &= \frac{\Gamma(\alpha)\pi^{-1}\sin\pi\alpha\rho}{p\Gamma(\alpha\rho + 1)\Gamma(\alpha\hat{\rho} + 1)} \\
&= \frac{\Gamma(\alpha)\pi^{-1}\sin\pi\alpha\hat{\rho}}{q\Gamma(\alpha\rho + 1)\Gamma(\alpha\hat{\rho} + 1)} \\
&= \frac{\Gamma(\alpha)[\sin\pi\alpha\rho + \sin\pi\alpha\hat{\rho}]}{\pi\Gamma(\alpha\rho + 1)\Gamma(\alpha\hat{\rho} + 1)},
\end{aligned}$$

where only the case $p > 0$ or $q > 0$ of the first two expressions above is adopted if $pq = 0$; if $pq \ne 0$ the two coincide, namely

$$p^{-1}\sin\pi\alpha\rho = q^{-1}\sin\pi\alpha\hat{\rho} = \sin\pi\alpha\rho + \sin\pi\alpha\hat{\rho} \quad (pq \ne 0),$$

as is implicit in the proof of the next result or directly derived (cf. [82, Appendix (A)]). Note that $\kappa = 0$ if and only if either $\alpha = 1$ and $\rho\hat{\rho} = 0$ or $\alpha = 2$.[5]

Lemma 6.5.1 (i) *For all* $0 < \alpha \le 2$, $\ell(x)\hat{\ell}(x)/[x^\alpha\mu(x)] \to 1/\kappa$, *in other words*

$$U_a(x)V_d(x) = \frac{\kappa + o(1)}{\mu(x)}.$$

(ii) *If* $\alpha = 2$, *then* $\ell^*(x)\hat{\ell}^*(x) \sim \int_0^x t\mu(t)\,dt$ *and, because of* $\int_0^x t\mu(t)\,dt \sim m(x) \sim L(x)/2$,

$$U_a(x)V_d(x) \sim x^2/m(x) \sim 2x^2/L(x).$$

(iii) *If* $\alpha = 1$ *and* $p \ne 1/2$ *(entailing* $\rho \in \{0, 1\}$ *so that* $\kappa = 0$*), then*

$$U_a(x)V_d(x) \sim \frac{x}{|A(x)|} \sim \begin{cases} \dfrac{x}{|p - q|\int_x^\infty \mu(t)\,dt} & (EX = 0), \\[3ex] \dfrac{x}{|p - q|\int_0^x \mu(t)\,dt} & (E|X| = \infty). \end{cases}$$

---

[5] In fact, $\kappa/(2 - \alpha) \to 1$ as $\alpha \uparrow 2$; if $\alpha = 1$, $\kappa/(\rho \wedge \hat{\rho}) \to 2$ as $\rho \wedge \hat{\rho} \to 0$.

For the identification of the asymptotic form of $U_a(x)V_d(x)$, explicit in terms of F, Lemma 6.5.1 covers all the cases of (AS) other than the case $\alpha = 2p = \rho \vee \hat\rho = 1$.

Proof (iii) follows immediately from Lemmas 6.3.1 and 6.4.3 (see also Remark 6.4.4). For the rest, we show (ii) first, which we need for the proof of (i).

PROOF OF (ii). Let $\alpha = 2$. Then $\ell = \ell^*$, $\hat\ell = \hat\ell^*$. Suppose that $EZ = E|\hat Z| = \infty$ so that $\ell^*(x) \wedge \hat\ell^*(x) \to \infty$, otherwise the proof below merely being simplified. Recalling (6.14) we have

$$\frac{P[-\hat Z \geq x]}{v^\circ} = \sum_{z=0}^{\infty} u_a(z)F(-x-z), \tag{6.50}$$

by which and its dual

$$
\begin{aligned}
(\ell\hat\ell)'(x) &= \ell^*(x)P[-\hat Z > x]/v^\circ + \hat\ell^*(x)P[Z > x] \\
&= \ell^*(x)\int_0^\infty \frac{\mu_-(x+y)}{\ell^*(y) \vee 1}\,dy\{1+o(1)\} \\
&\quad + \hat\ell^*(x)\int_1^\infty \frac{\mu_+(x+y)}{\hat\ell^*(y)}\,dy\{1+o(1)\}.
\end{aligned}
\tag{6.51}
$$

We claim that for any constant $\varepsilon \in (0,1)$,

$$\int_0^x \ell^*(t)\,dt \int_0^\infty \frac{\mu_-(t+y)}{\ell^*(y)\vee 1}\,dy \begin{cases} \leq \left[\varepsilon\int_0^x t\mu_-(t)\,dt + \int_0^x \eta_-(t)\,dt\right]\{1+o(1)\}, \\ \geq \int_0^x t\mu_-(t)\,dt\{1-\varepsilon+o(1)\}. \end{cases} \tag{6.52}$$

By splitting the inner integral of the LHS at εt and using the monotonicity of ℓ^* and μ_- it follows that

$$\ell^*(t)\int_0^\infty \frac{\mu_-(t+y)}{\ell^*(y)\vee 1}\,dy \leq [\varepsilon t\mu_-(t) + \eta_-(t)]\{1+o(1)\},$$

showing the first inequality of (6.52). On the other hand the LHS of (6.52) is greater than

$$
\begin{aligned}
\int_0^x \ell^*(t)\,dt \int_0^{x-t} \frac{\mu_-(t+y)}{\ell^*(y)\vee 1}\,dy &= \int_0^x \mu_-(w)\,dw \int_0^w \frac{\ell^*(w-y)}{\ell^*(y)\vee 1}\,dy \\
&\geq \int_0^x \mu_-(w)\,dw \int_0^{(1-\varepsilon)w} \{1+o(1)\}\,dy,
\end{aligned}
$$

which yields the second inequality of (6.52).

Put $L^*(x) = \int_0^x t\mu(t)\,dt$. Then (AS) (with $\alpha = 2$) is equivalent to the slow variation of L^* and in this case $m(x) = \int_0^x \eta(t)\,dt \sim L^*(x)$ (cf. Section A.1.1)), which entails $x\eta(x) = o(L^*(x))$. By (6.52) and its dual, which follows because of symmetric roles of the two tails of F, integrating the second expression of $(\ell\hat{\ell})'(x)$ in (6.51) leads to

$$\ell(x)\hat{\ell}(x) \begin{cases} \leq \{1 + \varepsilon + o(1)\}L^*(x), \\ \geq \{1 - \varepsilon + o(1)\}L^*(x), \end{cases}$$

which shows the formula of (ii) since ε is arbitrary.

PROOF OF (i). We have only to consider the case $0 < \rho < 1$ of $\alpha < 2$, since for $\rho = 1$ the result follows from Lemma 6.3.4 and for $\rho = 0$ by duality, and since for $\alpha = 2$, $x^2\mu(x) = o(\int_0^x t\mu(t)\,dt)$, so that the result follows from (ii). By symmetry, we may suppose $q > 0$ (entailing $\alpha\hat{\rho} < 1$). Denoting by $U_a\{dy\}$ the measure $dU_a(y)$ and writing the sum on the RHS of (6.50) as $\int_0^\infty \mu_-(x + xt)U_a\{x\,dt\}$ we see that

$$\frac{P[-\hat{Z} \geq x]}{v^\circ} \sim \frac{q\,U_a(x)L(x)}{x^\alpha} \int_0^\infty \frac{L(x + xt)}{L(x)(1+t)^\alpha} \cdot \frac{U_a\{x\,dt\}}{U_a(x)}.$$

Observe that $U_a\{x\,dt\}/U_a(x)$ weakly converges as $x \to \infty$ to the measure $\alpha\rho\,t^{\alpha\rho-1}\,dt$ on every finite interval and that $\int_M^\infty t^{-\alpha+\varepsilon}U_a\{x\,dt\}/U_a(x) \to 0$ as $M \to \infty$ uniformly in x for each $0 < \varepsilon < \alpha\hat{\rho}$. With the help of Potter's bound $L(x + xt)/L(x) = O((1 + t))^\varepsilon)$, valid for any $\varepsilon > 0$ (cf. [8]), we then see that the integral above converges to $\alpha\rho\int_0^\infty t^{\alpha\rho-1}(1 + t)^{-\alpha}\,dt = \alpha\rho B(\alpha\rho, \alpha\hat{\rho})$. We accordingly obtain

$$P[-\hat{Z} \geq x]/v^\circ \sim q\alpha\rho B(\alpha\rho, \alpha\hat{\rho})x^{-\alpha\hat{\rho}}L(x)/\ell(x). \tag{6.53}$$

Thus by (6.13) (the dual of (6.46)) and the identity $\Gamma(1 - t)\Gamma(1 + t) = \pi t/\sin t\pi$ $(|t| < 1)$,

$$V_d(x) \sim \frac{\pi^{-1}\sin\pi\alpha\hat{\rho}}{\alpha^2\rho\hat{\rho}\,B(\alpha\rho, \alpha\hat{\rho})} \cdot \frac{x^{\alpha\hat{\rho}}\ell(x)}{qL(x)} = \frac{\Gamma(\alpha)\pi^{-1}\sin\pi\alpha\hat{\rho}}{\Gamma(\alpha\rho + 1)\Gamma(\alpha\hat{\rho} + 1)} \cdot \frac{x^{\alpha\hat{\rho}}\ell(x)}{qL(x)}, \tag{6.54}$$

which combined with $U_a(x) \sim x^{\alpha\rho}/\ell(x)$ shows the asserted convergence.

The proof of Lemma 6.5.1 is complete. □

Remark 6.5.2 In the above, (6.53) is shown under the condition $q\rho\hat{\rho} > 0$. Here we prove that (6.53) holds (under (AS) with $\alpha < 2$) unless $q = \hat{\rho} = 0$ with the understanding that $sB(s, t) = 1$ for $s = 0, t > 0$ and $B(s, 0) = \infty$ for $s > 0$, and $\sim$ is interpreted according to our convention if the constant factor of its right-hand side equals zero or infinity. The proof is carried out in the following three cases

(a) $\alpha = \rho = 1 > p$;
(b) either $\alpha = \hat{\rho} = 1$ or $\alpha < 1 = q$;
(c) $\alpha > 1 = p$.

Suppose (a) holds. Then $\ell = \ell^*$ and $P[-\hat{Z} > x]/v^\circ \sim \ell_\sharp(x) \gg x\mu_-(x)/\ell^*(x)$. Since $q > 0 = \hat{\rho}$ (entailing $B(\alpha\rho, \alpha\hat{\rho}) = \infty$), this shows (6.53). If (b) holds, then $\rho = 0$, $\ell = \hat{\ell}_\sharp$ and $U_a(x) \sim 1/\hat{\ell}_\sharp(x)$, and by the representation $P[-\hat{Z} > x]/v^\circ = \sum_{y=0}^\infty u_a(y)\mu_-(x + y)$ we deduce that

$$P[-\hat{Z} > x]/v^\circ = \begin{cases} U_a(x)\mu_-(x)\{1 + o(1)\} \sim qx^{-\alpha}L(x)/\ell(x) & \text{if } q > 0, \\ o(\mu(x)/\ell(x)) & \text{if } q = 0, \end{cases}$$

showing (6.53). As for the case (c) we have $\alpha\rho = 1$, $\ell = \ell^*$, and $u_a(x) \sim 1/\ell^*(x)$, and see that $\sum_{y=0}^{\infty} u_a(y)\mu_-(x+y) \ll U_a(x)\mu(x) \sim x^{-\alpha\hat{\rho}}L(x)/\ell(x)$ since $\mu_-/\mu \to 0$.

If the sequence (c_n) is determined as in (2.17) then the asymptotic form of $U_a V_d$ given in Lemma 6.5.1 may also be expressed as

$$\frac{U_a(c_n)V_d(c_n)}{n/c_\sharp} \longrightarrow \kappa \text{ or } 2 \quad \text{according as } 0 < \alpha < 2 \text{ or } \alpha = 2 \qquad (6.55)$$

and is known if $0 < \rho < 1$ [23, Eq(15)], [88, Eq(15, 31)] apart from the explicit expression of the limit value.

6.6 Proof of Proposition 6.1.3

This section assumes that (AS) is valid, as mentioned before.

Lemma 6.6.1 *Suppose* $\rho > 0$. *Then for any* $\varepsilon > 0$, *there exists a constant* $M > 0$ *such that for all sufficiently large* R,

$$E_x\left[V_d(S_{\sigma(R,\infty)}); Z(R) \geq MR, \Lambda_R\right] \leq \varepsilon V_d(x) \quad (0 \leq x < R),$$

and, if $\alpha\rho = 1$,

$$\lim_{R\to\infty} \sup_{0\leq x < R} \frac{1}{V_d(x)} E_x\left[V_d(S_{\sigma(R,\infty)}); Z(R) \geq \varepsilon R, \Lambda_R\right] = 0.$$

Proof As in the proof of Lemma 6.3.6 but with $V_d(x) \sim x^{\alpha(1-\rho)}/\hat{\ell}(x)$ in place of $V_d(x) \sim 1/\ell_\sharp(x)$, one deduces that if $\rho > 0$, for all sufficiently large R,

$$\sum_{w\geq R+MR} p(w-z)V_d(w) \leq C_* \frac{V_d(R)\mu(R)}{M^{\alpha\rho}} \quad (0 \leq z < R) \qquad (6.56)$$

for some constant C_* which one can take to be $2/\alpha\rho$. Using Lemmas 6.2.1 and 6.5.1(i) in turn one then sees that for any $M > 0$

$$\frac{1}{V_d(x)} E_x\left[V_d(S_{\sigma(R,\infty)}); Z(R) \geq MR, \Lambda_R\right] \qquad (6.57)$$

$$= \frac{1}{V_d(x)} \sum_{z=0}^{R} \sum_{w\geq(M+1)R} g_{B(R)}(x,z)V_d(w)p(w-z)$$

$$\leq C_* \frac{U_a(R)V_d(R)\mu(R)}{M^{\alpha\rho}} = C_* \frac{\kappa + o(1)}{M^{\alpha\rho}}$$

$(B(R) := \mathbb{R} \setminus [0, R])$, so that the first half of the lemma is ensured by taking

$$M = [2C_*\kappa/\varepsilon]^{1/\alpha\rho}. \tag{6.58}$$

If $\alpha\rho = 1$, we have either $\kappa = 0$ or $1 < \alpha < 2$ with $p = 0$ and the second half follows from (6.57), for if $p = 0$, $\mu(R)$ can be replaced by $o(\mu(R))$ in (6.56) and thereafter.
□

Lemma 6.6.2 *Suppose $\rho > 0$. Then there exists a constant $\theta > 0$ such that $P_x(\Lambda_R) \geq \theta V_d(x)/V_d(R)$ for $0 \leq x < R$ and for R large enough. If $\alpha \leq 1$, then θ can be chosen so as to approach unity as $\rho \to 1$, and for any $\delta > 0$, for $x \geq \delta R$,*

$$P_x(\Lambda_R) \to 1 \quad as \quad R \to \infty \quad and \quad \rho \to 1 \ in \ this \ order.$$

Proof Because of the identity (6.11) (due to the optional stopping theorem), Lemma 6.6.1 shows that for any $\varepsilon > 0$ we can choose a constant $M > 0$ so that

$$(1 - \varepsilon)V_d(x) \leq E_x \left[V_d(S_{\sigma(R,\infty)}); Z(R) < MR \,\middle|\, \Lambda_R \right] P_x(\Lambda_R).$$

Since the conditional expectation on the RHS is less than $V_d(R + MR)$, which is asymptotically equivalent to $(1 + M)^{\alpha\hat{\rho}}V_d(R)$, we have the lower bound of $P_x(\Lambda_R)$ with $\theta = (1 - \varepsilon)(1 + M)^{-\alpha\hat{\rho}}$. Noting that M may be determined by (6.58) and that as $\rho \to 1$ (possible only when $\alpha \leq 1$), $\theta \to 1 - \varepsilon$, it follows that $\theta \to 1$ since ε may be arbitrarily small. Since we also have $V_d(x)/V_d(R) \sim (x/R)^{\alpha\hat{\rho}} \to 1$ (uniformly for $\delta R < x < R$), the second half of the lemma follows.
□

Lemma 6.6.3 *Suppose $0 < (\alpha \vee 1)\rho < 1$. Then for any $\delta < 1$, uniformly for $0 \leq x < \delta R$, as $R \to \infty$ and $\varepsilon \downarrow 0$ (interchangeably),*

$$P_x \left[Z(R) \leq \varepsilon R \,\middle|\, \Lambda_R \right] \to 0.$$

Proof By the restriction on $(\alpha \vee 1)\rho$ it follows that $\alpha < 2$ and

$$\begin{cases} \text{(a) } p > 0 \text{ so that } \mu_+ \text{ is regularly varying;} \\ \text{(b) } P[Z > x] \text{ is regularly varying with index } \in (-1, 0); \\ \text{(c) } V_d(x)/V_d(R) \leq CP_x(\Lambda_R) \text{ and } \forall \varepsilon > 0, \ \inf_{R \geq 1} \inf_{x \geq \varepsilon R} P_x(\Lambda_R) > 0. \end{cases} \tag{6.59}$$

The reasoning for (a) has already been given, while (b) and (c) are due to (6.47) and Lemma 6.6.2, respectively. By the generalised arcsine law (cf. Theorem A.2.3) (b) implies that $P[Z(r) < \varepsilon r] \to 0$ as $\varepsilon \to 0$ uniformly for $r \geq 1$. This, in particular, shows the asserted convergence of the lemma to be valid for $\frac{1}{4}R \leq x < \delta R$, therein $P_x(\Lambda_R)$ being bounded away from zero by (c).

For $x < \frac{1}{4}R$, take a constant $0 < h < 1/4$ arbitrarily and let $\mathcal{E}_{\varepsilon,R}$ stand for the event $\{(1 - h)R \leq S_{\sigma[R/4,\infty)} < (1 + \varepsilon)R\}$. It holds that

$$\begin{cases} \text{the event } \Lambda_R \cap \{Z(R) < \varepsilon R\} \text{ is contained in} \\ \Lambda_{R/4} \cap \left[\{S_{\sigma[R/4,\infty)} < (1 - h)R, Z(R) < \varepsilon R\} \cup \mathcal{E}_{\varepsilon,R} \right], \end{cases} \tag{6.60}$$

and that

$$P_x(\Lambda_{R/4} \cap \mathcal{E}_{\varepsilon,R}) \le \sum_{0 \le y < R/4} g_\Omega(x, \tfrac{1}{4}R - y) P[(1-h)R \le y + X < (1+\varepsilon)R]. \quad (6.61)$$

Put

$$\omega_R(h, \varepsilon) = \sup_{y < R/4} \frac{P[(1-h)R - y \le X < (1+\varepsilon)R - y]}{P[X \ge (1-h)R - y]}.$$

One sees that $\omega_R(h, \varepsilon) \le \sup_{z < R/2} P[z \le X < z + (h+\varepsilon)z]/P[X \ge z]$, hence the regular variation of μ_+ entails that $\omega_R(h, \varepsilon) \to 0$ as $h + \varepsilon \to 0$ uniformly for $R \ge 1$. Since on the RHS of (6.61) the probability under the summation sign is less than $\omega_R(h, \varepsilon)\mu_+(R/2)$, by Lemmas 6.2.1 and 6.5.1 the sum is evaluated to be at most a constant multiple of

$$\omega_R(h, \varepsilon)\mu_+(\tfrac{1}{2}R)V_d(x)U_a(\tfrac{1}{2}R) \le \omega_R(h, \varepsilon)\frac{V_d(x)}{V_d(\tfrac{1}{2}R)}\{\kappa + o(1)\}.$$

By (6.59c) and regular variation of V_d, $V_d(x)/V_d(\tfrac{1}{2}R) \le CP_x(\Lambda_R)$. Hence

$$P_x\left[\Lambda_{R/4} \cap \mathcal{E}_{\varepsilon,R}\right] / P_x(\Lambda_R) \le C_2 \omega_R(h, \varepsilon) \quad \text{for} \quad 0 \le x < R/4. \quad (6.62)$$

Now by (6.60), $P_x[Z(R) \le \varepsilon R \,|\, \Lambda_R]$ is less than

$$C_2\omega_R(h, \varepsilon) + \sup_{R/4 \le y < (1-h)R} P_y[Z(R) < \varepsilon R]\frac{P_x(\Lambda_{R/4})}{P_x(\Lambda_R)}.$$

By (6.59c) again, $P_x(\Lambda_{R/4})/P_x(\Lambda_R) = 1/P_x[\Lambda_R \,|\, \Lambda_{R/4}] < C$. By what we have observed above, for any $\varepsilon_0 > 0$, one can choose $h > 0$ so that the first term is less than ε_0 for all $0 < \varepsilon < h$ and all sufficiently large R, while for h thus chosen the second term is less than ε_0 for ε small enough and all R large enough. Thus the convergence asserted in the lemma follows.   □

Lemma 6.6.4 *Let $\alpha \le 1$. For any $\delta < 1$,*

$$\sup_{0 \le x < \delta R} \frac{P_x(\Lambda_R)V_d(R)}{V_d(x)} \to 0 \quad \text{as } R \to \infty \text{ and } \rho \to 0 \text{ in this order;}$$

if $\rho = 0$, the above supremum approaches zero as $R \to \infty$.

Proof The assertion for $\rho = 0$ is a particular case of Lemma 6.4.7. The proof of the rest is similar to that of Lemma 6.4.7. Let $\alpha \le 1$. Suppose $0 < \rho < 1$. Since for $\tfrac{1}{3}R \le x < \delta R$, $V_d(x)/V_d(R)$ is bounded away from zero and $P_x(\Lambda_R) \to 0$ as $R \to \infty$, $\rho \to 0$ in this order because of the dual of the last assertion of Lemma 6.6.2, we have only to consider the case $x < \tfrac{1}{3}R$. Putting $R' = \lfloor \tfrac{1}{3}R \rfloor$ and $R'' = 2R'$, let $J_1(R)$ and $J_2(R)$ be as in the proof of Lemma 6.4.7, so that $P_x(\Lambda_R) = J_1 + J_2$. Using Lemmas 6.2.1 and 6.5.1(i), we have

$$J_1(R)\frac{V_{\mathrm{d}}(R)}{V_{\mathrm{d}}(x)} \le V_{\mathrm{d}}(R)U_{\mathrm{a}}(R)\mu_+(R') \le 4p\kappa \qquad (6.63)$$

for R large enough. As before we see that as $R \to \infty$ and $\rho \to 0$ in turn,

$$J_2(R)\frac{V_{\mathrm{d}}(R)}{V_{\mathrm{d}}(x)} \le \frac{V_{\mathrm{d}}(R)}{V_{\mathrm{d}}(R')}\sum_{z=R'}^{R''} P_x\left[S_{\sigma(R',\infty)} = z\,\middle|\,\Lambda_{R'}\right]P_z(\Lambda_R) \to 0,$$

for, in this limit, $P_z(\Lambda_R) \to 0$ uniformly for $R' \le z \le R''$. Since $p\kappa \to 0$ as $\rho \to 0$, this together with (6.63) shows the first half of the lemma. □

Proof (of Proposition 6.1.3) Let $0 < (\alpha \vee 1)\rho < 1$. The lower bound of $P_x(\Lambda_R)$ asserted in (i) follows from Lemma 6.6.2. For the proof of the upper bound, let $1/2 < \delta < 1$. Lemma 6.6.3 entails

$$P_x\left[Z(R) > \varepsilon R\,\middle|\,\Lambda_R\right] \ge 1/2 \qquad (x < \delta R) \qquad (6.64)$$

for some $\varepsilon > 0$. Writing the equality (6.11) as

$$\frac{V_{\mathrm{d}}(x)}{P_x(\Lambda_R)} = \sum_{y=R+1}^{\infty} P_x[S_{\sigma(R,\infty)} = y\,|\,\Lambda_R]V_{\mathrm{d}}(y),$$

one deduces from (6.64) that the sum above is larger than $[V_{\mathrm{d}}(R) + V_{\mathrm{d}}((1+\varepsilon)R)]/2$. By (6.49) $\limsup V_{\mathrm{d}}((1+\varepsilon)R)/V_{\mathrm{d}}(R) = (1+\varepsilon)^{\alpha\hat{\rho}} > 1$, since $(\alpha \vee 1)\rho < 1$ entails $\hat{\rho} > 0$. Hence $P_x(\Lambda_R) \le \left(2/[1 + (1+\varepsilon)^{\alpha\hat{\rho}}]\right)V_{\mathrm{d}}(x)/V_{\mathrm{d}}(R)\{1 + o(1)\}$. This verifies the upper bound.

(ii) follows from Lemmas 6.6.2 and 6.6.4. □

6.7 Note on the Over- and Undershoot Distributions I

Recall that condition (C3), saying both u_{a} and V_{d} are s.v., holds if and only if $P_0[Z(R) < \varepsilon R] \to 1$ and $P_0[\hat{Z}(R) < -R/\varepsilon] \to 1$ $(R \to \infty)$ for any $\varepsilon > 0$. With the help of a few results from what has been obtained in the present chapter, we can read off exact modes of these convergences in terms of V_{d} or ℓ^* in the following proposition, where (AS) is not supposed. Put

$$N^+(R) = \inf\{n \ge 0 : S_{n+1} > R\}, \quad N^-(R) = \inf\{n \ge 0 : S_{n+1} < -R\}$$

and $\hat{Z}(R) = S_{N^-(R)+1} + R$. Their derivations are based on the following identities

$$P_0[Z(R) = x] = \sum_{y=0}^{R} u_{\mathrm{a}}(y)P[Z = x + R - y], \qquad (6.65)$$

$$P_0[S_{N^+(R)} = R - x] = g_\Omega(x, R)\mu_+(x) \qquad (6.66)$$

(since $g_{(R,\infty)}(0, R - x) = g_\Omega(x, R)$), and their duals. Let δ be a constant arbitrarily chosen from $(0, 1)$.

Proposition 6.7.1 (i) *Uniformly for* $1 \leq x \leq R$,[6]

(a) *if u_a is s.v.,* $P_0[Z(R) \leq x] = [\ell^*(x)/\ell^*(R)] + o(1)$ *as* $R \to \infty$,[7]
(b) *under (C3),* $P_0[S_{N^+(R)} \geq R - x] \sim \ell^*(x)/\ell^*(R)$ *as* $x \to \infty$, *and uniformly for* $0 \leq x < \delta R$,

$$P_0[S_{N^+(R)} = R - x] \sim \frac{V_d(x)\mu_+(x)}{\int_0^R V_d(y)\mu_+(y)\,dy} \qquad as \ R \to \infty.$$

(ii) *Uniformly for* $x \geq 1$, *as* $R \to \infty$

(a) *if V_d is s.v.,* $P_0[\hat{Z}(R) \leq -x] \sim V_d(R)/V_d(x \vee R)$,
(b) *under (C3),* $P_0[S_{N^-(R)} \geq -R + x] \sim V_d(R)/V_d(x \vee R)$ *and*

$$P_0[S_{N^-(R)} = -R + x] \sim \frac{\mu_-(x)/\ell^*(x)}{\int_R^\infty \mu_-(y)\,dy/\ell^*(y)} \qquad uniformly \ for \ x > R/\delta.$$

[(ii.b) entails that for any $M_R \gg R$ such that $V_d(M_R)/V_d(R) \to 1$, both $-\hat{Z}(R)/M_R$ and $S_{N^-(R)}/M_R$ diverge to $+\infty$ in probability.]

Proof By (6.65), $P_0[Z(R) > x] = \sum_{y=0}^R u_a(y)P[Z > x + R - y]$. If u_a is s.v., then $u_a \sim 1/\ell^*$, and, recalling the definition of ℓ^*, one sees that $xP[Z > x]/\ell^*(x) \to 0$ ($x \to \infty$), so that the above sum over $y < R/2$, being less than $U_a(R)P[Z \geq R/2]$, tends to zero. Hence

$$P_0[Z(R) > x] = o(1) + \frac{1 + o(1)}{\ell^*(R)}\left[\ell^*(x + \tfrac{1}{2}R) - \ell^*(x)\right] = 1 - \frac{\ell^*(x)}{\ell^*(R)} + o(1),$$

showing (a) of (i). For the proof of (ii.a) let V_d be s.v. Then $P[\hat{Z} < -x] \sim v^\circ/V_d(x)$, so that from the dual of (6.65) one deduces that uniformly for $x > R$,

$$P_0[\hat{Z}(R) < -x] = \sum_{y=0}^R \frac{v_d(y)}{V_d(x + R - y)}\{1 + o(1)\} \sim \frac{V_d(R)}{V_d(x)}.$$

As for (i.b), under (C3) we have $g_\Omega(x, R) \sim V_d(x)/\ell^*(R)$ for $x \leq \delta R$, and hence, by (6.66), $P_0[S_{N^+(R)} = R - x] = V_d(x)\mu_+(x)/\ell^*(R)$, which together yield the equivalences of (i.b), since $\sum_0^x V_d(y)\mu_+(y) \sim \ell^*(x)$ (Lemma 6.4.1) and the probability of the LHS of the first equivalence is increasing in x.

For the proof of (ii.b), let (C3) hold and observe that for each $0 < \delta < 1$,

$$P_0[S_{N^-(R)} = -R + x] = g_\Omega(R, x)\mu_-(x) \sim V_d(R)\mu_-(x)/\ell^*(x), \quad for \ x > R/\delta,$$

---

[6] To be precise, x must be restricted to integers if $\lim \ell^*(x) = EZ < \infty$.
[7] Virtually the same formula (together with closely related results) is obtained by Erickson [28, Eq(9.3)] under the condition: $xP[Z > x]$ is s.v.

where the equality is the dual of (6.66) and the asymptotic equivalence is uniform in x. Since $1/V_{\mathrm{d}}(x) \sim \ell_{\sharp}(x) = \int_x^{\infty} \mu_-(t)\,dt/\ell^*(t)$, one can accordingly conclude both relations of (ii.b). $\qquad \square$

Chapter 7
The Two-Sided Exit Problem for Relatively Stable Walks

This chapter is a continuation of Chapter 6. We use the same notation as therein. As in Chapter 6, we shall be primarily concerned with the event

$$\Lambda_R = \{\sigma_{(R,\infty)} < T\},$$

$(T = \sigma_\Omega, \Omega = (-\infty, -1])$. In Chapter 6 we obtained several sufficient conditions in order for (6.1) to hold, namely

$$P_x(\Lambda_R) \sim V_d(x)/V_d(R) \quad \text{uniformly for } 0 \le x < R \text{ as } R \to \infty.$$

If the r.w. S is p.r.s. (positively relatively stable), then one such condition – condition (C3) in Theorem 6.1.1 – is fulfilled (cf. Remark 6.1.2), hence (6.1) holds. At the same time, in Chapter 6, we also showed that if S is n.r.s., then $P_x(\Lambda_R) = o\left(V_d(x)/V_d(R)\right)$ uniformly for $0 \le x < \delta R$ ($0 < \delta < 1$). In this chapter, we obtain the precise asymptotic form of $P_x(\Lambda_R)$ in the case when S is n.r.s. under some additional regularity condition on the positive tail of F that is satisfied at least when F is in the domain of attraction of a stable law with exponent 1, $EX = 0$ and $\limsup m_-(x)/m(x) < 1/2$. The derivation of the asymptotic form of $P_0(\Lambda_R)$ is accompanied by those of the renewal sequence $u_a(x)$ and the Green function $g_\Omega(x, y)$. Indeed, the gist of the problem is to obtain the asymptotic form of $u_a(x)$, which yields estimates of $g_\Omega(x, y)$ sufficient for the derivation of the asymptotics of $P_x(\Lambda_R)$.

We suppose $\sigma^2 = \infty$, and for later citations, we bring in the conditions

(PRS) $A(x)/x\mu(x) \to \infty \quad$ as $\quad x \to \infty,$

(NRS) $A(x)/x\mu(x) \to -\infty \quad$ as $\quad x \to \infty$

$(A(x) = \int_0^x [\mu_+ - \mu_-(t)]\, dt)$. As mentioned previously, condition (PRS) holds if and only if S is p.r.s. We shall say F is p.r.s. (n.r.s., recurrent or transient) if so is the r.w. S. If F is p.r.s. (n.r.s.) both u_a and V_d (both v_d and U_a) are s.v. at infinity. When F is transient, the case of interest as much as the case of recurrence in this chapter, the Green kernel $G(x)$ defined by

K. Uchiyama, *Potential Functions of Random Walks in Z with Infinite Variance*,
Lecture Notes in Mathematics 2338, https://doi.org/10.1007/978-3-031-41020-8_7

$$G(x) = \sum_{n=0}^{\infty} P_0[S_n = x], \qquad (7.1)$$

plays a significant role. Under the relative stability condition, F is transient if and only if

$$\int_{x_0}^{\infty} \frac{\mu(t)}{A^2(t)} \, dt < \infty$$

and in this case $|A(x)| \to \infty$ (cf. [83]).

We shall also seek the asymptotic estimates of $P_x[\sigma_R < T]$ or $P_x[\sigma_0 < \sigma_{(R,\infty)}]$ which are not only of interest in themselves but are sometimes crucial for estimating $P_x(\Lambda_R)$. If F is n.r.s., the comparison of $P_0[\sigma_R < T]$ with $P_0(\Lambda_R)$ leads to the determination of the asymptotic form of the renewal sequence – as well as these two probabilities – under some regularity condition on the positive tail of F. Applying the obtained asymptotic estimates of $P_x[\sigma_R < T]$ and $P_x(\Lambda_R)$, we shall derive the asymptotic behaviour as $R \to \infty$ of the probability that R is ever hit by S, conditioned to avoid the negative half-line forever.

7.1 Statements of the Main Results

The following result (together with its dual) is fundamental for later arguments.

Theorem 7.1.1 *If F is recurrent and p.r.s., then $a(x)$, $x > 0$, is s.v. at infinity and as $y \to \infty$,*

$$v_{\mathrm{d}}(y) = o\left(a(y)/U_{\mathrm{a}}(y)\right), \quad \text{and}$$

$$g_\Omega(x, y) = a(x) - a(x - y) + o\left(a(x)\right) \qquad \text{uniformly for } x > y/2,$$

in particular $g_\Omega(y, y) \sim a(y)$; and for each constant $\delta < 1$

$$g_\Omega(x, y) \qquad\qquad\qquad\qquad\qquad\qquad\qquad\qquad\qquad (7.2)$$

$$= \begin{cases} \dfrac{V_{\mathrm{d}}(x)}{V_{\mathrm{d}}(y)} \left[a(y) - a(-y) + o(a(y))\right] \ (y \to \infty) & \text{uniformly for } 0 \le x \le \delta y, \\[2ex] o\left(\dfrac{U_{\mathrm{a}}(y)}{U_{\mathrm{a}}(x)} a(x)\right) \ (x \to \infty) & \text{uniformly for } 0 \le y < \delta x. \end{cases}$$

[For an alternative asymptotic form in the first case of (7.2) see (7.5) below.]

Remark 7.1.2 (a) The essential ingredients of Theorem 7.1.1 will be derived under a condition weaker than (PRS). (See Proposition 7.3.1 and Remark 7.3.2(a).) Under an additional regularity condition on F the estimate of $v_{\mathrm{d}}(x)$ and the second estimate in (7.2) are refined in the dual form in Theorems 7.1.4 and 7.1.6 below (see the comments following Corollary 7.1.8).

(b) Under (PRS) some asymptotic estimates of $a(x)$ and $G(x)$ as $x \to \pm\infty$ are obtained in Section 4.1 and [83], respectively; in particular it follows that as $x \to \infty$,

$$\frac{1}{A(x)} \sim a(x) - a(-x). \tag{7.3}$$

Note that by definition, $a(x) - a(-x) = G(x) - G(-x)$ if F is transient. (See (7.36) and (7.90) for more about a and G, respectively.)

(c) In view of the identity $g_\Omega(x, y) = \hat{g}_\Omega(y, x)$, the dual statement of Theorem 7.1.1 may read as follows: If F is recurrent and n.r.s., then as $x \to \infty$

$$g_\Omega(x, y) = a(-y) - a(x - y) + o\left(a(-y)\right) \quad \text{uniformly for } y > x/2, \tag{7.4}$$

in particular $g_\Omega(x, x) \sim a(-x)$; and

$$g_\Omega(x, y)$$

$$= \begin{cases} o\left(\dfrac{V_d(x)}{V_d(y)} a(-y)\right) \ (y \to \infty) & \text{uniformly for } 0 \le x < \delta y, \\[3mm] \dfrac{U_a(y)}{U_a(x)} \left[a(-x) - a(x) + o(a(-x))\right] \ (x \to \infty) & \text{uniformly for } 0 \le y \le \delta x. \end{cases}$$

(d) If both u_a and V_d are s.v., then by Spitzer's formula (2.6)

$$g_\Omega(x, y) \sim \frac{V_d(x)}{\ell^*(y)} \sim \frac{V_d(x) U_a(y)}{y} \quad \text{for } 0 \le x < \delta y. \tag{7.5}$$

Comparing this to (7.2) (with $x = y/2$) and using (7.3) one sees that if F is recurrent and p.r.s, then

$$V_d(x) U_a(x)/x = 1/A(x) + o(a(x)).$$

For transient F, we shall give in Section 7.5 a formula analogous to (7.2), and according to it, the corresponding formula becomes

$$V_d(x) U_a(x)/x = 1/A(x) + o(G(x)).$$

As to the estimates of $g_\Omega(x, y)$, the result corresponding to Theorem 7.1.1 for the transient walk is much cheaper because G is bounded – we shall give it in Section 7.5 as Lemma 7.5.1 (in the dual setting, i.e., for an n.r.s. walk), whereas the estimation of u_a (given in Theorem 7.1.6 below not for v_a because of the dual setting) is more costly than for the recurrent walk. Here we state the well-established fact that if F is transient, then $0 < G(0) = 1/P_0[\sigma_0 = \infty] < \infty$ and

$$g_\Omega(x, x) \longrightarrow G(0). \tag{7.6}$$

(See Section A.5.3 for the proof of the latter assertion.)

If F is n.r.s., then $P_x(\Lambda_R) \to 0$ $(0 \leq x < \delta R)$, and the exact estimation of $P_x(\Lambda_R)$ seems demanding to obtain in general. However, if the positive and negative tails of F are not balanced in the sense that

$$\begin{cases} \lim\sup_{x\to\infty} a(x)/a(-x) < 1 & \text{if } F \text{ is recurrent,} \\ \lim\sup_{x\to\infty} G(x)/G(-x) < 1 & \text{if } F \text{ is transient} \end{cases} \tag{7.7}$$

and if the positive tail of F satisfies an appropriate regularity condition, then we can compute the precise asymptotic form of $g_\Omega(x,y)$ for $0 \leq x < \delta y$ (which is lacking in the second formula in (c) of Remark 7.1.2), and thereby obtain that of $P_x(\Lambda_R)$ for F that is n.r.s. We need to assume

$$\exists \lambda > 1, \quad \lim\sup \frac{\mu_+(\lambda t)}{\mu_+(t)} < 1. \tag{7.8}$$

Proposition 7.1.3 *Suppose that F is n.r.s. Then*

$$P_0(\Lambda_R) \geq v^\circ U_a(R)\mu_+(R)\{1 + o(1)\}. \tag{7.9}$$

Let (7.8) hold in addition. Then as $x \to \infty$

$$u_a(x) \geq \frac{U_a(x)\mu_+(x)}{-A(x)}\{1 + o(1)\} \tag{7.10}$$

and under (7.7), $U_a(x)V_d(x) \sim -x/A(x)$ and

$$\begin{cases} u_a(x) = o(U_a(x)/x) \quad \text{and} \quad \exists c > 0, \ P_0[\sigma_R < T]/P_0(\Lambda_R) > c, \\ \qquad\qquad\qquad\qquad\qquad\qquad\qquad\qquad\qquad \text{if } F \text{ is recurrent}; \\ u_a(x) = o\left(\frac{|A(x)|U_a(x)}{x}\right) \quad \text{and} \quad \frac{P_0[\sigma_R < T]}{P_0(\Lambda_R)} \geq \frac{1+o(1)}{|A(R)|G(0)}, \\ \qquad\qquad\qquad\qquad\qquad\qquad\qquad\qquad\qquad \text{if } F \text{ is transient.} \end{cases}$$

In the next theorem, we obtain precise asymptotic forms of $P_0(\Lambda_R)$ and $u_a(x)$ in the case when F is recurrent by assuming, in addition to (NRS), (7.7) and (7.8), the 'continuity' condition

$$\lim_{\delta \uparrow 1} \lim\sup_{x\to\infty} \mu_+(\delta x)/\mu_+(x) = 1. \tag{7.11}$$

Theorem 7.1.4 *If F is recurrent and n.r.s. and if (7.11), (7.7) and (7.8) hold, then*

$$u_a(x) \sim \frac{U_a(x)\mu_+(x)}{-A(x)}, \tag{7.12}$$

$$P_0(\Lambda_R)/v^\circ \sim U_a(R)\mu_+(R), \tag{7.13}$$

and for each $\delta < 1$, uniformly for $0 \leq x < \delta y$ as $y \to \infty$,

$$g_\Omega(x, y) \sim \frac{V_d(x)U_a(y)}{|A(y)|(x+1)} \int_{y-x-1}^y \mu_+(t)\, dt; \quad \text{and} \tag{7.14}$$

$$\frac{P_x[\sigma_y < T]}{P_x[\Lambda_y]} \sim \frac{a(-y) - a(y)}{a(-y)}. \tag{7.15}$$

Remark 7.1.5 (a) Under (NRS), condition (7.7) follows if we suppose

$$\begin{cases} \limsup_{x \to \infty} m_-(x)/m(x) < 1/2 & \text{if } EX = 0, \\ \limsup_{x \to \infty} A_+(x)/A(x) < 1/2 & \text{if } F \text{ is transient,} \end{cases}$$

where $A_\pm(x) = \int_0^x \mu_\pm(t)\, dt$ $(x \geq 0)$ [note $A(x) = \eta_-(x) - \eta_+(x)$ if $EX = 0$]. (See Section 7.7, where related matters are discussed in the dual setting.)

(b) Suppose that $EX = 0$, $\limsup \mu_-(x)/\mu(x) < \frac{1}{2}$ and the asymptotic 'continuity' condition (7.11) holds. Then the assumption of Theorem 7.1.4 holds if

$$\mu_+(x) \asymp L(x)/x \quad (x \to \infty) \quad \text{for some s.v. } L, \tag{7.16}$$

as is readily verified (the converse is not true).

If F is transient one may expect the formulae parallel to those given in Theorem 7.1.4 to be true on an ad hoc basis, but the analysis runs into a difficulty. The next theorem ensures them under the following condition, more restrictive than the continuity condition (7.11):

$$\exists C_* > 0, \quad p(x) \leq C_* \frac{\mu_+(x)}{x} \quad (x \geq 1). \tag{7.17}$$

Theorem 7.1.6 *Let F be transient and n.r.s. Suppose that (7.17) holds in addition to (7.7) and (7.8). Then the formulae (7.12) to (7.15) are valid, where (7.15) is more naturally written as*

$$\frac{P_x[\sigma_y < T]}{P_x[\Lambda_y]} \sim \frac{G(-y) - G(y)}{G(0)}. \tag{7.18}$$

Remark 7.1.7 Let $E|X| = \infty$. Then by Erickson's criterion (5.22), under (NRS) our hypothesis of S being oscillatory is expressed as

$$\int_1^\infty \frac{\mu_+(t)}{A_-(t)}\, dt = \infty.$$

Because of the identity $P_x[\sigma_y < T] = g_\Omega(x, y)/g_\Omega(y, y)$ $(y \neq x)$, combining (7.6), (7.14) and (7.4) with the equivalence (7.15) or (7.18) leads to the following

Corollary 7.1.8 *If the assumption of Theorem 7.1.4 or that of Theorem 7.1.6 holds (according as F is recurrent or transient), then uniformly for $0 \leq x < \delta R$,*

$$P_x(\Lambda_R) \sim |A(R)| g_\Omega(x, R) \tag{7.19}$$

$$\sim \frac{V_d(x)U_a(R)}{x+1} \int_{R-x-1}^R \mu_+(t)\, dt \quad \text{as } R \to \infty.$$

In the dual setting, Theorem 7.1.4 is paraphrased as follows. If F is recurrent and p.r.s., (7.11) and (7.8) hold with μ_- in place of μ_+ and $\limsup a(-x)/a(x) < 1$, then

$$1 - P_0(\Lambda_y) \sim V_d(y)\mu_-(y) \sim v_d(y)A(y) \quad (y \to \infty), \tag{7.20}$$

$$P_x[\sigma_0 < \sigma_{(R,\infty)}] \sim g_\Omega(R, R - x)/a(R) \quad \text{uniformly for } 0 \le x \le R$$

and for each $\varepsilon > 0$,

$$g_\Omega(x, y) \sim \frac{U_a(y)V_d(x)}{A(x)(y+1)} \int_{x-y-1}^x \mu_-(t)\, dt \quad (x \to \infty) \tag{7.21}$$

$$\text{uniformly for } 0 \le y < (1 - \varepsilon)x,$$

$$\frac{P_x[\sigma_0 < \sigma_{(R,\infty)}]}{P_x[T < \sigma_{(R,\infty)}]} \sim \frac{a(R) - a(-R)}{a(R)} \quad (R \to \infty) \text{ uniformly for } \varepsilon R < x \le R,$$

and

$$1 - P_x(\Lambda_R) \sim A(R)g_\Omega(R, R - x) \sim \frac{U_a(R - x)V_d(R)}{R - x + 1} \int_{x-1}^R \mu_-(t)\, dt \tag{7.22}$$

$$\text{as } R \to \infty \text{ uniformly for } \varepsilon R < x \le R;$$

in particular by (7.21) as $x \to \infty$

$$P_x[\sigma_y < T] \asymp \frac{yV_d(x)\mu_-(x)}{V_d(y)A(x)} = o\left(\frac{y/V_d(y)}{x/V_d(x)}\right) \quad \text{uniformly for } 1 \le y < \delta x,$$

where for the last formula we have used $V_d(y)U_a(y) \asymp ya(y)$ (see **L(3.1)** of the next section and (7.39)). It would be straightforward to state the dual results of Theorem 7.1.6.

Remark 7.1.9 In [84, Eq(2.22)] it is shown that if $EZ < \infty$, then

$$1 - P_x(\Lambda_R) \sim [V_d(R) - V_d(x)]/V_d(R) \quad \text{as } R - x \to \infty \text{ for } x \ge 0,$$

which, giving an exact asymptotic for $0 \le x \le \delta R$, partially complements (7.22).

From the estimates of u_a and $g_\Omega(x, y)$ of Theorems 7.1.4 and 7.1.6 we can derive some exact asymptotic estimates of

$$P_x[S_{\sigma(R,\infty)-1} = y \,|\, \Lambda_R], \tag{7.23}$$

the conditional probability of S exiting the interval $[0, R]$ through y, given Λ_R. We shall carry out the derivation in Section 7.6. The result obtained will lead to the evaluation of the distribution of the overshoot $Z(R)$ (Proposition 7.6.4).

We conclude this section by giving the asymptotic forms of $g_{B(R)}(x, y)$, the results being essentially consequences of what we have stated above.

Theorem 7.1.10 *Let F be n.r.s. Put $B(R) = \mathbb{R} \setminus [0, R]$ and take a constant $\frac{1}{2} < \delta < 1$ arbitrarily and put $\varepsilon = 1 - \delta$.*

(i) *As $y \to \infty$, uniformly for $\frac{1}{2}x < y \le x < \delta R$,*

$$g_{B(R)}(x, y) = a(-y) - a(x - y) + o(a(-y)), \tag{7.24}$$

in particular $g_{B(R)}(y, y) \sim a(-y)$ uniformly for $R > y/\delta$.

(ii) *If either (7.7) holds or $p(-x)x = O(\mu_-(x))$, then as $x \to \infty$,*

$$g_{B(R)}(x, y) \sim \frac{U_a(y)U_a(R - x)}{U_a(R)\hat{\ell}^*(x)} \qquad \text{uniformly for } 0 \le y < \delta x, \ R \ge x. \tag{7.25}$$

(iii) *Suppose that either the assumption of Theorem 7.1.4 or that of Theorem 7.1.6 holds and $x\mu_+(x)$ is s.v. Then*

$$g_{B(R)}(x, y) = \frac{V_d(x)yu_a(y)}{(x + 1)}\left[\log\frac{1 - (x + 1)/R}{1 - (x + 1)/y} + o(1)\right] \qquad (R \to \infty)$$

uniformly for $0 \le x < \delta y$ and $\varepsilon R < y < \delta R$.
$$\tag{7.26}$$

[Recall $u_a(y) \sim \mu_+(y)U_a(y)/|A(y)|$ in (7.26) according to Theorem 7.1.6.]

By $g_{B(R)}(x, y) = \hat{g}_{B(R)}(y, x)$ we have the dual statement of Theorem 7.1.10 (recall $\hat{g}_{B(R)}$ designates the dual of $g_{B(R)}$). If the duals of assumptions (ii) and (iii) are valid, then

$$g_{B(R)}(x, y) \sim \frac{V_d(x)V_d(R - y)}{V_d(R)\ell^*(y)} \qquad (y \to \infty) \text{ for } 0 \le x < \delta y, \ R \ge y, \tag{7.27}$$

$$g_{B(R)}(x, y) = \frac{xv_d(x)U_a(y)}{(y + 1)}\left[\log\frac{1 - y/R}{1 - y/x} + o(1)\right] \qquad (R \to \infty) \tag{7.28}$$

uniformly for $0 \le y < \delta x$ and $\varepsilon R < x < \delta R$

instead of (7.25) and (7.26), respectively. By Feller's duality, one can bring out some cases of x, y for the validity of (7.24) to (7.28) not meeting the restrictions specified therein: an instance is afforded by (7.29) below (take $x > \delta R$).

Proof (of Theorem 7.1.10) We show the duals of (i) and (ii). Let F be p.r.s. Then by Lemma 5.7.3 $g_{B(R)}(x, y) \sim g_\Omega(x, y)$ for $0 \le x \le y < \delta R$, which combined with (7.4) yields the dual of (7.24). This and (7.5) together show that if $0 \le x < \delta y$ and $y < \delta R$, $g_{B(R)}(x, y) \sim V_d(x)/\ell^*(y)$, which conforms to (7.27). Because of the identity $g_{B(R)}(x, y) = g_{B(R)}(R - y, R - x)$ it follows that as $R - x \to \infty$,

$$g_{B(R)}(x, y) \sim V_d(R - y)/\ell^*(R - x) \quad \text{for } R - y < \delta(R - x) \text{ and } x > \varepsilon R, \tag{7.29}$$

which also conforms to (7.27) and shows it for $y \ge \delta R$ and $\varepsilon R < x < \frac{1}{2}R$. Now one notices that (7.27) is verified under the additional restriction $x \wedge (R - y) < \varepsilon R$, provided F is merely p.r.s.

Put $R' = \lfloor \frac{1}{2} R \rfloor$. Then (7.27) follows if we can show that

$$g_{B(R)}(x, y) \sim P_x(\Lambda_{R'})g_{B(R)}(R', y) \quad \text{for } 0 \le x < \varepsilon y \text{ and } \delta R \le y \le R. \quad (7.30)$$

Indeed, using $P_x(\Lambda_{R'}) \sim V_d(x)/V_d(R')$, valid under (PRS), one infers from (7.29) that the RHS is asymptotically equivalent to that of (7.27). Since for any $\varepsilon_1 > 0$, $P_x[Z(R') > \varepsilon_1 R \,|\, \Lambda_R'] \to 0$, in view of the representation

$$g_{B(R)}(x, y) = \sum_z P_x \left[Z(R') = z; \Lambda_R' \right] g_{B(R)}(R' + z, y)$$

one may expect (7.30) to be true. However, $\sup_z g_{B(R)}(R' + z, y)/g_{B(R)}(R', y)$ tends to infinity, so the problem is non-trivial and delicate without the additional assumption made for (7.27) in the theorem. We postpone the proof of (7.30) to Section 8.6 (given after the proof of Lemma 8.6.2, where we encounter a similar problem).

The proof of (iii) rests on Theorems 7.1.4 and 7.1.6. Suppose that the assumption of Theorem 7.1.4 holds and $x\mu_+(x)$ is s.v. We apply Lemma 7.6.1 proved in Section 7.6, which says that uniformly for $0 \le x < \delta R, 0 \le y \le R$,

$$g_{B(R)}(x, y) = g_\Omega(x, y) - [U_a(y)/U_a(R)] g_\Omega(x, R)\{1 + o(1)\}. \quad (7.31)$$

Substitution of the asymptotic form of $g_\Omega(x, \cdot)$ in (7.14) leads to

$$g_{B(R)}(x, y) = \frac{V_d(x)U_a(y)}{x+1} \left[\frac{1+o(1)}{|A(y)|} \int_{y-x-1}^y \mu_+(t)\, dt - \frac{1+o(1)}{|A(R)|} \int_{R-x-1}^R \mu_+(t)\, dt \right].$$

Noting that $|A(y)|$ is s.v. and $U_a(y)/|A(y)| \sim u_a(y)/\mu_+(y)$, we can easily identify the RHS with that of (7.26) for $0 \le x < \delta y$ and $\varepsilon R < y < \delta R$. The proof under the assumption of Theorem 7.1.6 is similar.   □

Section 7.2 states some of the results from Chapter 6 that are fundamental in the later discussions. The proof of Theorem 7.1.1 will be given in Section 7.3. In Section 7.4, we shall prove Proposition 7.1.3 (in the case when F is recurrent) and Theorem 7.1.4 after showing miscellaneous lemmas in preparation for their proofs. Proposition 7.1.3 (in the case $E|X| = \infty$) and Theorem 7.1.6 will be proved in Section 7.5.

7.2 Basic Facts from Chapter 6

We shall use the results obtained in Chapter 6, some of which we collect below. First of all recall (6.12), which says

$$P_x(\Lambda_R) \le V_d(x)/V_d(R),$$

and that (6.1) holds, i.e., $P_x(\Lambda_R) \sim V_d(x)/V_d(R)$, under

(C3) both $V_d(x)$ and $xU_a(x)$ are s.v. as $x \to \infty$.

For the convenience of later citations we write down the dual of (C3):

$(\widehat{C3})$ both $xV_d(x)$ and $U_a(x)$ are s.v. as $x \to \infty$.

The condition (C3) follows from (PRS) and $(\widehat{C3})$ from (NRS), as mentioned previously. Put, for $t \geq 0$,

$$\ell^*(t) = \int_0^t P[Z > s]\,ds \quad \text{and} \quad \hat{\ell}^*(t) = \frac{1}{v^\circ} \int_0^t P[-\hat{Z} > s]\,ds$$

(as in Chapter 6), and

$$\ell_{\sharp}(t) = \int_t^\infty \frac{\mu_-(s)}{\ell^*(s)}\,ds \quad \text{and} \quad \hat{\ell}_{\sharp}(t) = \int_t^\infty \frac{\mu_+(s)}{\hat{\ell}^*(s)}\,ds, \tag{7.32}$$

respectively, in case (C3) and in case $(\widehat{C3})$.

The ladder height Z is r.s. if and only if $xU_a(x)$ is s.v. (Theorem A.2.3) and if this is the case, u_a is s.v. and $u_a(x) \sim 1/\ell^*(x)$ (cf. (5.28)). The following are of repeated use.

L(2.1) For $x \geq 0$ and $R \geq 1$, $\displaystyle\sum_{y=0}^R g_\Omega(x, y) < V_d(x)U_a(R)$.

L(3.1) Under (C3), ℓ^* and $\ell_{\sharp}$ are s.v., $u_a(x) \sim 1/\ell^*(x)$ and $V_d(x) \sim 1/\ell_{\sharp}(x)$.
By the duality this entails that under $(\widehat{C3})$, $\hat{\ell}^*$ and $\hat{\ell}_{\sharp}$ are s.v.,

$$v_d(x) \sim 1/\hat{\ell}^*(x) \quad \text{and} \quad U_a(x) \sim 1/\hat{\ell}_{\sharp}(x). \tag{7.33}$$

L(3.4) If either (C3) or $(\widehat{C3})$ holds, then $V_d(x)U_a(x)\mu(x) \to 0$.

L(3.6) If (C3) holds, then for each $\varepsilon > 0$, $P_x[Z(R) > \varepsilon R \,|\, \Lambda_R] \to 0$ as $R \to \infty$ uniformly for $0 \leq x < R$.

L(4.7) If either $(\widehat{C3})$ or (AS) with $\alpha < 1 = \hat{\rho}$ holds, then

$$P_x(\Lambda_R) = o\,(V_d(x)/V_d(R)) \quad (R \to \infty) \text{ uniformly for } 0 \leq x < \delta R.$$

These results follow from Lemmas 6.2.1, 6.3.1, 6.3.4, 6.3.6 and 6.4.7, respectively.

7.3 Proof of Theorem 7.1.1 and Relevant Results

Here we shall suppose that F is recurrent. Consider the Green function of S killed as it hits $(-\infty, 0]$, instead of $\Omega = (-\infty, -1]$. We make this choice for convenience in applying (2.15), which yields the relation (7.34) below. Recall (2.3) and (5.2):

$$g(x, y) = g_{\{0\}}(x, y) - \delta_{x,0} = a(x) + a(-y) - a(x - y)$$

and use (2.15) to see that for $x \geq 1$,

$$E_x[a(S_{\sigma_{(-\infty,0]}})] = \begin{cases} a(x) & \text{if } EZ = \infty, \\ o\,(a(x)) & (x \to \infty) \quad \text{if } EZ < \infty; \end{cases} \tag{7.34}$$

$$g_{(-\infty,0]}(x,y) = g_{\{0\}}(x,y) - E_x[g(S_{\sigma_{(-\infty,0]}},y)], \tag{7.35}$$

which take less simple forms for $E_x[a(S_T)]$ and $g_\Omega(x,y)$.

The following conditions, assumed in the next proposition, are all satisfied if either (PRS) or (AS) with $\alpha = 1$ and $\rho > 0$ holds (see Remark 7.3.2(a),(c) below):

(1) $a(x)$ is almost increasing and $a(-x)/a(x)$ is bounded as $x \to \infty$;

(2) $\displaystyle \sup_{z':z<z'\leq 2z} \frac{|a(-z') - a(-z)|}{a(z)} \longrightarrow 0$ as $z \to \infty$;

(3) $P_x[S_T > -\varepsilon x] \longrightarrow 0$ as $x \to \infty$ and $\varepsilon \downarrow 0$ in this order.

Proposition 7.3.1 *Suppose that the conditions (1) to (3) above hold. Then for each $\varepsilon > 0$, $\limsup_{x \to \infty} a(-x)/a(\varepsilon x) \leq 1$ and as $x \to \infty$:*

(i) $g_\Omega(x,y) = a(x) - a(x-y) + o\,(a(x))$ *uniformly for $x > \varepsilon y > 0$,*
    in particular, $g_\Omega(x,y) \sim a(x)$ if $a(x-y) = o(a(x))$ and $x > \varepsilon y > 0$;

(ii) $g_\Omega(x,y) = a(x) - a(-x) + o\,(a(x))$ *uniformly for $\varepsilon y \leq x \leq (1-\varepsilon)y$.*

Remark 7.3.2 (a) Let (PRS) hold. Then, according to Corollary 4.1.2,

$$\begin{cases} a(x) \text{ is s.v.; } \quad a(x) - a(-x) \sim 1/A(x)\,(> 0); \\ a(x) \sim \displaystyle\int_{x_0}^x \frac{\mu_-(t)}{A^2(t)}\,dt, \quad a(-x) = \displaystyle\int_{x_0}^x \frac{\mu_+(t)}{A^2(t)}\,dt + o(a(x)), \end{cases} \tag{7.36}$$

as $x \to \infty$. Hence, by (PRS), for $x_0 < z < z' \leq 2z$

$$a(-z') - a(-z) = \int_z^{z'} \frac{\mu_+(t)}{A^2(t)}\,dt + o(a(z)) = o\left(\frac{z'-z}{A(z)z}\right) + o(a(z)),$$

and one sees that (2) is satisfied. We also know that by **L(3.1)** $V_d(x)$ is s.v., or what amounts to the same thing,

$$\forall M > 1, \quad \lim P_x[S_T < -Mx] = 1 \tag{7.37}$$

(cf. Theorem A.2.3), in particular (3) is satisfied. (1) is evident from (7.36).

(b) Let F be recurrent and p.r.s. By virtue of the Spitzer's formula (2.6) we know $g_\Omega(x,y) \sim V_d(x)/\ell^*(y)$ for $0 \leq x < \delta y$ ($0 < \delta < 1$), which, in conjunction with (7.36), $V_d(x) \sim 1/\ell_\sharp(x)$ and the second half of Proposition 7.3.1, leads to the equivalence relations

$$a(x) \asymp \frac{1}{A(x)} \iff \limsup \frac{a(-x)}{a(x)} < 1 \iff a(x) \asymp \frac{1}{\ell^*(x)\ell_\sharp(x)}, \tag{7.38}$$

as well as $1/\ell^*(x)\ell_\sharp(x) \sim g_\Omega(x, x/\delta) = a(x) - a(-x) + o(a(x))$, so that each of the conditions in (7.38) implies

$$A(x) \sim \ell^*(x)\ell_\sharp(x) \qquad (7.39)$$

and that $\lim a(-x)/a(x) = 1$ if and only if $\ell^*(x)\ell_\sharp(x)a(x) \to \infty$.

(c) Suppose that F satisfies (AS) with $\alpha = 1$ and $\rho > 0$. Then conditions (1) to (3) are satisfied. Indeed, for $\rho = 1$, (PRS) is satisfied, while for $0 < \rho < 1$,

$$a(x) \sim a(-x) \sim c_\rho [1/L]_*(x) \qquad (7.40)$$

$([1/L]_*(x) = \int_1^x [tL(t)]^{-1}\, dt)$ with a certain constant $c_\rho > 0$ (Proposition 4.2.1(iv)), entailing (1) and (2); moreover, V_d is regularly varying with index $\hat\rho \in (0, 1)$, so that (3) is satisfied owing to the generalised arcsin law ([31, p. 374], Theorem A.2.3).Thus Proposition 7.3.1 is applicable. For $0 < \rho < 1$, (ii) can be refined to

$$g_\Omega(x, y) = o(a(y)) \quad (y \to \infty) \quad \text{uniformly for } 0 \le x < (1 - \varepsilon)y. \qquad (7.41)$$

For $x > \varepsilon y$, this is immediate from (7.40) and (i). For $0 < x \ll y$, one has only to use $g_\Omega(x, y) = \sum_z P_x[S_{\sigma[\varepsilon y, \infty)} = z, \Lambda_{\varepsilon y}]g_\Omega(z, y)$. Indeed, by Proposition 6.1.3 the sum restricted to $z < 2y$ is $o(a(y))$. Since $a(z) - a(z - y) = o(a(z))$ for $z \ge 2y$, using $H^{-x}_{[0,\infty)}(a) = a(-x)$ (valid if $E|\hat{Z}| = \infty$), one sees that the sum over $z \ge 2y$ is also $o(a(y))$. Hence (7.41) holds true.

The following corollary follows immediately from Proposition 7.3.1 and Remark 7.3.2(a,c) because of

$$P_x[\sigma_y < T] = \frac{g_\Omega(x, y)}{g_\Omega(y, y)} \quad (x \ne y). \qquad (7.42)$$

Corollary 7.3.3 *Suppose that F is recurrent and satisfies either (PRS) or (AS) with $\hat\rho < \alpha = 1$. Then, for any $0 < \delta < 1$ and $\varepsilon > 0$, as $y \to \infty$,*

$$P_x[\sigma_y < T] = \begin{cases} \dfrac{a(x) - a(-x)}{a(x)} + o(1) & \text{uniformly for } \varepsilon y \le x < \delta y, \\ o(1) & \text{uniformly for } x > y/\delta; \end{cases}$$

in particular under (AS) with $\hat\rho < \alpha = 1 = 2p$, $P_x[\sigma_y < T] \to 0$ as $y \to \infty$ uniformly for $x \notin [\delta y, y/\delta]$.
[For a transient r.w., $a(x) = G(0) - G(-x)$, so that $a(x) - a(-x) = G(x) - G(-x)$. If a transient F is p.r.s., by the estimate of $G(x)$ given in [83] (see (7.90) and Lemma 7.5.1 – given in the dual setting – of this chapter) one sees that a formula analogous to the above holds but for $R/\delta < x < R/\varepsilon$ (resp. $x < \delta R$) instead of $\varepsilon R \le x < \delta R$ (resp. $x > R/\delta$).]

Proof One has only to examine the last statement in the case $x < \varepsilon y$, which is verified similarly to the corresponding part of (7.41). [See Proposition 8.3.1 for a sharper result.]   □

The proofs of Proposition 7.3.1 and Theorem 7.1.1 will be given after the following two lemmas.

Lemma 7.3.4 *If (1) and (2) hold, then for any $M \geq 1$,*

$$\sup\left\{\frac{|a(-x - z) - a(-z)|}{a^\dagger(x) + a(-z)} : 0 \leq x < Mz\right\} \longrightarrow 0 \qquad (z \to \infty). \tag{7.43}$$

Proof From (1) and (2) it follows that

$$|a(-x - z) - a(-z)|/a(x) \to 0 \quad (z \to \infty) \quad \text{uniformly for } z \leq x < Mz. \tag{7.44}$$

Pick $\varepsilon > 0$ arbitrarily and choose z_0 – possible under (2) – so that

$$|a(-x - z) - a(-z)| < \varepsilon^2 a(z) \quad \text{whenever } z \geq z_0 \text{ and } 0 \leq x < z. \tag{7.45}$$

Let $0 < x \leq z$ and $z > z_0$. Then if $a(-z) > \varepsilon a(z)$, $|a(-x - z) - a(-z)| < \varepsilon^2 a(z) < \varepsilon a(-z)$. On the other hand, if $a(-z) \leq \varepsilon a(z)$, by the inequalities

$$-\frac{a(-z)}{a(z)}a(-x) \leq a(-x - z) - a(-z) \leq \frac{a(-x - z)}{a(z + x)}a(x)$$

(cf. Section 5.1), condition (1) entails that $|a(-x - z) - a(-z)| \leq \varepsilon C a(x)$, where we have also used the bound (7.45) to have

$$a(-x - z)/a(z + x) \leq C'[a(-z) + \varepsilon^2 a(z)]/a(z) < 2C'\varepsilon.$$

Thus we have $|a(-x - z) - a(-z)| \leq \varepsilon C[a(x) + a(-z)]$, showing the supremum in (7.43) restricted to $0 < x \leq z$ tends to zero. Combined with (7.44) this concludes the proof. $\qquad\qquad\square$

Lemma 7.3.5 *If (1) to (3) hold, then for any $\varepsilon > 0$, as $y \to \infty$,*

$$E_x[a(S_T - y)] = E_x[a(S_T)] + o\,(a(x)) \quad \text{uniformly for } x > \varepsilon y. \tag{7.46}$$

Proof By (3), for any $\varepsilon > 0$ and $\varepsilon_1 > 0$ we can choose $M > 1$ and $M' > 1$ so that

$$P_x[S_T \geq -y/M] < \varepsilon_1 \quad \text{if } x > \varepsilon y, y > M'. \tag{7.47}$$

Supposing (1) and (2) to hold we apply Lemma 7.3.4 to see that as $y \to \infty$,

$$a(-y - z) = a(-z) + o\,(a(-z) + a(y)) \quad \text{uniformly for } z > y/M,$$

hence

$$E_x[a(S_T - y); S_T < -y/M] = E_x[a(S_T); S_T < -y/M]\{1 + o(1)\} + o(a(y))$$
$$= E_x[a(S_T)]\{1 + o(1)\} + O\,(\varepsilon_1 a(y)) + o(a(y)).$$

Here (7.47) as well as (1) is used for the second equality, and by the same reasoning (with the help of (2)), the left-most expression is written as $E_x[a(S_T - y)] + O(\varepsilon_1 a(y))$. Noting $E_x[a(S_T)] \sim E_x[a(S_{\sigma_{(-\infty,0]}})] \le Ca^\dagger(x)$, we can conclude that

$$E_x[a(S_T - y)] = E_x[a(S_T)] + o\,(a(x \vee y)) + O(\varepsilon_1 a(y)),$$

which shows (7.46), ε_1 being arbitrary (note that $a(y) \le Ca(x)/\varepsilon$ because of sub-additivity of a). $\qquad\square$

Proof (of Proposition 7.3.1) Note that for the asymptotic estimates of our concern under P_x, T and $\sigma_{(-\infty,0]}$ may be interchangeable as $x \to \infty$. Then, applying (7.35) and Lemma 7.3.5 in turn one sees that as $y \to \infty$

$$g_{(-\infty,0]}(x, y) = a(x) - a(x - y) - \left(E_x[a(S_{(-\infty,0]})] - E_x[a(S_{(-\infty,0]} - y)]\right)$$
$$= a(x) - a(x - y) + o(a(x))$$

uniformly for $x > \varepsilon y$, showing the formula (i) of the proposition. The assumption (2) entails $a(x - y) - a(-x) = o(a(y))$ for $\varepsilon R \le x < (1 - \varepsilon)y$, where, because of $a(nx) \le na(x)$ ($n = 2, 3, \ldots$), we may replace $o(a(y))$ by $o(a(x))$. Hence (ii) follows from (i). Taking $y = 2x$ in (i) it follows that $\limsup a(-x)/a(x) \le 1$. $\qquad\square$

Proof (of Theorem 7.1.1) Taking (6.1) into account, it follows that under (PRS) for any $0 < \delta < 1$

$$g_\Omega(x, y) = \begin{cases} P_x(\Lambda_y)\,[a(y) - a(-y) + o(a(y))] & \text{uniformly for } 0 \le x < \delta y, \\ o(a(y)) & \text{uniformly for } x > y/\delta. \end{cases}$$

Indeed, if $x \ge \frac{1}{4}y$, this readily follows from Proposition 7.3.1, the slow variation of V_d and (7.36), while uniformly for $0 \le x < \frac{1}{4}y$ we have

$$g_\Omega(x, y) = \sum_{\frac{1}{4}y \le w < \frac{1}{2}y} P_x[S_{\sigma(\frac{1}{4}y,\infty)} = w, \Lambda_{\frac{1}{4}y}]g_\Omega(w, y) + o\left(P_x(\Lambda_{\frac{1}{4}y})g_\Omega(y, y)\right)$$
$$= P_x(\Lambda_y)[a(y) - a(-y) + o(a(y))],$$

by **L(3.6)** and what we have just shown. Thus the first formula of (7.2) obtains. From the dual of **L(4.7)** (applied with $x = 0$) one deduces that

$$v_d(x) = o\,(a(x)/U_a(x)) \tag{7.48}$$

(in our setting the dual of (7.48) is easier to deal with: see (7.52)). The second formula of (7.2) follows from (7.48) in view of Spitzer's formula for $g_\Omega(x, y)$. $\qquad\square$

For later usage, here we state the following result – corresponding to Corollary 7.3.3 for the n.r.s. walk – that follows immediately from the dual of Theorem 7.1.1 (stated in Remark 7.1.2(c)):

If F is n.r.s and recurrent, then as $y \to \infty$,

$$P_x[\sigma_y < T] \qquad\qquad\qquad\qquad\qquad\qquad\qquad (7.49)$$

$$= \begin{cases} o\left(V_d(x)/V_d(y)\right) & \text{uniformly for } 0 \le x < \delta y, \\ \dfrac{U_a(y)}{U_a(x)}\left[\dfrac{a(-x) - a(x)}{a(-y)} + o\left(\dfrac{a(-x)}{a(-y)}\right)\right] & \text{uniformly for } x > y/\delta. \end{cases}$$

By Lemma 6.2.2(ii), the second case gives a lower bound for $x \in (y, y/\delta)$.

7.4 Proof of Proposition 7.1.3 (for F Recurrent) and Theorem 7.1.4

This section consists of three subsections. In the first one we obtain some basic estimates of $P_0(\Lambda_R)$ and $u_a(x)$ under (NRS) and prove Proposition 7.1.3 in the case when F is recurrent. In the second we derive the asymptotic forms of $P_0(\Lambda_R)$ and $u_a(x)$ asserted in Theorem 7.1.4. The rest of Theorem 7.1.4 is proved in the third subsection.

Let $B(R) = \mathbb{R} \setminus [0, R]$ as before and define

$$N(R) = \sigma_{B(R)} - 1, \qquad\qquad\qquad\qquad\qquad (7.50)$$

the first epoch at which the r.w. S leaves the interval $[0, R]$. Throughout this section we assume that (NRS) holds, i.e., $x\mu(x)/A(x) \to -\infty$ ($x \to \infty$). Recall that this entails $P_0[S_n > 0] \to 0$, $V_d(x) \sim x/\hat{\ell}^*(x)$ and $U_a(x) \sim 1/\hat{\ell}_\sharp(x)$.

7.4.1 Preliminary Estimates of u_a and the Proof of Proposition 7.1.3 in the Case When F is Recurrent

In this subsection, we are mainly concerned with the case when F is recurrent, but some of the results are also valid for transient F. Note that the function $a(x)$ is always well defined and approaches the constant $G(0)$ as $|x| \to \infty$ if F is transient.

If F is n.r.s., then $g_\Omega(y, y) \sim a(-y)$ (by Theorem 7.1.1) so that as $y \to \infty$,

$$P_0[\sigma_y < T] \sim g_\Omega(0, y)/a(-y) = v^\circ u_a(y)/a(-y), \qquad (7.51)$$

hence by **L(27)** (applied with $x = 0$),

$$u_a(y) = o\left(a(-y)/V_d(y)\right). \qquad\qquad\qquad (7.52)$$

Lemma 7.4.1 *Let F be recurrent and n.r.s. and suppose* $\limsup a(x)/a(-x) < 1$. *Then*

$$u_a(y) = o(U_a(y)/y). \qquad\qquad\qquad\qquad (7.53)$$

Proof Under $\limsup a(x)/a(-x) < 1$, we have

$$a(-x) \asymp 1/A(x) \sim V_{\mathrm{d}}(x)U_{\mathrm{a}}(x)/x$$

owing to (7.39) of Remark 7.3.2(b), so that (7.53) follows from (7.52).   □

Lemma 7.4.2 *Suppose (NRS) holds.*

(i) $P_0(\Lambda_R) \geq U_{\mathrm{a}}(R)\mu_+(R)\{v^\circ + o(1)\}$.

(ii) *If μ_+ is of dominated variation, then for each $0 < \varepsilon < 1$,*

$$P_0\left[\varepsilon R \leq S_{N(R)} < (1-\varepsilon)R \,\middle|\, \Lambda_R\right] \to 0 \ \text{ and}$$
$$P_0\left[S_{N(R)} < \varepsilon R, \Lambda_R\right] \asymp U_{\mathrm{a}}(R)\mu_+(R).$$

[Here a non-increasing f is of dominated variation if $\liminf f(2x)/f(x) > 0$.]

Proof We have the trivial inequality $g_{B(R)}(x, y) \leq g_\Omega(x, y)$. We need to find an appropriate lower bound of $g_{B(R)}(x, y)$. To this end we use the identity

$$g_{B(R)}(x, y) = g_\Omega(x, y) - E_x[g_\Omega(S_{\sigma(R,\infty)}, y); \Lambda_R] \quad (0 \leq x, y \leq R). \quad (7.54)$$

By Spitzer's formula (2.6) we see that for $z \geq R$, $1 \leq y < \delta R$ ($\delta < 1$),

$$g_\Omega(z, y) \leq U_{\mathrm{a}}(y)/\hat{\ell}^*(R)\{1 + o(1)\},$$

so that for $0 \leq x, y < \delta R$,

$$g_{B(R)}(x, y) \geq g_\Omega(x, y) - U_{\mathrm{a}}(y)P_x(\Lambda_R)/\hat{\ell}^*(R)\{1 + o(1)\}. \quad (7.55)$$

Since U_{a} is s.v., $R/\hat{\ell}^*(R) \sim V_{\mathrm{d}}(R)$ and $P_0(\Lambda_R)V_{\mathrm{d}}(R) \to 0$ by virtue of **L(4.7)**, letting $x = 0$ in (7.55) and summing both sides of it lead to

$$\sum_{y=0}^{R/2} g_{B(R)}(0, y) = \sum_{y=0}^{R/2} g_\Omega(0, y) + o(U_{\mathrm{a}}(R)) \sim v^\circ U_{\mathrm{a}}(R). \quad (7.56)$$

Hence

$$P_0(\Lambda_R) \geq \sum_{y=0}^{R/2} g_{B(R)}(0, y)\mu_+(R - y) \geq U_{\mathrm{a}}(R)\mu_+(R)\{v^\circ + o(1)\}, \quad (7.57)$$

showing (i).

Let $0 < \varepsilon < \frac{1}{2}$. The first probability in (ii) is less than $\sum_{y=\varepsilon R}^{(1-\varepsilon)R} u_{\mathrm{a}}(y)\mu_+(R - y)$. After summing by parts, this sum may be expressed as

$$[U_{\mathrm{a}}(y)\mu_+(R - y)]_{y=\varepsilon R}^{(1-\varepsilon)R} - \int_{y=\varepsilon R}^{(1-\varepsilon)R} U_{\mathrm{a}}(t)d\mu_+(R - y) \quad (7.58)$$

apart from the error term of a smaller order of magnitude than $U_a(R)\mu_+(\varepsilon R)$. Because of the slow variation of U_a, the above difference is $o\,(U_a(R)\mu_+(\varepsilon R))$. By (i) we, therefore, obtain the first relation of (ii), provided that μ_+ is of dominated variation. The second relation of (ii) follows from the first and (7.56).   □

Lemma 7.4.2(ii) says that, given Λ_R, the conditional law of $S_{N(R)}/R$, the position of departure of the scaled r.w. S_n/R from the interval $[0,1]$, tends to concentrate near the boundary. If the positive tail of F satisfies the continuity condition (7.11), we shall see that such a concentration should be expected to occur only about the lower boundary (see Lemma 7.4.6); otherwise, this may be not true.

Lemma 7.4.3 *Suppose* $\limsup \mu_+(\lambda x)/\mu_+(x) < 1$ *for some* $\lambda > 1$. *Then*

$$\sup_{0 \le x \le R} P_x[Z(R) > MR \mid \Lambda_R] \to 0 \quad as \ M \to \infty \ uniformly \ for \ R. \qquad (7.59)$$

Proof For any integer $M > 1$, writing $M' = M + 1$ we have

$$P_x[Z(R) > MR \mid \Lambda_R] \le \frac{\sum_{y=0}^{R} g_{B(R)}(x,y)\mu_+(M'R - y)}{\sum_{y=0}^{R} g_{B(R)}(x,y)\mu_+(R - y)} \le \frac{\mu_+(MR)}{\mu_+(R)},$$

of which the last expression approaches zero as $M \to \infty$ uniformly in R under the supposition of the lemma.   □

Proof (of Proposition 7.1.3 (recurrent case)) Suppose that F is recurrent and the assumption of Proposition 7.1.3 holds, namely

$$\begin{cases} \text{(a)} & F \text{ is n.r.s.;} \\ \text{(b)} & F \text{ is recurrent and } \limsup_{x\to\infty} a(x)/a(-x) < 1; \\ \text{(c)} & \exists \lambda > 1, \ \limsup \mu_+(\lambda x)/\mu_+(x) < 1. \end{cases} \qquad (7.60)$$

Then Lemmas 7.4.2(i) and 7.4.3 are applicable, of which the former one gives the lower bound of $P_0(\Lambda_R)$ asserted in Proposition 7.1.3. By (7.51) we have

$$v^\circ u_a(R) \sim a(-R)P_0[\sigma_R < T] \ge -P_0(\Lambda_R)/A(R)\{1 + o(1).\} \qquad (7.61)$$

Here, for the inequality we have employed, in turn, (7.59), (b) of (7.60), the second half of (7.49) and $a(-x) - a(x) \sim -1/A(x)$. Since $a(-R) \asymp 1/A(R)$ because of (b), we have $P_0[\sigma_R < T] \ge cP_0(\Lambda_R)$ for some constant $c > 0$, showing the required lower bound of u_a. The rest of the assertions follow from Remark 7.1.2(d) (or (7.39)) and Lemma 7.4.1.   □

The following lemma is used in the next subsection.

Lemma 7.4.4 *Suppose that (7.60) holds.*

(i) *For any $1/2 \leq \delta < 1$, as $y \to \infty$*

$$
g_\Omega(w, y) \begin{cases} = V_d(w) \times o\left(U_a(y)/y\right) & 0 \leq w < \delta y, \\ \leq a(-y)\{1 + o(1)\} \leq C/|A(y)| & \delta y \leq w \leq y/\delta, \\ \sim U_a(y)/\hat{\ell}^*(w) & w > y/\delta > 0. \end{cases} \tag{7.62}
$$

(ii) $\displaystyle \sup_{y \geq 0} \sum_{w=0}^\infty g_\Omega(w, y)\mu_+(w) < \infty.$

Proof Let F be recurrent. By (2.6), Spitzer's representation of $g_\Omega(w, y)$, one can easily deduce (i) with the help of (7.53) and Theorem 7.1.1 (see (7.67) below). For the convenience of later citations, we note that (i) entails that for some constant C

$$
g_\Omega(w, y) \begin{cases} \leq CV_d(w)U_a(y)/y & 0 \leq w < y/2, \\ \leq CU_a(y)/\hat{\ell}^*(w) & w \geq 2y > 0. \end{cases} \tag{7.63}
$$

By (7.62), $\sum_{w=0}^\infty g_\Omega(w, y)\mu_+(w)$ is bounded above by a constant multiple of

$$
\frac{U_a(y)}{y} \sum_{w=0}^{y/2} V_d(w)\mu_+(w) + \frac{y\mu_+(y/2)}{|A(y)|} + U_a(y) \sum_{w=2y}^\infty \frac{1}{\hat{\ell}^*(w)}\mu_+(w) \tag{7.64}
$$

for $y \geq x_0$. The first two terms above approach zero as $y \to \infty$, for by **L(3.4)** $V_d(w)\mu_+(w) = o(1/U_a(w))$, while the last term tends to unity since the second sum equals $\hat{\ell}_\sharp(2y) \sim 1/U_a(y)$. Thus (ii) follows.   □

7.4.2 Asymptotic Forms of u_a and $P_0(\Lambda_R)$

Throughout this subsection we suppose (7.60) holds, so that the results obtained in the preceding subsection are applicable; in particular, we have the bound on $g_\Omega(x, y)$ in the last lemma, as well as the following bounds

$$
\begin{cases} u_a(y) = o\left(U_a(y)/y\right); \text{ and} \\ P_0(\Lambda_y) \asymp P_0[\sigma_y < T] \sim v^\circ u_a(y)/a(-y). \end{cases} \tag{7.65}
$$

Recalling $B(R) = \mathbb{R} \setminus [0, R]$, one sees that

$$
P_x(\Lambda_R) = \sum_{w=0}^R g_{B(R)}(x, R - w)\mu_+(w).
$$

For each $r = 1, 2, \ldots$, $g_{B(r)}(x, r - y)$ is symmetric in $x, y \in B(r)$:

$$
g_{B(r)}(x, r - y) = g_{B(r)}(y, r - x), \tag{7.66}
$$

since both sides are equal to $g_{-B(r)}(-r + y, -x)$. Recall that under (NRS), $\hat{Z}$ is r.s., $v_d(x) \sim 1/\hat{\ell}^*(x)$ and $U_a(x) \sim 1/\hat{\ell}_\sharp(x)$, where $\hat{\ell}_\sharp(t) = \int_t^\infty \mu_+(t)\, dt/\hat{\ell}^*(t)$.

By the dual of Theorem 7.1.1 and Remark 7.3.2(b) it follows that under (7.60),

$$a(-x) - a(x) \sim \frac{-1}{A(x)} \sim \frac{1}{\hat{\ell}^*(x)\hat{\ell}_\sharp(x)} \sim \frac{V_d(x)U_a(x)}{x}. \qquad (7.67)$$

The next lemma is crucial for proving Theorem 7.1.4. Recall that $N(R) = \sigma_{(R,\infty)} - 1$ and that the continuity condition (7.11), stronger than dominated variation, states

$$\lim_{\delta \uparrow 1} \limsup_{x \to \infty} \mu_+(\delta x)/\mu_+(x) = 1.$$

Lemma 7.4.5 *Suppose that (7.60) holds and μ_+ varies dominatedly. Then for some constant C,*

$$P_0\left[S_{N(R)} \geq \tfrac{1}{8}R, \Lambda_R\right] \leq CU_a(R)\mu_+(R) + o\left(P_0(\Lambda_{R/2})\right), \qquad (7.68)$$

and if one further supposes the continuity condition (7.11),

$$P_0\left[S_{N(R)} \geq \tfrac{1}{8}R, \Lambda_R\right] = o\left(U_a(R)\mu_+(R)\right) + o\left(P_0(\Lambda_{R/2})\right). \qquad (7.69)$$

Proof We use the representation

$$P_0\left[S_{N(R)} \geq \tfrac{1}{8}R, \Lambda_R\right] = \sum_{0 \leq w \leq R/8} g_{B(R)}(0, R - w)\mu_+(w). \qquad (7.70)$$

Splitting the r.w. paths by the landing points, y say, when S (started at the origin) exits $[0, \tfrac{1}{2}R]$, we obtain for $0 \leq w < R/8$,

$$g_{B(R)}(0, R - w) = \sum_{R/2 < y \leq R} P_0[S_{\sigma_{B(R/2)}} = y]g_{B(R)}(y, R - w).$$

Hence

$$g_{B(R)}(0, R - w) = \sum_{z=0}^{R/2} \sum_{R/2 < y \leq R} g_{B(\frac{1}{2}R)}(0, z)p(y - z)g_{B(R)}(y, R - w), \qquad (7.71)$$

where $p(x) = P[X = x]$. Taking $0 < \varepsilon < 1/8$ arbitrarily we decompose the double sum into the following three parts:

$$I_\varepsilon(w) = \sum_{z=0}^{R/2} \sum_{(1-\varepsilon)R < y \leq R}, \qquad II_\varepsilon(w) = \sum_{z=0}^{R/4} \sum_{\frac{1}{2}R < y \leq (1-\varepsilon)R}$$

$$\text{and} \quad III_\varepsilon(w) = \sum_{\frac{1}{4}R < z \leq \frac{1}{2}R} \sum_{\frac{1}{2}R < y \leq (1-\varepsilon)R}.$$

First we evaluate $II_\varepsilon(w)$. Employing the symmetry (7.66) we see that

$$g_{B(R)}(y, R - w) = g_{B(R)}(w, R - y) \leq g_\Omega(w, R - y). \tag{7.72}$$

Since $g_\Omega(w, R - y) \leq CV_d(w)U_a(R)/(\varepsilon R)$ in its range of y by virtue of (7.63), we infer that

$$\frac{II_\varepsilon(w)}{U_a(R)\mu_+(R)} \leq \frac{CV_d(w) \sum_{z=0}^{R/4} g_{B(\frac{1}{2}R)}(0, z)\mu_+(\frac{1}{4}R)}{\varepsilon R\mu_+(R)} \tag{7.73}$$

$$\leq \varepsilon^{-1} CV_d(w)U_a(R)/R.$$

By L(3.4), $V_d(w)\mu_+(w) = o\left(1/U_a(w)\right)$, so that

$$\int_0^R V_d(w)\mu_+(w)\, dw = o\left(\frac{R}{U_a(R)}\right). \tag{7.74}$$

Hence

$$\frac{1}{U_a(R)\mu_+(R)} \int_0^R II_\varepsilon(w)\mu_+(w)\, dw \longrightarrow 0. \tag{7.75}$$

By the first case of (7.62), we have as above

$$III_\varepsilon(w) = \left[V_d(w) \times o\left(\frac{U_a(R)}{R}\right)\right] \sum_{\frac{1}{4}R<z\leq\frac{1}{2}R} \sum_{\frac{1}{2}R<y\leq(1-\varepsilon)R} g_{B(\frac{1}{2}R)}(0, z)p(y - z).$$

Making the change of variables $z' = \frac{1}{2}R - z$ and $y' = y - \frac{1}{2}R$ shows that the above double sum is less than

$$\sum_{0\leq z'<\frac{1}{2}R} g_{B(\frac{1}{2}R)}(0, \tfrac{1}{2}R - z')\mu_+(z') = P_0(\Lambda_{R/2}).$$

Hence applying (7.74) we find

$$\frac{1}{U_a(R)\mu_+(R)} \int_0^R III_\varepsilon(w)\mu_+(w)\, dw = o\left(P_0(\Lambda_{R/2})\right). \tag{7.76}$$

Because of (7.75) and (7.76) as well as Lemma 7.4.2(i), to verify (7.68) it suffices to show that for a fixed $\varepsilon > 0$,

$$\frac{1}{U_a(R)\mu_+(R)} \sum_{w=0}^{R/8} I_\varepsilon(w)\mu_+(w) \leq C. \tag{7.77}$$

Since $g_{B(R)}(0, z) < v^\circ u_a(z)$ and $\sum_{w=0}^R \mu_+(w)g_{B(R)}(y, R - w) = P_y(\Lambda_R)$, the above sum is less than

$$\sum_{z=0}^{R/2} \sum_{y=(1-\varepsilon)R}^R p(y - z)u_a(z)P_y(\Lambda_R), \tag{7.78}$$

which is at most a constant multiple of $U_a(R)\mu_+(R)$. Thus (7.77) follows. Since the sum $\sum_{y=(1-\varepsilon)R}^{R} p(y-z)P_y(\Lambda_R)$ is less than $\mu_+(R-\varepsilon R-z) - \mu_+(R-z-1)$, the continuity condition (7.11) assures that uniformly for $z \le R/2$, as $\varepsilon \to 0$ the double sum (7.78) is of a smaller order of magnitude than $U_a(R)\mu_+(R)$, so that

$$\frac{1}{U_a(R)\mu_+(R)} \sum_{w=0}^{R} I_\varepsilon(w)\mu_+(w) \to 0 \quad \text{as } R \to \infty \text{ and } \varepsilon \downarrow 0 \text{ in this order,} \quad (7.79)$$

showing (7.69).   □

Lemma 7.4.6 *Suppose that (7.60) holds and μ_+ varies dominatedly. Then*

$$P_0(\Lambda_R) \asymp U_a(R)\mu_+(R) \quad \text{and} \quad u_a(x) \asymp U_a(x)\mu_+(x)a(-x), \quad (7.80)$$

and if one further supposes the continuity condition (7.11) to hold,

$$P_0[S_{N(R)} \ge \tfrac{1}{8}R \,|\, \Lambda_R] \longrightarrow 0. \quad (7.81)$$

Proof Lemma 7.4.2(ii) and 7.4.5 together show that

$$P_0(\Lambda_R) \le CU_a(R)\mu_+(R) + o\left(P_0(\Lambda_{R/2})\right).$$

On putting $\lambda(R) = P_0(\Lambda_R)/[CU_a(R)\mu_+(R)]$, the dominated variation of μ_+ allows us to rewrite this inequality as

$$\lambda(R) \le 1 + o\left(\lambda(R/2)\right).$$

One can choose $R_0 > 0$ so that $\lambda(R) \le 1 + \tfrac{1}{2}\lambda(R/2)$ if $R \ge R_0$, which yields

$$\lambda(R) \le 1 + 2^{-1}\left[1 + 2^{-1}\lambda(R/4)\right] \le \ldots \le 2 + 2^{-n}\lambda(2^{-n}R)$$

as far as $2^{-n}R \ge R_0$. Taking $n = n(R)$ such that $2^{-n}R \ge R_0 > 2^{-n-1}R$, one obtains that $P_0(\Lambda_R) = O\left(U_a(R)\mu_+(R)\right)$, which shows (7.80). This entails $P_0(\Lambda_{R/2}) \asymp P_0(\Lambda_R)$, hence (7.81) follows from the second half of Lemma 7.4.5.   □

Remark 7.4.7 The continuity condition (7.11), used at the end of the proof of Lemma 7.4.5, cannot be removed for $P_0(\Lambda_R)/v^\circ \sim U_a(R)\mu_+(R)$ to hold [note that the contribution to the sum (7.78) from $z < \varepsilon R$ bears significance for any $\varepsilon > 0$].

As a consequence of Lemmas 7.4.2 and 7.4.6 we obtain

Lemma 7.4.8 *Suppose that (7.60) and (7.11) hold. Then*

(i) *for each $\varepsilon > 0$, as $R \to \infty$*

$$P_0[S_{N(R)} \ge \varepsilon R \,|\, \Lambda_R] \to 0, \quad (7.82)$$

and $P_0[\varepsilon R \le Z(R) \le \varepsilon^{-1}R \,|\, \Lambda_R] \to 1$ as $\varepsilon \downarrow 0$ uniformly in $R > 1$;

(ii) $P_0(\Lambda_R)/v^\circ \sim U_a(R)\mu_+(R)$ and $\dfrac{P_0[\sigma_R < T]}{P_0(\Lambda_R)} \sim \dfrac{a(-R) - a(R)}{a(-R)}$;

(iii) $u_a(y) \sim \dfrac{U_a(y)\mu_+(y)}{-A(y)}$.

Proof The first convergence of (i) follows from Lemmas 7.4.2(ii) and 7.4.6, and it implies the second one – owing to (7.11). By (7.82) we have

$$P_0(\Lambda_R) \sim P_0[S_{N(R)} \le \varepsilon R, \Lambda_R] \sim v^\circ \int_0^{\varepsilon R} u_a(t)\mu_+(R - t)\,dt.$$

The last integral is between $U_a(\varepsilon R)\mu_+(R - \varepsilon R)$ and $U_a(\varepsilon R)\mu_+(R)$, hence it may be written as $U_a(R)\mu_+(R)\{1 + o_\varepsilon(1)\}$ as $R \to \infty$ and $\varepsilon \downarrow 0$ owing to (7.11) – since U_a is s.v., showing the first relation of (ii). By the second half of (i),

$$P_0[\sigma_R < T] = P_0(\Lambda_R)\left[o_\varepsilon(1) + \sum_{y=\varepsilon R}^{R/\varepsilon} P_0[Z(R) = y \mid \Lambda_R]P_{R+y}[\sigma_R < T]\right],$$

where $o_\varepsilon(1) \to 0$ as $\varepsilon \downarrow 0$. By (7.49) the second probability under the summation sign is asymptotically equivalent to $[a(-R) - a(R)]/a(-R)$, and we have the second relation of (ii). By $-1/A(y) \sim a(-y) - a(y)$ and the second relation of (ii)

$$P_0[\sigma_y < T] \sim P_0(\Lambda_y)/|A(y)a(-y)|,$$

while the probability on the LHS $\sim v^\circ u_a(y)/a(-y)$ – since $g_\Omega(y, y) \sim a(-y)$. Combined with the first relation of (ii) this yields (iii). $\qquad\square$

7.4.3 Proof of Theorem 7.1.4.

By virtue of Lemma 7.4.8, Theorem 7.1.4 follows if we can show the following

Proposition 7.4.9 *Suppose that (7.60) and (7.11) hold. Then for each $\delta < 1$, uniformly for $0 \le x < \delta y$, as $y \to \infty$,*

$$g_\Omega(x, y) \sim \frac{V_d(x)U_a(y)}{|A(y)|(x + 1)} \sum_{k=0}^{x} \mu_+(y - k), \tag{7.83}$$

$$P_x(\Lambda_y) \sim g_\Omega(x, y)|A(y)|, \tag{7.84}$$

and $\forall \varepsilon > 0, \; P_x[S_{N(y)} \ge x + \varepsilon y \mid \Lambda_y] \to 0.$

Lemma 7.4.10 *Under the same assumption as in Proposition 7.4.9, the equivalence (7.83) holds for each $\delta < 1$.*

Proof Substituting the asymptotic form of u_a of Lemma 7.4.8 into (2.6) one has

$$g_\Omega(x, y) \sim \frac{U_a(y)}{-A(y)} \sum_{k=0}^{x} v_d(k)\mu_+(y - x + k) \quad \text{uniformly for } 0 \leq x < \delta y.$$

By virtue of (7.11), this shows (7.83) is valid for each x fixed. We must verify that the above sum may be replaced by $[1/\hat{\ell}^*(x)] \sum_{k=0}^{x} \mu_+(y - k)$ when $x \to \infty$. However, observing that the sum restricted to $k < \varepsilon x$ is less than $\varepsilon x \mu_+(y - x)/\hat{\ell}^*(x)$, which is at most $C\varepsilon$ times the remaining sum, we can easily perform the verification. □

From (7.83) and $g_\Omega(x, x) \sim a(-x)$ it follows that uniformly for $0 \leq x < \delta R$, as $R \to \infty$,

$$P_x[\sigma_R < T] \sim \frac{U_a(R)V_d(x)}{-a(-R)A(R)(x + 1)} \int_{R-x-1}^{R} \mu_+(t)\, dt. \tag{7.85}$$

Lemma 7.4.11 *Suppose the assumption in Proposition 7.4.9 holds. Then for each $0 < \delta < 1$, uniformly for $0 \leq x < \delta R$,*

$$P_x\left[S_{N(R)} \geq \tfrac{1}{2}(1 + \delta)R \,\big|\, \Lambda_R\right] \to 0.$$

Proof Put $R' = \lfloor \tfrac{1}{2}(1+\delta)R \rfloor$, $R'' = R - R'$ and $Q_R(x) = P_x[S_{N(R)} > R', \Lambda_R]$. Then

$$\frac{Q_R(x)}{P_x(\Lambda_{R'})} = \sum_{k=1}^{R''} P_x\left[Z(R') = k \,\big|\, \Lambda_{R'}\right] Q_R(k).$$

Since for each $\varepsilon > 0$, $Q_R(k) < P_k(\Lambda_R) \to 0$ uniformly for $R' \leq k \leq R - \varepsilon R''$ and $P_x(\Lambda_{R'}) < P_x[\sigma_R < T] \asymp P_x(\Lambda_R) < P_x(\Lambda_{R'})$ according to the second half of Lemma 7.4.3, we have only to show that as $\varepsilon \downarrow 0$,

$$\limsup_{R \to \infty} \sup_{0 \leq x < \delta R} \frac{1}{P_x(\Lambda_R)} \sum_{y=0}^{\varepsilon R''} P_x\left[Z(R') = R'' - y \,\big|\, \Lambda_{R'}\right] Q_R(R - y) \to 0. \tag{7.86}$$

As in the proof of Lemma 7.4.5 we see that the sum above is dominated by

$$\sum_{y=0}^{\varepsilon R''} \sum_{z=0}^{R'} g_\Omega(x, z) p(R - y - z) \sum_{w=0}^{\infty} g_\Omega(w, y)\mu_+(w), \tag{7.87}$$

which, by Lemma 7.4.4(ii), is at most a constant multiple of

$$\sum_{z=0}^{R'} g_\Omega(x, z)[F(R - z) - F(R - \varepsilon R'' - z)].$$

By **L(2.1)** $\sum_{z=0}^{R'} g_\Omega(x, z) \leq V_d(x)U(R)$. Hence, by (7.11) the above sum is of a smaller order of magnitude than $V_d(x)U_a(R) \times \mu_+(R) \asymp P_x(\Lambda_R)$ as $R \to \infty$ and $\varepsilon \downarrow 0$ in this order. Thus we have (7.86), as desired. □

Proof (of Proposition 7.4.9 and Theorem 7.1.4) Let $0 < \varepsilon < 1/2$. By the asymptotic form of u_a in Lemma 7.4.8(iii), it follows that

$$g_\Omega(x, z) \le CV_d(x)u_a(z) \quad \text{for } z \ge x + \varepsilon x.$$

As in the proof of Lemma 7.4.2(ii) we infer that

$$P_x[x + \varepsilon y < S_{N(y)} < (1 - \varepsilon)y \,|\, \Lambda_y] \longrightarrow 0 \quad \text{uniformly for } 0 \le x < \delta y.$$

Hence the last formula of Proposition 7.4.9 follows from Lemma 7.4.11.

As before, from what we have just shown and the continuity condition (7.11), we deduce that uniformly for $0 \le x < \delta y$ (see Remark 7.1.2(c)),

$$P_x[\varepsilon y < Z(y) < y/\varepsilon \,|\, \Lambda_y] \longrightarrow 1 \quad \text{as } y \to \infty \text{ and } \varepsilon \downarrow 0 \text{ in this order.} \quad (7.88)$$

Hence $P_x[\sigma_y < T \,|\, \Lambda_y] \sim P_{2y}[\sigma_y < T] \sim [a(-y) - a(y)]/a(-y)$ as $y \to \infty$ uniformly for $0 \le x < \delta y$. Rewriting these relations, we obtain

$$P_x(\Lambda_y) \sim -A(y)a(-y)P_x[\sigma_y < T] \sim -A(y)g_\Omega(x, y). \quad (7.89)$$

The first relation is the same as that giving the asymptotic form of $P_x[\sigma_y < T]$ in Theorem 7.1.4, and the second implies the asymptotic form of $P_x(\Lambda_y)$ in (7.84). This finishes the proof of Proposition 7.4.9 (hence of Theorem 7.1.4). $\qquad\qquad \square$

7.5 Proof of Proposition 7.1.3 (for F Transient) and Theorem 7.1.6

In this section we suppose that F is transient and n.r.s. According to [83], we then have

$$\begin{cases} \text{(a)} & A(x) \to -\infty, \text{ and } G(-x) - G(x) \sim -1/A(x), \\ \text{(b)} & A(x) \text{ and } G(-x) \text{ are s.v.,} \\ \text{(c)} & G(-x) \sim \displaystyle\int_x^\infty \frac{\mu_-(t)}{A^2(t)}\, dt, \quad G(x) = \displaystyle\int_x^\infty \frac{\mu_+(t)}{A^2(t)}\, dt + o(G(-x)), \end{cases} \quad (7.90)$$

as $x \to \infty$. [The latter half of (a) is not stated in [83], but obtained in the proof of Theorem of [83] (see Eq(57), Eq(30) and Section 5.3 of [83]); it also follows from Theorem 4.1.1(i) of the present treatise – its proof applies to the transient walks.]

Lemma 7.5.1 *If F is transient and n.r.s., then $g_\Omega(x, x) \to G(0)$, and for any $M > 1$, as $y \to \infty$,*

$$g_\Omega(x, y) = G(y - x) - G(y) + o(G(-y)) \quad \text{uniformly for } 0 \le x < My.$$

Proof Under (NRS) $\hat{Z}$ is r.s., so that $P_x[S_T < -\varepsilon x] \to 0$ for each $\varepsilon > 0$. Hence the result is immediate from the identity $g_\Omega(x, y) = G(y - x) - E_x[G(y - S_T)]$. $\qquad \square$

From Lemma 7.5.1 (and (7.90)) it follows that for each $0 < \delta < 1$, as $y \to \infty$,

$$g_\Omega(x, y) = \begin{cases} G(-y) - G(y) + o(G(-y)) & y/\delta \leq x < My, \\ o(G(-y)) & x < \delta y, \\ G(y - x) - G(x) + o\,(G(-y)) & \delta y < x \leq y/\delta \end{cases} \tag{7.91}$$

uniformly for $0 \leq x < My$. Observing $\int_0^z G(t)\,dt = z\,[G(z) + o(G(-z))]\,(z \to \infty)$[1] and $G(\lambda y) - G(y) = o(G(-y))$ for each $\lambda > 0$, one obtains

$$\sum_{z:|y-z|<\varepsilon y} g_\Omega(z, y) = \varepsilon y\,\{[G(-y) - G(y)] + o\,(G(-y))\}, \tag{7.92}$$

for each $0 < \varepsilon < 1/2$. In general, the estimate of $g_\Omega(x, y)$ in (7.91) is not exact for $x \leq y/\delta$ unless $|x - y|/y$ is sufficiently small, but if $x/y \asymp 1$, (7.92) may provide a bound that suffices for the present purpose.

Proof (of Proposition 7.1.3 (for F transient)) By Lemma 7.4.3 we have

$$P_0[Z(R) > MR \mid \Lambda_R] \longrightarrow 0 \quad \text{as } R \wedge M \to \infty$$

under (7.8), while under (7.7), by Lemma 7.5.1

$$P_z[\sigma_R < T] \geq [G(-R) - G(R)]\,/G(0)\{1 + o(1)\} \quad \text{for } R < z < MR, M > 1.$$

It therefore follows that

$$P_0[\sigma_R < T \mid \Lambda_R] \geq [G(-R) - G(R)]/G(0)\{1 + o(1)\}.$$

Hence by (7.90a),

$$v^\circ u_a(R) = G(0)P_0[\sigma_R < T] \geq -P_0(\Lambda_R)/A(R)\{1 + o(1)\}. \tag{7.93}$$

The lower bound (7.9) of $P_0(\Lambda_R)$ follows from Lemma 7.4.2, which together with (7.93) yields that of $u_a(x)$. As in Remark 7.3.2(b), observe that $|A(x)| \sim \hat{\ell}^*(x)\hat{\ell}_\#(x)$ in view of the first case of (7.91) and **L(3.1)**. The upper bound of $u_a(x)$ is obtained by $P_0(\Lambda_R) = o(1/V_d(R))$.   □

For the proof of Theorem 7.1.6, we need to show

$$P_0[\sigma_R < T; \Lambda_R] \sim P_0[\sigma_R < T] = \frac{v^\circ u_a(R)}{g_\Omega(R, R)} \sim \frac{G(-R) - G(R)}{G(0)}P_0(\Lambda_R). \tag{7.94}$$

To this end, we shall employ the identity

$$P_0[\sigma_R < T \mid \Lambda_R] = \sum_{y=1}^{\infty} P_0[Z(R) = y \mid \Lambda_R]\frac{g_\Omega(R + y, R)}{g_\Omega(R, R)}. \tag{7.95}$$

---

[1] If $\varphi(x)$ is a positive and bounded function of $x \geq 0$, and $f(x) := \int_x^\infty \varphi(y)\,dy$ is s.v., then $\int_0^x y\varphi(y)\,dy = o(f(x))\,(x \to \infty)$.

Our task will be reduced to showing $E_0\left[G(-Z(R)); Z(R) < \varepsilon R \,|\, \Lambda_R\right] /G(-R) \to 0$ as $R \to \infty$ and $\varepsilon \downarrow 0$, where the crux of the matter will lie in the verification of the bound $E_0\left[G(-Z(R)) \,|\, \Lambda_R\right] = O(G(-R))$ given in Lemma 7.5.2 below.

In the rest of this section, let the assumption of Theorem 7.1.6 be valid, namely

$$
\begin{cases}
\text{(a)} & F \text{ is n.r.s.;} \\
\text{(b)} & F \text{ is transient and } \limsup_{x\to\infty} G(x)/G(-x) < 1; \\
\text{(c)} & \exists \lambda > 1, \; \limsup \mu_+(\lambda t)/\mu_+(t) < 1; \\
\text{(d)} & \exists C_* > 0, \; p(x) \le C_*\mu_+(x)/x.
\end{cases}
\tag{7.96}
$$

Note that under (d), μ_+ varies dominatedly and satisfies the continuity condition (7.11). As noted above, $-A(x) \sim \ell^*(x)\ell_\sharp(x)$ in view of the first case of (7.91). The dual of Lemma 6.4.9 accordingly yields that under (7.96),

$$
u_a(y) \sim U_a(y)\mu_+(y)/|A(y)|,
\tag{7.97}
$$

since (a), (c), and (d) together entail the dual of the assumption of Lemma 6.4.9.

Lemma 7.5.2 *Suppose (7.96) holds. Then for some constant C,*

$$
E_0[G(-Z(R)) \,|\, \Lambda_R] \le CG(-R).
$$

Proof We have only to estimate $E_0\left[G(-Z(R)), Z(R \le R \,|\, \Lambda_R\right]$, which we write as

$$
\frac{1}{P_0(\Lambda_R)} \sum_{w=0}^{R} \sum_{z=0}^{R} g_{B(R)}(0, R - w)p(w + z)G(-z).
$$

By virtue of (7.96d), the trivial bound $\mu_+(w + z) \le \mu_+(w)$ yields that the contribution of the outer sum restricted to $w \ge R/5$ is at most

$$
\frac{1}{P_0(\Lambda_R)} \sum_{w=R/5}^{R} g_{B(R)}(0, R - w)\frac{\mu_+(w)}{R/5} \sum_{z=0}^{R} G(-z) \le G(-R)\{5 + o(1)\}.
\tag{7.98}
$$

We show that the contribution of the remaining sum is $o\left(G(-R)\right)$. To this end, we proceed as in the proof of Lemma 7.4.5. Let $I_\varepsilon(w), II_\varepsilon(w)$ and $III_\varepsilon(w)$ be as therein. We may take $\varepsilon = 1/4$ and drop the subscript ε from them for the present purpose. What is to be shown may be paraphrased as

$$
\frac{1}{P_0(\Lambda_R)} \sum_{w=0}^{R/5} \sum_{z=0}^{R} [I(w) + II(w) + III(w)] \, p(w + z)G(-z) = o\left(G(-R)\right).
\tag{7.99}
$$

First of all we note that combining (7.8) and (7.96d) leads to

$$\sum_{z=0}^{\infty} p(w+z)G(-z) \le C_* \frac{\mu_+(w)}{w} \sum_{z=0}^{w} G(-z) + \mu_+(w)G(-w) \qquad (7.100)$$

$$\le C'\mu_+(w)G(-w) \quad (w \ge 1).$$

Recall

$$g_{B(R)}(0, R-w) = \sum_{z=0}^{R/2} \sum_{R/2<y\le R} g_{B(R/2)}(0,z)p(y-z)g_{B(R)}(y, R-w)$$

and that $I(w)$, $II(w)$ and $III(w)$ were defined as the contributions to this sum from the ranges,

$$0 \le z \le \tfrac{1}{2}R, \tfrac{3}{4}R < y \le R; \quad 0 \le z \le \tfrac{1}{4}R, \tfrac{1}{2}R < y \le \tfrac{3}{4}R$$
$$\text{and} \quad \tfrac{1}{4}R < z \le \tfrac{1}{2}R, \tfrac{1}{2}R < y \le \tfrac{3}{4}R,$$

respectively. By (7.96d) it follows that for $w \le R/5$,

$$I(w) \le \frac{C_* U_a(R)\mu_+(\tfrac{1}{4}R)}{R/4} \sum_{y=\frac{3}{4}R}^{R} g_\Omega(w, R-y) \le \frac{C'U_a^2(R)\mu_+(R)}{R}V_d(w),$$

where for the latter inequality **L(2.1)** is used. Using in turn $P_0(\Lambda_x)/v^\circ \ge U_a(x)\mu_+(x)\{1 + o(1)\}$, (7.100) and $V_d(x)\mu_+(x) = o(1/U_a(x))$, we accordingly obtain

$$\frac{1}{P_0(\Lambda_R)} \sum_{w=0}^{R/5} \sum_{z=0}^{\infty} I(w)p(w+z)G(-z)$$

$$\le \frac{CU_a(R)}{R} \sum_{w=0}^{R/5} V_d(w)\mu_+(w)G(-w) = o\left(G(-R)\right).$$

One can derive the same bound for $II(w)$ in a similar way.

As for $III(w)$, we apply (7.97) at two places to see that for $w \le R/5$,

$$III(w) \le \sum_{z=R/4}^{R/2} \sum_{y=R/2}^{3R/4} u_a(z)p(y-z)g_\Omega(w, R-y)$$

$$\le C\left[\frac{U_a(R)\mu_+(R)}{A(R)}\right]^2 V_d(w) \sum_{z=R/4}^{R/2} \sum_{y=R/2}^{3R/4} p(y-z)$$

$$\le C'\left[\frac{U_a(R)\mu_+(R)}{A(R)}\right]^2 V_d(w)A_+(R).$$

As above we have $\sum_{w=0}^{R} V_{\mathrm{d}}(w)\mu_{+}(w)G(-w) \ll RG(-R)/U_{\mathrm{a}}(R)$, so that

$$\frac{1}{U_{\mathrm{a}}(R)\mu_{+}(R)} \sum_{w=0}^{R/5} III(w)\mu_{+}(w)G(-w) \ll \frac{R\mu_{+}(R)G(-R)}{A(R)} = o\left(G(-R)\right).$$

Thus (7.99) is verified as required. □

Lemma 7.5.3 *Suppose (7.96) holds. Then for each $\varepsilon > 0$,*

(i) $E_{0}[G(-Z(R)), Z(R) < \varepsilon R \mid \Lambda_{R}] \leq 5\{\varepsilon + o(1)\}G(-R);$ *and*
(ii) $P_{0}[S_{N(R)} > \varepsilon R \mid \Lambda_{R}] \to 0.$

Proof In the proof of Lemma 7.5.2 we have shown

$$\frac{1}{P_{0}(\Lambda_{R})} \sum_{w=0}^{R/5} \sum_{z=0}^{R} g_{\Omega}(0, R - w)p(w + z)G(-z) = o\left(G(-R)\right). \tag{7.101}$$

By the same reasoning as was advanced for the bound (7.98), we also have

$$\frac{1}{P_{0}(\Lambda_{R})} \sum_{w=R/5}^{R} \sum_{z=0}^{\varepsilon R} g_{B(R)}(0, R - w)p(w + z)G(-z) \leq \{5 + o(1)\}\varepsilon G(-R),$$

for any $\varepsilon > 0$. These together show (i). Since by (7.9),

$$\frac{1}{P_{0}(\Lambda_{R})} \sum_{\frac{1}{5}R < w \leq R - \varepsilon R} g_{B(R)}(0, R - w)\mu_{+}(w) = \frac{o(U(R)\mu_{+}(R))}{P_{0}(\Lambda_{R})} \to 0,$$

noting $G(-x)$ is asymptotically decreasing, one can easily deduce (ii) from (7.101).□

Lemma 7.5.4 *Suppose (7.96) holds. Then for each $\delta < 1$ there exists a constant C such that*

$$P_{x}(\Lambda_{R}) \leq CV_{\mathrm{d}}(x)U_{\mathrm{a}}(R)\mu_{+}(R) \quad (0 \leq x < \delta R). \tag{7.102}$$

Proof Let $0 \leq x < \delta R$ for $\delta < 1$ and $\varepsilon = \frac{1}{2}(1 - \delta)$. By **L(2.1)**,

$$\sum_{y=0}^{(1-\varepsilon)R} g_{\Omega}(x, y)\mu_{+}(R - y) \leq V_{\mathrm{d}}(x)U_{\mathrm{a}}(R)\mu_{+}(\varepsilon R). \tag{7.103}$$

For $\frac{1}{8}R \leq x < \delta R$, using the asymptotic form of u_{a} in (7.97) and **L(3.4)**, one infers

$$\sum_{w=0}^{\varepsilon R} g_{B(R)}(x, R - w)\mu_{+}(w) \leq \sum_{w=0}^{\varepsilon R} g_{\Omega}(w, R - x)\mu_{+}(w) = o\left(R\mu_{+}(R)/|A(R)|\right)$$

$$= o\left(V_{\mathrm{d}}(x)U_{\mathrm{a}}(R)\mu_{+}(R)\right).$$

For $x < \frac{1}{8}R$ we proceed as in the proof of Lemma 7.4.5, based on the representation

$$P_x(\Lambda_R) = \sum_{w=0}^{R} \sum_{0 \le z < R/2} \sum_{y=R/2}^{R} g_{B(R/2)}(x,z)p(y-z)g_{B(R)}(y,R-w)\mu_+(w).$$

Because of (7.103) we may only consider the sum restricted to $w < R/5$. The triple sum restricted to $\frac{1}{4}R \le z < \frac{1}{2}R \le y \le \frac{3}{4}R$, $w < R/5$ is of smaller order of magnitude than

$$V_d(x)\left[\frac{U_a(R)\mu_+(R)}{A(R)}\right]^2 A_+(R)\frac{R}{U_a(R)} = o\left(V_d(x)U_a(R)\mu_+(R)\right).$$

Similarly, the rest of the sum is evaluated to be of a smaller order of magnitude than $V_d(x)U_a(R)\mu_+(R)$. This concludes (7.102). $\qquad\qquad\square$

Lemma 7.5.5 *Suppose (7.96) holds. Then for each $\delta < 1$, uniformly for $0 \le x < \delta R$, as $R \to \infty$ and $\varepsilon \downarrow 0$ in this order,*

(i) $E_x[G(-Z(R)), Z(R) < \varepsilon R \,|\, \Lambda_R]/G(-R) \to 0$, *and*
(ii) $P_x[S_{N(R)} > \frac{1}{2}(1+\delta)R \,|\, \Lambda_R] \to 0$ *and* $P_x[Z(R) < \varepsilon R \,|\, \Lambda_R] \to 0$.

Proof The results are immediate from Lemmas 7.5.3 and 7.4.2 if $x = 0$. As for general x, by the asymptotic form of u_a given in (7.97), we have the same asymptotic form of $g_\Omega(x,y)$ as given in Lemma 7.4.10, and this together with the bound of $P_x(\Lambda_R)$ in the preceding lemma enables us to follow the proof of Lemma 7.4.11 to show both (i) and (ii). $\qquad\qquad\square$

Proof (of Theorem 7.1.6) By virtue of Lemmas 7.5.1 and 7.5.5(ii), from the identity $P_x[\sigma_R < T \,|\, \Lambda_R] = E_x[g_\Omega(R + Z(R), R) \,|\, \Lambda_R]/g_\Omega(R, R)$ we deduce

$$P_x(\Lambda_R) \sim \frac{G(0)P_x[\sigma_R < T]}{G(-R) - G(R)} \sim -A(R)g_\Omega(x, R) \qquad\qquad (7.104)$$

$$\text{uniformly for } 0 \le x < \delta R.$$

With this as well as (7.97) we can follow the proof of Theorem 7.1.4 to show the rest of the results of Theorem 7.1.6. $\qquad\qquad\square$

7.6 Estimation of $P_x\left[S_{N(R)} = y \,\middle|\, \Lambda_R\right]$ and the Overshoots

In this section, we suppose

either the assumption of Theorem 7.1.4 or that of Theorem 7.1.6 holds $\quad$ (7.105)

(entailing (NRS)) and compute the conditional probability of S exiting $B(R)$ through $y \in B(R)$, given Λ_R and $S_0 = x \in B(R)$. Denote it by $q_R(x, y)$:

$$q_R(x, y) := P_x\left[S_{N(R)} = y \,\middle|\, \Lambda_R\right].$$

The results are stated in Proposition 7.6.3 after proving two lemmas.

Let $1/2 < \delta < 1$. In Sections 7.4 and 7.5, we have shown that uniformly for $0 \le x < \delta y$, as $y \to \infty$,

$$P_x(\Lambda_y) \sim -A(y)g_\Omega(x,y), \quad P_0(\Lambda_y)/v^\circ \sim -u_a(y)A(y) \sim U_a(y)\mu_+(y), \quad (7.106)$$

$$g_\Omega(x,y) \sim \frac{U_a(y)V_d(x)}{|A(y)|(x+1)} \sum_{k=0}^{x} \mu_+(y-k). \qquad (7.107)$$

From the expressions of $g_\Omega(x,y)$ in (7.4) and Lemma 7.5.1 (with the help of the slow variation of $A(x)$) one can easily deduce that for each $0 < \varepsilon < 1/2$, as $x \to \infty$

$$\sum_{z:(1-\varepsilon)x<z<x} g_\Omega(x,z) \sim \frac{\varepsilon x}{|A(x)|} \quad \text{and} \quad \sum_{z:x\le z<\varepsilon^{-1}x} g_\Omega(x,z) = o\left(\frac{x}{|A(x)|}\right). \qquad (7.108)$$

Lemma 7.6.1 *Under (7.105), uniformly for $0 \le x < \delta R$, $0 \le y < R$, as $R \to \infty$,*

$$g_{B(R)}(x,y) = g_\Omega(x,y) - \frac{U_a(y)}{U_a(R)}\,g_\Omega(x,R)\{1+o(1)\}. \qquad (7.109)$$

[This refines Lemma 5.7.3, the latter giving $g_{B(R)}(x,y) \sim g_\Omega(x,y)$ $(y \le x < \delta R)$; note that $g_\Omega(x,y) \asymp U_a(y)/\hat{\ell}^*(x) \sim 1/|A(x)|$ for $\frac{1}{2}x < y \le x$.]

Proof Since $\lim_{\varepsilon\downarrow 0}\inf_R P_x[\varepsilon R < Z(R) < R/\varepsilon\,|\,\Lambda_R] = 1$ ((7.88), Lemma 7.5.5(ii)), by the dual of (7.5) (i.e., $g_\Omega(x,y) \sim U_a(y)/\hat{\ell}^*(x)$ $(y < \delta x)$), we deduce first that uniformly for $0 \le x < \delta R$, $0 \le y < R$,

$$E_x\left[g_\Omega(S_{\sigma(R,\infty)},y)\,|\,\Lambda_R\right] \sim U_a(y)/\hat{\ell}^*(R),$$

and then, using $-A(R)/\hat{\ell}^*(R) \sim 1/U_a(R)$ and (7.106), that

$$E_x\left[g_\Omega(S_{\sigma(R,\infty)},y),\ \Lambda_R\right] \sim g_\Omega(x,R)U_a(y)/U_a(R).$$

Thus (7.109) follows. $\qquad\qquad\qquad\qquad\qquad\qquad\qquad\qquad\qquad\qquad\qquad\qquad\quad\square$

By Lemma 7.6.1 and (7.106), we see that uniformly for $0 \le x < \delta R$, $0 \le y \le R$,

$$q_R(x,y) = \frac{g_{B(R)}(x,y)}{P_x(\Lambda_R)}\mu_+(R-y) \qquad (7.110)$$

$$= \left[\frac{g_\Omega(x,y)}{g_\Omega(x,R)}\{1+o(1)\} - \frac{U_a(y)}{U_a(R)}\right]\frac{\mu_+(R-y)}{-A(R)}$$

(since, on the RHS of (7.109), the first term must be larger than the second).

Before stating our results for the general case, we deal with the case $x = 0$, which may be worth considering separately.

Lemma 7.6.2 *Under (7.105), as $R \to \infty$,*

$$q_R(0,y) = \begin{cases} \dfrac{u_a(y)}{U_a(R)}\{1 + o(1)\} & as \quad y/R \downarrow 0, \\[2mm] o\left(\mu_+(R-y)/|A(R)|\right) & as \quad y/R \uparrow 1, \end{cases} \tag{7.111}$$

and if $x\mu_+(x)$ is s.v. in addition, then for each $\varepsilon > 0$, the first formula of (7.111) holds uniformly for $0 \le y < (1-\varepsilon)R$.

Proof Taking $x = 0$ in (7.110) and substituting $g_\Omega(0,y) = v°u_a(y)$ into its last expression we obtain

$$q_R(0,y) = \frac{u_a(y)\mu_+(R-y)}{u_a(R)|A(R)|}\left[1 - \frac{u_a(R)/U_a(R)}{u_a(y)/U_a(y)} + o(1)\right] \qquad (0 \le y < R).$$

By using $u_a(x) \sim U_a(x)\mu_+(x)/|A(x)|$ we can further rewrite this as

$$q_R(0,y) = \frac{u_a(y)\mu_+(R-y)}{U_a(R)\mu_+(R)}\left[1 - \frac{\mu_+(R)/A(R)}{\mu_+(y)/A(y)} + o(1)\right],$$

which shows the second case of (7.111). If $y/R \to 0$, the ratio inside the large square brackets tends to zero due to (7.8) (see footnote 4 on page 122). Hence the first case follows. For $\varepsilon R < y < (1-\varepsilon)R$, we then see that if $x\mu_+(x)$ is s.v.,

$$q_R(0,y) \sim \frac{u_a(y)\mu_+(R-y)}{U_a(R)\mu_+(R)}\left[1 - \frac{\mu_+(R)}{\mu_+(y)}\right] \sim \frac{u_a(y)}{U_a(y)}.$$

The proof is complete.   □

Taking $0 < \varepsilon < 1/2$, we put $\delta = 1 - \varepsilon$ and let x, y be such that

$$0 \le x \wedge y \le x \vee y < \delta R$$

in the rest of this section. Substituting (7.107) and/or the dual of (7.5) into (7.110) and using $\mu_+(y)/|A(y)| \sim u_a(y)/U_a(y) = o(1/y)$, we obtain the following proposition in a similar way to the proof above.

Proposition 7.6.3 *Suppose condition (7.105) holds. Let ε, x and y be as above.*

(i) *If $x \vee y < 1/\varepsilon$, then as $R \to \infty$,*

$$q_R(x,y) \sim \frac{g_\Omega(x,y)}{V_d(x)U_a(R)}.$$

(ii) *Uniformly for $y < \delta x$, as $x \to \infty$,*

$$g_{B(R)}(x,y) \sim g_\Omega(x,y) \sim V_d(x)U_a(y)/x, \quad and$$

$$q_R(x,y) \sim \frac{U_a(y)\mu_+(R-y)}{U_a(R)\int_{R-x-1}^{R}\mu_+(t)\,dt} \asymp \frac{U_a(y)}{U_a(R)x}.$$

(iii) *Uniformly for $0 \leq x < \delta y$, as $y \to \infty$,*

$$
q_R(x,y) \sim \begin{cases} \dfrac{\mu_+(R-y)}{-A(R)}\left(\dfrac{\int_{y-x-1}^{y}\mu_+(t)\,dt}{\int_{R-x-1}^{R}\mu_+(t)\,dt} - 1 + o(1)\right) = o\left(\dfrac{1}{R}\right) & \text{for } y > \varepsilon R, \\[3ex] \dfrac{u_a(y)}{U_a(R)} \cdot \dfrac{\int_{y-x-1}^{y}\mu_+(t)\,dt}{(x+1)\mu_+(y)} \asymp \dfrac{u_a(y)}{U_a(R)} & \text{as } y/R \to 0. \end{cases}
$$

For $x \geq \varepsilon R$, (ii) provides a nice description of $q_R(x,y)$. Below let $x < \varepsilon R$. From the above estimates, with the help of Corollary 7.1.8 and (7.108), we infer first that as $R \to \infty$,

$$
q_R(x,y) \begin{cases} = o\,(1/R) & \text{for } x < \delta y,\, y > \varepsilon R, \\ \asymp u_a(y)/U_a(R) & \text{if } x < \delta y,\, y/R \to 0, \\ \asymp \dfrac{g_\Omega(x,y)}{V_d(x)U_a(R)} & \text{if } \delta x < y \leq x/\delta, \\ \asymp U_a(y)/[U_a(R)x] & \text{if } y \leq \delta x, \end{cases} \tag{7.112}
$$

and then that for each constant $0 < \varepsilon_1 \leq 1$, as $x \to \infty$,

$$
\varepsilon_1 \frac{U_a(x)}{CU_a(R)} \leq \sum_{(1-\varepsilon_1)x < z \leq x} q_R(x,z) \leq \varepsilon_1 \frac{CU_a(x)}{U_a(R)};
$$

$$
\sum_{x < z \leq x/\varepsilon_1} q_R(x,z) = o\left(\frac{U_a(x)}{U_a(R)}\right); \tag{7.113}
$$

$$
\sum_{n \leq z < R} q_R(x,z) \leq C'\left(1 - \frac{U_a(n)}{U_a(R)}\right) + o(1) \qquad (2x < n < R),
$$

where C is some positive constant independent of ε_1, and $o(1)$ in the last formula is uniform in n.

Let $(n_R)_{R=1}^{\infty}$ be any sequence of positive integers such that $U_a(n_R) \sim U_a(R)$ and $n_R/R \to 0$. Then, employing (7.112) and (7.113), one sees that the mass of the conditional distribution $P_x\left[S_{N(R)} \in \cdot \mid \Lambda_R\right]$ tends to concentrate on the set of y satisfying

$$
\begin{aligned} (1+x)/\varepsilon_1 < y < n_R \quad & \text{if } U_a(x)/U_a(R) \to 0, \text{ and} \\ 0 < y < x, \quad & \text{if } U_a(x)/U_a(R) \to 1, \end{aligned}
$$

where ε_1 may be any positive number. This suggests that the conditional walk drifts to the right in the former case and to the left in the latter up to the epoch of exiting $B(R)$. If $\varepsilon < U_a(x)/U_a(R) < \delta$, the mass is distributed on both sides of x.

The following proposition is a consequence of the results obtained above.

Proposition 7.6.4 *Suppose that either the assumption of Theorem 7.1.4 or that of Theorem 7.1.6 holds. Suppose, in addition, that $x\mu_+(x)$ is s.v. Then for each $\varepsilon > 0$, uniformly for $y \geq 0$ and $0 \leq x < (1 - \varepsilon)R$, as $R \to \infty$,*

$$P_x\left[Z(R) \leq y \,\middle|\, \Lambda_R\right] = 1 - \frac{\log[1 - (R+y)^{-1}(x+1)]}{\log[1 - R^{-1}(1+x)]} + o(1);$$

in particular, as $x/R \to \xi \in (0,1)$ and $y/R \to \eta \geq 0$,

$$P_x\left[Z(R) \leq y \,\middle|\, \Lambda_R\right] \longrightarrow \frac{\xi}{-\log(1-\xi)} \int_0^\eta \frac{dt}{(1+t)(1-\xi+t)}.$$

Proof Let $\varepsilon_0 > 0$ and n_R be as above. Observe

$$P_x\left[Z(R) = y' \,\middle|\, \Lambda_R\right] = \sum_{z=0}^R \frac{g_{B(R)}(x,z)p(y'+R-z)}{P_x(\Lambda_R)} = \sum_{z=0}^R q_R(x,z)\frac{p(y'+R-z)}{\mu_+(R-z)}.$$

Then by the last inequality of (7.113) we see that uniformly for $x > n_R$, the contribution of the last sum restricted to $z > x$ to the sum of the LHS over $y' \geq 1$ tends to zero, so that

$$P_x\left[Z(R) \leq y \,\middle|\, \Lambda_R\right] = \sum_{z=0}^x \frac{g_{B(R)}(x,z)}{P_x(\Lambda_R)}P[R-z < X \leq R-z+y] + o(1).$$

Observe that Lemma 5.7.3 (together with (7.63), valid also for transient F) allows us to replace $g_{B(R)}(x,z)$ by $U_a(z)/\hat{\ell}^*(x)$. Then we apply (7.106) and (7.107) (with $y = R$) to infer that the above sum is asymptotically equivalent to

$$\frac{1}{U_a(R) \int_{R-x}^R \mu_+(t)\, dt} \sum_{z=0}^x U_a(z)\left[\mu_+(R-z) - \mu_+(R+y-z)\right].$$

Now, using $\mu_+(t) \sim L_+(x)/x$, we see that for $x > n_R$,

$$\sum_{z=0}^x U_a(z)\mu_+(R-z) \sim U_a(R)L_+(R) \int_{R-x}^R \frac{dt}{t} = -U_a(R)L_+(R) \log\left[1 - R^{-1}x\right],$$

and similarly $\sum_{z=0}^x U_a(y)\mu_+(R+y-z) \sim -U_a(R)L_+(R) \log[1 - (R+y)^{-1}x]$. Thus we obtain the formula of Proposition 7.6.4 in the case $x > n_R$.

For $x \leq n_R$, by (7.113) again, we see that

$$P_x\left[Z(R) \leq y \,\middle|\, \Lambda_R\right] = \sum_{z=0}^{2n_R} \frac{g_{B(R)}(x,z)}{P_x(\Lambda_R)}P[R-z < X \leq R-z+y] + o(1).$$

In view of (7.109) and Lemma 6.2.1 it follows that $\sum_{z=0}^{2n_R} g_{B(R)}(x,z) \sim V_d(x)U_a(n_R)$. Since, by (7.11), $\mu_+(R-z) - \mu_+(R+y-z) = \mu_+(R) - \mu_+(R+y) + o(\mu_+(R))$ for

$0 \le z \le 2n_R$, we therefore obtain

$$P_x\left[Z(R) \le y \mid \Lambda_R\right] = \frac{1}{\mu_+(R)}\left[\mu_+(R) - \mu_+(R+y)\right] + o(1) = \frac{y}{R+y} + o(1).$$

Now the asserted formula of the proposition follows immediately. □

Remark 7.6.5 In case (PRS) with

$$(*) \quad \limsup a(-x)/a(x) < 1 \quad \text{or} \quad \limsup G(-x)/G(x) < 1$$

depending on whether $E|X|$ is finite or infinite, below, we derive the estimates of $P_x[S_{N(R)} = y \mid \Lambda_R]$ corresponding to the results given above.

Let $0 < \varepsilon < 1/2 < \delta < 1 - \varepsilon$ and $0 \le x \le R$. By **L(2.1)** we have

$$\sum_{y=0}^{\delta R} g_\Omega(x,y)\mu_+(R-y) \le V_d(x)U_a(\delta R)\mu_+((1-\delta)R)$$

and by $P_x(\Lambda_R) \le V_d(x)/V_d(R)$,

$$P_x[S_{N(R)} < \delta R \mid \Lambda_R] \le \frac{\sum_{y=0}^{\delta R} g_\Omega(x,y)\mu_+(R-y)}{P_x(\Lambda_R)}$$

$$\le V_d(R)U_a(R)\mu_+((1-\delta)R).$$

Owing to **L(3.4)** it therefore follows that if (C3) holds, then

$$P_x[S_{N(R)} < \delta R \mid \Lambda_R] \longrightarrow 0,$$

which says that the conditional distribution of $S_{N(R)}$ tends to concentrate in the interval contained in $[\delta R, R]$ for any $\delta < 1$. (According to Theorem 6.1.1 and Remark 6.3.5 the same is true under $\lim m_+(x)/m(x) = 0$ or the condition (AS) with $\alpha = 2$; note that V_a is sub-additive.)

Now suppose that (PRS) and $(*)$ hold. Then $A(x) \sim \ell^*(x)\ell_\sharp(x)$ and $g_\Omega(x,y) = o(1/A(x))$ for $x > y/\delta$ (by the second formula of (7.2) of Theorem 7.1.1 and the dual of Lemma 7.5.1). For $0 \le w < \varepsilon R$ and $x \le \delta R$, observe that

$$E_w[g_\Omega(S_{\sigma(R,\infty)}, R-x) ; \Lambda_R] \le P_w(\Lambda_R) \sup_{y>R} g_\Omega(y, R-x) + o(\bar{a}(w)/R),$$

where we have used the bound $\sum_{z>2R} p(z-y)/A(z) = o(1/R)$ valid uniformly for $0 \le y \le R$. Since $P_w(\Lambda_R) \sim V_d(w)/V_d(R) \sim V_d(w)A(R)/\ell^*(R)$, the RHS is $o(V_d(w)/\ell^*(R))$, hence by the identity (7.54) one sees that uniformly for $0 \le w < \varepsilon R < x \le \delta R$,

$$g_{B(R)}(x, R-w) \sim g_\Omega(w, R-x) \sim V_d(w)/\ell^*(R).$$

On writing $g_{B(R)}(x, R - w)$ as $\sum_{z=R/2}^{R} P_x[S_{\sigma_{B(R/2)}} = z] g_{B(R)}(z, R - w)$ $(w < \tfrac{1}{2}R)$, the above equivalence is extended to $0 \le x \le \varepsilon R$ so that uniformly for $0 \le x \le \delta R, 0 \le w < \varepsilon R$,

$$g_{B(R)}(x, R - w) \sim P_x(\Lambda_R) V_d(w)/\ell^*(R). \qquad (7.114)$$

Hence one can conclude that uniformly for $0 \le x \le \delta R$ and $(1 - \varepsilon)R < y \le R$,

$$P_x[S_{N(R)} = y \,|\, \Lambda_R] = g_{B(R)}(x, y)\mu_+(R - y)/P_x(\Lambda_R)$$

$$\sim \frac{1}{\ell^*(R)} V_d(w)\mu_+(w), \quad w := R - y. \qquad (7.115)$$

Note that the sum of the last expression over $0 \le w < n_R$ tends to unity whenever $\ell^*(n_R) \sim \ell^*(R)$ by virtue of Lemma 6.4.1, and, on comparing the result of Proposition 6.7.1(ib), the conditioning on Λ_R has little influence on the distribution of $S_{N(R)}$.

7.7 Conditions Sufficient for (7.7) or $\ell^*(x)\ell_\sharp(x) \sim A(x)$

In this section, we suppose F is p.r.s. Since this entails (C3), from Lemma 6.4.3 it follows that (6.37) holds, so that

$$\text{Eq(6.37):} \qquad \ell^*(t)\ell_\sharp(t) = A(t) + o\,(A_+(x))$$

and if $EX = 0$, both η_- and η are s.v. If the positive and negative tails of F are not balanced in the sense that

$$\begin{cases} \limsup \eta_+(x)/\eta_-(x) < 1 & \text{if } EX = 0, \\ \limsup A_-(x)/A_+(x) < 1 & \text{if } E|X| = \infty, \end{cases} \qquad (7.116)$$

then (6.37) shows (7.39), namely,

$$\ell^*(x)\ell_\sharp(x) \sim A(x).$$

Be warned that if F is recurrent and $E|X| = \infty$, then $\limsup A_-(x)/A_+(x) = 1$, hence the second case of (7.116) is impossible; if $\lim A_-(x)/A_+(x) = 1$ in addition, then $a(-x)/a(x) \to 1$, as is easily deduced from (4.3).

We observe that the following implications hold

$$(7.116) \implies \begin{cases} \limsup a(-x)/a(x) < 1 & \text{if } EX = 0 \\ \limsup G(-x)/G(x) < 1 & \text{if } F \text{ is transient} \end{cases}$$

$$\implies \ell^*(x)\ell_\sharp(x) \sim A(x).$$

If $EX = 0$, then $A = \eta_- - \eta_+$ and by Corollary 4.1.2 $a(x) \sim \int_{x_0}^x \left[\mu_-(y)/A^2(y)\right] dy \asymp \int_{x_0}^x \left[\mu_-(y)/\eta_-^2(y)\right] dy \sim 1/\eta_-(x) \asymp 1/A(x)$ under (7.116), hence by Theorem 4.1.1 $a(x) \asymp a(x) - a(-x)$, showing the first implication; for F transient, one can proceed analogously based on (7.90). As for the second implication, see Remark 7.3.2(b) and the remark given right after (7.96).

If $EX = 0$, then, since both η and η_- are s.v., it follows that

$$\limsup \frac{m_+(x)}{m(x)} < \frac{1}{2} \iff \limsup \frac{\eta_+(x)}{\eta(x)} < \frac{1}{2} \implies \limsup \frac{a(-x)}{a(x)} < 1,$$

and from (6.38) – paraphrased as $\ell^*(x)\ell_\sharp(x) = \eta_-(x) - \eta_+(x) + o(\eta_+(x))$ – we also infer that $\limsup \ell^*(x)\ell_\sharp(x)/\eta_-(x) \le 1$ and

$$\liminf \ell^*(x)\ell_\sharp(x)/\eta_-(x) > 0 \iff \limsup m_+(x)/m(x) < 1/2.$$

Remark 7.7.1 If $EX = 0$ and (AP) holds with $\alpha = 1 = 2p$ and $\rho = 1$ (entailing that F is p.r.s.), then

$$\limsup \frac{A(x)V_d(x)U_a(x)}{x} \ge 1 \quad \text{and} \quad \liminf \frac{A(x)V_d(x)U_a(x)}{x} \le 1,$$

owing to Lemma 6.4.5 and Theorem 7.1.1 (see Remark 6.4.6). However, under the same condition above, it is unclear whether the above ratio under lim sup/lim inf is bounded away from 0 in general, nor is it away from ∞, to say nothing of its convergence.

Chapter 8
Absorption Problems for Asymptotically Stable Random Walks

Throughout this chapter we assume the asymptotic stability condition:

$$(AS) \begin{cases} \text{(a)} & X \text{ is attracted to a stable law of exponent } 0 < \alpha \le 2 \\ \text{(b)} & \text{there exists} \quad \rho := \lim P_0[S_n > 0] \end{cases}$$

(the same condition as introduced in Section 2.3), unless otherwise stated explicitly. We are concerned with the hitting probabilities of a point y by the r.w. S started from x and killed as it enters the half line Ω or the set $B(R) = \mathbb{R} \setminus [0, R]$ for a large number R. We are interested in how they depend on x, y, or R. The problem links closely to the estimates of the Green function of the killed walk. The Green function contains the complete information of the hitting probabilities, while the exact estimates of the latter are often crucial for computing the former.

We continue to use the same notation as in Sections 2.3 and 6.1, in particular as in (2.16) we let an s.v. function L be such that $E[X^2; |X| < x] \sim L(x)$ ($\alpha = 2$) and

$$\mu_+(x) \sim pL(x)x^{-\alpha} \quad \text{and} \quad \mu_-(x) \sim qL(x)x^{-\alpha} \quad (0 < \alpha < 2)$$

for some constant $p = 1 - q \in [0, 1]$.

Under (AS), there exist s.v. functions ℓ and $\hat{\ell}$ such that (6.49) holds, i.e.,

$$U_a(x) \sim x^{\alpha\rho}/\ell(x) \quad \text{and} \quad V_d(x) \sim x^{\alpha\hat{\rho}}/\hat{\ell}(x),$$

as mentioned at the beginning of Section 6.5, in which we have seen that

$$U_a(x)V_d(x)\mu(x) \longrightarrow \kappa = \frac{\pi^{-1} \sin \pi\alpha\hat{\rho}}{q\alpha^2 \rho\hat{\rho}B(\alpha\rho, \alpha\hat{\rho})} \qquad (q > 0)$$

(see Lemma 6.5.1) and if $p\rho < 1$ and $0 < \alpha < 2$,

$$P[-\hat{Z} \ge x]/v° \sim q\alpha\rho B(\alpha\rho, \alpha\hat{\rho})x^{-\alpha\hat{\rho}} L(x)/\ell(x), \tag{8.1}$$

K. Uchiyama, *Potential Functions of Random Walks in Z with Infinite Variance*, Lecture Notes in Mathematics 2338, https://doi.org/10.1007/978-3-031-41020-8_8

where $B(s,t) = \Gamma(s)\Gamma(t)/\Gamma(s+t)$, with the understanding $sB(s,t) = 1$ if $s = 0, t > 0$ (see (6.53) and Remark 6.5.2). Recall that we may and do take $\ell = \ell^*$ or $\ell_\sharp$ (defined in (6.48)) according as $\alpha\rho = 1$ or 0, and analogously for $\hat\ell$ by duality.

We are also interested in the conditional probability $P_x[\sigma_R < T \mid \Lambda_R]$. For its evaluation we shall compute the Green function $g_\Omega(x,y)$ in the case $0 < \rho < 1$; if $\rho\hat\rho = 0$ we have obtained relevant results in Chapter 7.

According to [71, Theorem 23.1], if $\sigma^2 < \infty$ and $EX = 0$, as $R \to \infty$,

$$\frac{1}{2}\sigma^2 g_{B(R)}(\xi R, \eta R) \longrightarrow \xi \wedge \eta - \xi\eta \qquad \text{uniformly for } 0 \le \xi, \eta \le 1. \qquad (8.2)$$

The case $\sigma^2 = \infty$ is studied by Kesten [40], where he derived an explicit analytic expression of the limit of $g_{B(R)}(x,y)R^{1-\alpha}$ as $x/R \to \xi, y/R \to \eta$ ($\xi, \eta \in (0,1)$) under the condition that $\psi(t) = \psi(-t)$ and $\lim[1-\psi(t)]/|t|^\alpha = Q(> 0)$ ($1 < \alpha \le 2$). His argument is based on his result concerning $P_x[S_{\sigma_{B(R)}} = \cdot]$ and $g_\Omega(x,y)$. In this section, we follow his derivation and, using Rogozin's result (2.28) and our estimates of $g_\Omega(x,y)$, generalise Kesten's expression of the scaling limit of $g_{B(R)}$ to the r.w. satisfying (AS) with $1 \le \alpha \le 2$ (Proposition 8.5.3).

In Section 8.1, we shall derive asymptotic forms of u_a and v_d that yield those of $g_\Omega(x,y)$ (Theorem 8.2.1), which will be used to show the other results stated therein and which will be fundamental for the derivation of the asymptotic form of $g_{B(R)}(x,y)$ (given in Section 8.4). The results on $P_x[\sigma_R < T \mid \Lambda_R]$ are stated and proved in Section 8.3.

8.1 Strong Renewal Theorems for the Ladder Height Processes

This section is devoted to getting a sufficient condition for the strong renewal theorem for U_d and/or V_d that may read, respectively,

$$\begin{cases} \text{(a)} \quad u_a(x) \sim \alpha\rho\, x^{\alpha\rho-1}/\ell(x), \\ \text{(b)} \quad v_d(x) \sim \alpha\hat\rho\, x^{\alpha\hat\rho-1}/\hat\ell(x). \end{cases} \qquad (8.3)$$

It is known [33], [28] that if $1/2 < \alpha\hat\rho < 1$, (b) follows from $V_d(x) \sim x^{\alpha\hat\rho}/\hat\ell(x)$ without any extra assumption and similarly for $U_a(x)$ when $1/2 < \alpha\rho < 1$ (cf. [8]).

The following result, crucial throughout this chapter, rests on the recent result by Caravenna and Doney [14].

Lemma 8.1.1 *Suppose that $\rho\hat\rho > 0$. Then if $\alpha(1 + \hat\rho) > 1$, (a) implies (b) and if $\alpha(1+\rho) > 1$, (b) implies (a); in particular if $\alpha(\rho \vee \hat\rho) > 1/2$ and $\alpha(1+\rho \wedge \hat\rho) > 1$, both (a) and (b) hold.*

Proof We have only to prove the implication (a) $\Rightarrow$ (b) under $\alpha(1 + \hat\rho) > 1$, $\hat\rho \le \rho$; the converse is its dual and (a) holds under $\alpha\rho > 1/2$. First of all we recall that if $\alpha\hat\rho = 1$, then $\hat{Z}$ is r.s. and (b) follows (cf. Section A.1.2, [83]). This especially ensures the result for $\alpha = 2$. Below we may and do suppose $\alpha < 2$.

Let (a) hold. We may suppose $\alpha\hat{\rho} \leq 1/2$; otherwise, (b) is trivial. According to [14] (b) holds if

$$\lim_{\varepsilon\downarrow 0} \limsup_{x\to\infty} xP[-\hat{Z} \geq x] \sum_{z=1}^{\varepsilon x} \frac{P[-\hat{Z} = x - z]}{z(P[-\hat{Z} \geq z])^2} = 0 \tag{8.4}$$

(the converse also holds). By the conditions $\hat{\rho} > 0$ and $\alpha\hat{\rho} \leq 1/2$ it follows that $q\hat{\rho} > 0$ (see the table in Section 2.6), so that (8.1) holds with the constant factor positive. Because of (6.14) and (8.1), namely

$$\frac{P[-\hat{Z} = x]}{v^\circ} = \sum_{y=0}^{\infty} u_a(y)p(-x-y) \quad \text{and} \quad P[-\hat{Z} \geq x] \sim \frac{CL(x)}{x^{\alpha\hat{\rho}}\ell(x)}$$

(with $C > 0$), it follows from (a) that the sum in (8.4) is dominated by a constant multiple of

$$J_\varepsilon(x) := \sum_{z=1}^{\lfloor \varepsilon x \rfloor} \frac{1}{z}\left(\frac{z^{\alpha\hat{\rho}}\ell(z)}{L(z)}\right)^2 \sum_{y=1}^{\infty} \frac{y^{\alpha\rho-1}}{\ell(y)}p(-x-y+z).$$

We decompose the repeated sum on the RHS into four parts:

$$I_\varepsilon(x) = \sum_{z=1}^{\lfloor \varepsilon x \rfloor}\sum_{y=z+1}^{\varepsilon x}, \quad II_\varepsilon(x) = \sum_{z=1}^{\lfloor \varepsilon x \rfloor}\sum_{y=1}^{z}, \quad III_\varepsilon(x) = \sum_{z=1}^{\lfloor \varepsilon x \rfloor}\sum_{y=\varepsilon x}^{x} \quad \text{and} \quad IV_\varepsilon(x) = \sum_{z=1}^{\lfloor \varepsilon x \rfloor}\sum_{y>x}.$$

Introducing $w = y - z$ and taking the summation over y first with w fixed, we deduce

$$I_\varepsilon(x) = \sum_{w=1}^{\lfloor \varepsilon x \rfloor - 1} p(-x-w) \sum_{y=w+1}^{\varepsilon x} \frac{y^{\alpha\rho-1}(y-w)^{2\alpha\hat{\rho}-1}}{\ell(y)}\left(\frac{\ell(y-w)}{L(y-w)}\right)^2.$$

We claim that the inner sum is at most a constant multiple of $(\varepsilon x)^{\alpha(1+\hat{\rho})-1}\ell(x)/L^2(x)$, provided $\alpha(1+\hat{\rho}) > 1$. For the proof we decompose it into three parts:

$$K_1(w) = 1(w < \varepsilon x/2)\sum_{y=w+1}^{2w}, \quad K_2(w) = 1(w < \varepsilon x/2)\sum_{y=2w+1}^{\varepsilon x}$$
$$\text{and} \quad K_3(w) = 1(\varepsilon x/2 \leq w < \lfloor \varepsilon x \rfloor)\sum_{y=w+1}^{\varepsilon x}.$$

Observing

$$K_1(w) \asymp \frac{w^{\alpha\rho-1}}{\ell(w)}\sum_{y=w+1}^{2w}(y-w)^{2\alpha\hat{\rho}-1}\left(\frac{\ell(y-w)}{L(y-w)}\right)^2 \sim \frac{w^{\alpha+\alpha\hat{\rho}-1}\ell(w)}{2\alpha\hat{\rho}L^2(w)}$$

and $y - w \geq \frac{1}{2}y$ for $y \geq 2w$ so that $K_2(w) \asymp \sum_{2w}^{\varepsilon x} y^{\alpha+\alpha\hat{\rho}-2}\ell(y)/L^2(y)$, we obtain the bound $K_1(w)+K_2(w) \leq C(\varepsilon x)^{\alpha+\alpha\hat{\rho}-1}\ell(x)/L^2(x)$ if $\alpha(1+\hat{\rho}) > 1$. It is easily seen that

$K_3(w)$ admits the same bound. Thus the claim is verified. Since $\sum_{w=1}^{\varepsilon x} p(-x - w) \le C(\varepsilon)L(x)x^{-\alpha}$ with $C(\varepsilon) \to 0$ as $x \to \infty$ and $\varepsilon \downarrow 0$ in this order, it follows that

$$I_\varepsilon(x) \le C(\varepsilon)\varepsilon^{\alpha(1+\hat\rho)-1}x^{\alpha\hat\rho-1}\ell(x)/L(x).$$

A proof similar to the above applies to $II_\varepsilon(x)$ to yield the same bound.

As for $IV_\varepsilon(x)$, we use Potter's bound $\ell(x)/\ell(y) \le C_1(x/y)^\delta$ $(y \ge x)$ with $\delta = \alpha\rho/2$ to see that the inner sum of the repeated sum defining $IV_\varepsilon(x)$ is at most

$$\frac{C_1 x^\delta}{\ell(x)} \sum_{y=x}^\infty y^{-\delta}y^{\alpha\rho-1}p(-x + z - y),$$

which, as is shown by summation by parts, is bounded above by $C_2\mu_-(x)x^{\alpha\rho-1}/\ell(x)$, so that

$$IV_\varepsilon(x) \le C_2\mu_-(x)\frac{x^{\alpha\rho-1}}{\ell(x)} \sum_{z=1}^{\varepsilon x} \frac{1}{z}\left(\frac{z^{\alpha\hat\rho}\ell(z)}{L(z)}\right)^2$$

$$\sim C_3\varepsilon^{2\alpha\hat\rho}x^{\alpha\hat\rho-1}\ell(x)/L(x)$$

with $C_3 = qC_2/(2\alpha\hat\rho)$. A similar summation by parts to the above leads to

$$III_\varepsilon(x) \le C_4\varepsilon^{\alpha+\alpha\hat\rho-1}x^{\alpha\hat\rho-1}\ell(x)/L(x).$$

Combined with the bound of $I_\varepsilon(x) + II_\varepsilon(x) + IV_\varepsilon(x)$ obtained above this gives

$$\limsup_{x\to\infty} xP[-\hat Z \ge x]J_\varepsilon(x) \le C_5\varepsilon^{\alpha+\alpha\hat\rho-1} \vee \varepsilon^{2\alpha\hat\rho}.$$

Since $\alpha + \alpha\hat\rho - 1 > 0$ and $\hat\rho > 0$, this shows (8.4). $\square$

Remark 8.1.2 In the proof above, the property of the positive tail of F is used only through those of the distributions of Z and $\hat Z$. Since the regular variation of $u_a(x)$ and $\mu_-(x)$ implies that of $P[-\hat Z > x]$, it accordingly follows – whether (AS) is true or not – that

If $U_a(x) \sim x^\beta/\ell(x)$, $\mu_-(x) \sim L_-(x)x^{-\alpha}$ with L_-, ℓ s.v., $1/2 < \beta \le 1$, $2\alpha > \beta+1$ and $0 < \alpha - \beta < 1$, then (8.3b) holds with $\alpha\hat\rho = \alpha - \beta$ and

$$\hat\ell(x) \sim \frac{\Gamma(\alpha - \beta + 1)\Gamma(\beta + 1)}{\Gamma(\alpha)\pi^{-1}\sin\alpha\hat\rho\pi} \cdot \frac{L_-(x)}{\ell(x)}.$$

If Z is r.s. in particular, then from the condition $\mu_-(x) \sim L_-(x)x^{-\alpha}$ with $1 < \alpha < 2$, it follows that $v_d(x) \sim \left[(\alpha - 1)\pi^{-1}\sin(\alpha - 1)\pi\right]x^{\alpha-2}\ell(x)/L_-(x).$

The last statement of Lemma 8.1.1 depends on the fact that if $(\rho \vee \hat\rho)\alpha > 1/2$, then either Z or $\hat Z$ admits the strong renewal theorem (i.e., (a) or (b) of (8.3) holds). For this reason, the case $\alpha = 2\rho = 1$ is excluded from it. On the other hand the first formula of (7.12) in Theorem 7.1.4 together with Remark 7.1.2(d) entails that if $\alpha = \rho = 1$, $EX = 0$ and $p < 1/2$, then v_d satisfies the 'strong renewal theorem (s.r.t.)' under (AS) in the form

$$v_d(x) \sim u_a(x)\mu_-(x)[V_d(x)]^2 \tag{8.5}$$

and similarly for $u_a(x)$ in the case $p > 1/2$ by duality. Taking this as well as (8.3) into account and recalling that $u_a(x) \sim 1/\ell^*(x)$ if $\alpha\rho = 1$, we have the 's.r.t.' for u_a and v_d under (AS) except for the following cases:

(1) $\rho\hat\rho > 0$, and either $\alpha(\rho \vee \hat\rho) \le 1/2$ or $\alpha(1 + \rho \wedge \hat\rho) \le 1$;
(2) $\rho\hat\rho = 0$ and either $p = 1/2$ or $E|X| = \infty$.

The next lemma, addressing the problem in these cases, shows the 's.r.t.' is still valid for U_a and/or V_d under some additional assumption on the tails of F.

Lemma 8.1.3 (i) *If $\alpha = 1$, the law of X is symmetric and $L(x)$ is almost decreasing, then (8.3) holds.*
(ii) *If $\rho > 0$ and $p(x) = O(\mu_+(x)/x)$ ($x \uparrow \infty$), then $u_a(x) \sim \alpha\rho x^{\alpha\rho-1}/\ell(x)$, while if $\hat\rho > 0$ and $p(-x) = O(\mu_-(x)/x)$ ($x \uparrow \infty$), then $v_d(x) \sim \alpha\hat\rho x^{-\alpha\hat\rho}/\hat\ell(x)$.*
(iii) *Let $\rho = 1$ and suppose that either $1/2 < \alpha \le 1$ or $p(x) = O(\mu_+(x)/x)$. Then $u_a(x) \sim \alpha x^{\alpha-1}/\ell(x)$. If $\alpha = 1$ and if, in addition, $p(-x) = O(\mu_-(x)/x)$ and $\limsup \mu_-(\lambda x)/\mu_-(x) < 1$ for some $\lambda > 1$ then v_d satisfies (8.5).*

Proof (i) follows from Theorem 1.5 of [14], which says that if $U_a(x) \sim \sqrt{x}/\ell(x)$ with s.v. ℓ and $1/\ell$ is almost increasing, then $u_a(x) \sim \frac{1}{2}U_a(x)/x$. For, if $\alpha = 1$ and the law of X is symmetric, then $U_a(x) \sim \sqrt{x/[v°\kappa L(x)]}$.

For the proof of the first half of (ii) we may suppose $\alpha\rho < 1$, which, together with $\rho > 0$, entails

(∗) $p > 0$ and $\lim P[Z \ge x]U_a(x)$ exists and is positive,

hence it suffices to show that under (∗) the following implications hold:

$$p(x) \le C\frac{\mu_+(x)}{x} \ (x \ge 1) \implies \limsup \frac{xP[Z = x]}{P[Z \ge x]} < \infty \implies \lim \frac{xu_a(x)}{U_a(x)} = \alpha\rho.$$

The left-most statement implies that $d(-\mu_+(y)/y) \le C_1\mu_+(y)\,dy/y^2$ and hence

$$P[Z = x] \le C\sum_{y=0}^{\infty} v_d(y)\frac{\mu_+(x+y)}{x+y} \le C'\frac{V_d(x)\mu(x)}{x},$$

which, combined with $U_a(x)V_d(x)\mu(x) \to \kappa$, shows the first implication. The second one is shown by Doney [21]. The second half of (ii) is the dual of the first. The proof of (ii) is finished.

The first half of (iii) has been verified — right above under its second supposition and by Appendix (A.2.1.1) (when $\alpha = 1$) and the general result mentioned immediately before Lemma 8.1.1 (when $\alpha < 1$) under the first supposition. The second half follows from Lemma 6.4.9. □

8.2 The Green Function $g_\Omega(x, y)$

Lemmas 8.1.1 and 8.1.3 allow us to compute the asymptotic form of $g_\Omega(x, y)$ for $\alpha \geq 1$ (as given in Theorem 8.2.1 below) unless $\alpha = 1$ and $\rho \in \{0, 1\}$, in which case we have obtained a partial result in Proposition 7.3.1. When $\alpha = 2\rho = 1$ we shall often need to use (i) and (ii) of Lemma 8.1.3; for ease of later reference we bring in the following condition:

$$\begin{cases} \alpha = 1, \ 0 < \rho < 1 \text{ and if } \rho = 1/2, \\ \text{either } the\ law\ of\ X\ is\ symmetric\ and\ L(x)\ is\ almost\ decreasing, \\ \text{or} \quad either\ p(x) = O\ (\mu_+(x)/x)\ or\ p(-x) = O\ (\mu_-(x)/x), \end{cases} \quad (8.6)$$

which, on combining Lemmas 8.1.1 and 8.1.3, entails (8.3).

Put for $\alpha \geq 1$,

$$\mathfrak{h}_\lambda(\zeta) = \mathfrak{h}_\lambda^{(\alpha)}(\zeta) = \int_0^1 t^{\lambda-1}(\zeta - 1 + t)^{\alpha-\lambda-1}\, dt \quad (0 < \lambda \leq 1, \zeta \geq 1).[1]$$

Note that $\mathfrak{h}_\lambda(\zeta) < \infty$ whenever $\alpha \vee \zeta > 1$, while if $\alpha = 1$

$$-\log\left(1 - \zeta^{-1}\right) < \mathfrak{h}_\lambda^{(1)}(\zeta) < -\log(\zeta - 1) + \lambda^{-1} \quad \text{for } 1 < \zeta \leq 2$$

(see (8.7) below for the lower bound); and that for all $1 \leq \alpha \leq 2$,

$$\mathfrak{h}_\lambda(\zeta) \sim \lambda^{-1}\zeta^{\alpha-\lambda-1} \quad \text{as} \quad \zeta \to \infty$$

and for $\alpha > 1$,

$$\mathfrak{h}_{\alpha-1}(\cdot) \equiv \mathfrak{h}_\lambda(1) = \frac{1}{\alpha - 1} \quad \text{and} \quad \mathfrak{h}_1(\zeta) = \frac{\zeta^{\alpha-1} - (\zeta - 1)^{\alpha-1}}{\alpha - 1}.$$

The change of variable $t \mapsto w = (\zeta-1)/(\zeta-1+t)$ yields the following representation

$$\mathfrak{h}_\lambda(\zeta) = (\zeta - 1)^{\alpha-1} \int_{1-1/\zeta}^1 w^{-\alpha}(1 - w)^{\lambda-1}\, dw \quad (\zeta > 1). \quad (8.7)$$

In the sequel we shall usually write $\mathfrak{h}_\lambda$ instead of $\mathfrak{h}_\lambda^{(\alpha)}$ if $\alpha > 1$ as above.

Theorem 8.2.1 (i) *Let $1 < \alpha \leq 2$. Then*

$$g_\Omega(x, y) \sim \begin{cases} \dfrac{\alpha^2 \rho\hat{\rho}\, V_{\mathrm{d}}(x)}{\ell(y)[x \vee 1]^{1-\alpha\rho}}\, \mathfrak{h}_{\alpha\hat{\rho}}\left(\dfrac{y}{x \vee 1}\right) & (y \to \infty) \ \ uniformly\ for\ 0 \leq x \leq y, \\[4mm] \dfrac{\alpha^2 \rho\hat{\rho}\, U_{\mathrm{a}}(y)}{\hat{\ell}(x)[y \vee 1]^{1-\alpha\hat{\rho}}}\, \mathfrak{h}_{\alpha\rho}\left(\dfrac{x}{y \vee 1}\right) & (x \to \infty) \ \ uniformly\ for\ 0 \leq y \leq x, \end{cases}$$

$$(8.8)$$

---

[1] $\mathfrak{h}_\lambda(1)$, defined by continuity, is understood to be equal to $(\alpha - 1)^{-1} = \infty$ for $\alpha = 1$.

in particular, as $x \to \infty$

$$g_\Omega(x, x) \sim \frac{\alpha^2 \rho \hat{\rho}}{\alpha - 1} \cdot \frac{V_d(x) U_a(x)}{x}. \tag{8.9}$$

(ii) *Let $\alpha = 1$ and $0 < \rho < 1$. If (8.6) holds, then for each $0 < \delta < 1$ the first and second equivalences of (8.8) hold uniformly for $0 \le x < \delta y$ and for $0 \le y < \delta x$, respectively; in particular as $(x/(x + y), y/(x + y)) \to (\xi, \eta) \in (0, 1)^2$ with $\xi \ne \eta$,*

$$g_\Omega(x, y) \sim \frac{\rho \hat{\rho} \, V_d(x) U_a(x)}{x} H(\xi, \eta) \sim \frac{2 \sin^2 \rho \pi}{\pi^2 x \mu(x)} H(\xi, \eta), \tag{8.10}$$

where

$$H(\xi, \eta) = \begin{cases} \mathfrak{h}_{\hat{\rho}}^{(1)}(\eta/\xi) & \xi < \eta, \\ \mathfrak{h}_{\rho}^{(1)}(\xi/\eta) & \xi > \eta. \end{cases}$$

If F is recurrent ((8.6) is not required), then as $x \to \infty$

$$g_\Omega(x, x) \sim a(x) \sim a(-x) \sim (2\pi^{-2} \sin^2 \rho \pi)[1/L]_*(x), \tag{8.11}$$

where $[1/L]_*(x) := \int_1^x [L(t)t]^{-1} \, dt \sim \int_1^x [t^2 \mu(t)]^{-1} \, dt$, and, for each $\varepsilon > 0$

$$g_\Omega(x, y) = o\left(g_\Omega(y, y)\right) \quad \text{as } y \to \infty \text{ uniformly for } x : |x - y| > \varepsilon y. \tag{8.12}$$

(iii) *Let F be transient. Then $g_\Omega(x, x) \sim a(x) \sim a(-x) \to 1/P[\sigma_0 = \infty]$ and $g_\Omega(x, y) \to 0$ as $|y - x| \to \infty$; and if $\alpha = 1$ and $0 < \rho < 1$, $\int_1^\infty [xL(x)]^{-1} \, dx < \infty$.*

Proof Let $1 < \alpha \le 2$ and $x \le y$ so that $u_a(y) \sim \alpha \rho y^{\alpha \rho - 1}/\ell(y)$ and $g_\Omega(x, y) = \sum_{k=0}^x v_d(k) u_a(y - x + k)$. For simplicity suppose $x \ge 1$. If $x/y \to 0$, then

$$g_\Omega(x, y) \sim \alpha \rho \, V_d(x) y^{\alpha \rho - 1}/\ell(y),$$

which we see agrees with the asserted formula by using $\mathfrak{h}_{\alpha \hat{\rho}}(\zeta) \sim (\alpha \hat{\rho})^{-1} \zeta^{\alpha \rho - 1}$ as $\zeta \to \infty$. For $y \asymp x$ by Lemma 8.1.1 the above sum divided by $\alpha^2 \rho \hat{\rho}$ is asymptotically equivalent to

$$\sum_{k=1}^x \frac{k^{\alpha \hat{\rho} - 1}(y - x + k)^{\alpha \rho - 1}}{\hat{\ell}(k)\ell(y - x + k)} \sim \frac{x^{\alpha - 1}}{\hat{\ell}(x)\ell(y)} \int_0^1 t^{\alpha \hat{\rho} - 1} \left(\frac{y}{x} - 1 + t\right)^{\alpha \rho - 1} dt$$

$$\sim \frac{V_d(x) \mathfrak{h}_{\alpha \hat{\rho}}(y/x)}{\ell(y) x^{1 - \alpha \rho}},$$

verifying the first formula of (8.8). The second one is dealt with in the same way. (i) has been proved.

Let $\alpha = 1$ and $0 < \rho < 1$. The first assertion of (ii) is verified in the same way as proof of (i) (use Lemma 8.1.3 in case $\rho = 1/2$); as for (8.10), note that

$$\kappa\rho\hat\rho = 2\pi^{-2}\sin^2\rho\pi \quad \text{if } \alpha = 1, 0 < \rho < 1 \tag{8.13}$$

and $\ell(x)\hat\ell(x) \sim \kappa^{-1}L(x)$. If F is recurrent, then, according to Proposition 4.2.1(iv), $a(x) \sim a(-x) \sim \kappa' \int_0^x [L(t)t]^{-1}\,dt$. Hence, (8.11) as well as (8.12) follows from Proposition 7.3.1 (see Remark 7.3.2(c) and Lemma 7.4.3). [Note that for each y, $\lim_{x\to\infty} g_\Omega(x, y) = 0$.]

The first half of (iii) follows from the standard result given in Section A.5.3, in particular, the fact that $\lim_{|x|\to\infty} a(x) = G(0) = 1/P_0[\sigma = \infty]$. For the second, see Section 6.3 of [83]. □

Remark 8.2.2 (a) With the constant $\kappa_{\alpha,\rho}$ given in (8.35) of Section 8.4 the equivalences (8.9) and (8.11) are written as

$$g_\Omega(x, x) \sim 2\kappa_{\alpha,\rho}\,\bar a(x), \tag{8.14}$$

if F is recurrent and $0 < \rho < 1$. For $\alpha > 1$ this follows from $V_d(x)U_a(x)\mu(x) \to \kappa$ (Lemma 6.5.1) and Proposition 4.2.1 (see (8.41)). The case $\alpha = 1$ is immediate.

(b) The equivalence (8.14) also follows from (8.3). Indeed, the latter entails $v_d(k)u_a(k) \sim \alpha^2\rho\hat\rho/[k^{2-\alpha}\ell(k)\hat\ell(k)]$, so that if F is recurrent, as $x \to \infty$

$$(*) \qquad g_\Omega(x, x) \sim \alpha^2\rho\hat\rho \int_1^x \frac{dt}{\ell(t)\hat\ell(t)t^{2-\alpha}}.$$

Lemma 6.5.1 together with Proposition 4.2.1 shows that the RHS of $(*)$ is asymptotically equivalent to $2\kappa_{\alpha,\rho}\bar a(x)$ (see (8.41)).

Remark 8.2.3 Let $(c_n)_{n=1}^\infty$ be the same norming sequence specified as in (2.17), or what amounts to the same thing,

$$\begin{aligned} n\mu(c_n) &\longrightarrow c_\sharp && (0 < \alpha < 2), \\ nL(c_n)/c_n^2 &\longrightarrow c_\sharp && (\alpha = 2). \end{aligned}$$

Let $Y = (Y(t))$ be the stable process obtained as the limit of

$$S_{\lfloor nt\rfloor}/c_n : \quad P[Y(1) \le \eta\,|\,Y(0) = \xi] = \lim P_{\lfloor \xi c_n\rfloor}[S_n/c_n \le \eta] \quad (\xi, \eta \in \mathbb{R}).^2$$

Denote by $G^Y_{(-\infty,0]}(\xi, \eta)$ $(\xi, \eta > 0)$ the Green kernel for Y killed upon entering $(-\infty, 0]$, so that

$$\int_0^a G^Y_{(-\infty,0]}(\xi, \eta)\,d\eta = \int_0^\infty P\left[Y(t) \le a; Y(s) > 0 \text{ for } s < t\,\middle|\,Y(0) = \xi\right]\,dt \quad (a > 0).$$

---

[2] If $\alpha \neq 1$, the characteristic function of $Y(1)$ is given by (2.18), while if $\alpha = 1$, $-\log E[e^{iY(1)\theta}\,|\,Y(0) = 0] = \frac{1}{2}\pi c_\sharp|\theta| + ic_\sharp b\theta$ (in other words $P[Y(1) \in \cdot\,|\,Y(0) = 0] = P[c_\sharp(Y^\circ + b) \in \cdot]$ with Y° and b given in Section 4.2.1 (see (4.31))). We consider Y to be defined on the same probability space as S.

According to Silverstein [75, Theorem 6(i)], $G^Y_{(-\infty,0]}(\xi,\eta)$ is given in the form

$$G^Y_{(-\infty,0]}(\xi,\eta) = \tilde{\kappa} \int_0^{\xi \wedge \eta} (\xi - u)^{\alpha\hat{\rho}-1}(\eta - u)^{\alpha\rho-1} \, du \qquad (8.15)$$

for some constant $\tilde{\kappa}$. Here, based on Theorem 8.2.1, we prove that if $1 \le \alpha \le 2$ and condition (8.6) holds when $\alpha = 1$, then

$$\lim_{n\to\infty} \frac{c_n}{n} g_\Omega(c_n\xi, c_n\eta) = G^Y_{(-\infty,0]}(\xi,\eta) \quad \text{for } \xi, \eta > 0, {}^3 \qquad (8.16)$$

and that (8.15) holds with

$$\tilde{\kappa} = \begin{cases} 2/c_\sharp & \text{if } \alpha = 2, \\ \alpha^2 \rho\hat{\rho} \kappa/c_\sharp & \text{if } 1 \le \alpha < 2. \end{cases} \qquad (8.17)$$

If one chooses $[\Gamma(1 - \alpha) \cos \frac{1}{2}\alpha\pi]^{-1}$ or $2/\pi$ for $c_\sharp$ according as $1 < \alpha < 2$ or $\alpha = 1$ so that $\Re\Phi(\theta) = |\theta|^\alpha$ in (2.18), then (8.17) reduces to

$$\tilde{\kappa} = \begin{cases} [\Gamma(\alpha\rho)\Gamma(\alpha\hat{\rho})]^{-1} \cos[(\rho - \frac{1}{2})\alpha\pi] & \text{if } 1 \le \alpha < 2 \\ \pi^{-1} \sin^2 \rho\pi & \text{if } \alpha = 1. \end{cases} \qquad (8.18)$$

We have, at least formally,

$$G^Y_{(-\infty,0]}(\xi,\eta) \, d\eta = \lim_{n\to\infty} \int_0^\infty P_{\lfloor c_n\xi\rfloor}\left[S_{\lfloor nt\rfloor} \in c_n \, d\eta, \lfloor nt\rfloor < T\right] dt.$$

The integral on the RHS may be written as

$$\frac{1}{n} \sum_{k\in c_n \, d\eta} g_\Omega(c_n\xi, k) \sim \frac{1}{n} c_n g_\Omega(c_n\xi, c_n\eta) \, d\eta,$$

and by the functional limit theorem for Y_n we see that

$$n^{-1} c_n \int_0^\eta g_\Omega(c_n\xi, c_n r) \, dr \to \int_0^\eta G^Y_{(-\infty,0]}(\xi, r) \, dr.$$

Hence to verify (8.16), it suffices to show that $n^{-1}c_n g_\Omega(c_n\xi, c_n r)$ is convergent and bounded locally uniformly for $r \ne \xi$.[4] This is immediate from Theorem 8.2.1. Noting that if $\xi < \eta$,

$$\xi^{\alpha-1} \mathfrak{h}_{\alpha\hat{\rho}}(\eta/\xi) = \int_0^\xi (\xi - u)^{\alpha\hat{\rho}-1}(\eta - u)^{\alpha\rho-1} \, du$$

and similarly when $\eta < \xi$, one can easily deduce (8.15) from (8.8).

---

[3] $g_\Omega(x, y)$ is considered to be extended to a step function on $[0, \infty)^2$.

[4] If $\alpha = 1$, then $n^{-1}c_n g_\Omega(x, x) \to \infty$ by (8.11) and r must be bounded away from ξ.

Let $1 < \alpha \leq 2$. Since $\mathfrak{h}_\lambda(\zeta) \sim \lambda^{-1}\zeta^{\alpha-\lambda-1}$ as $\zeta \to \infty$, Theorem 8.2.1(i) entails that

$$g_\Omega(x, y) \asymp \begin{cases} V_d(x)U_a(y)/y & \text{for } 0 \leq x \leq y, \\ V_d(x)U_a(y)/x & \text{for } 0 \leq y \leq x, \end{cases} \tag{8.19}$$

where $\asymp$ can be replaced by $\sim$ if $y/x \to \infty$ or 0.

Here we give one more result obtained by applying Lemmas 8.1.1 and 8.1.3.

Lemma 8.2.4 *Let* $1 \leq \alpha \leq 2$ *and* (8.6) *hold if* $\alpha = 1$. *Then for* $x, y \geq 0$,

$$g_{B(R)}(x, y) \sim g_\Omega(x, y) \quad as \quad (x \vee y)/R \to 0.$$

Proof By the upper bound of $P_x(\Lambda_R)$ in (6.12), it follows that

$$\sum_{k=1}^\infty P_x[Z(R) = k, \Lambda_R] g_\Omega(R + k, y) \leq \frac{V_d(x)}{V_d(R)} \sup_{k \geq 1} g_\Omega(R + k, y).$$

Let $y \leq x$. Then the above supremum is bounded by a constant multiple of $g_\Omega(x, y)$ in view of Spitzer's formula since v_d is asymptotically decreasing (by Lemmas 8.1.1 and 8.1.3), while $V_d(x)/V_d(R) \to 0$ as $x/R \to 0$. Because of (7.54) it accordingly follows that $g_{B(R)}(x, y) \sim g_\Omega(x, y)$. For $x < y$, $g_{B(R)}(x, y) = \hat{g}_{B(R)}(y, x) \sim \hat{g}_\Omega(y, x) = g_\Omega(x, y)$ since what has just been verified applies to the dual walk. $\square$

8.3 Asymptotics of $P_x[\sigma_R < T \mid \Lambda_R]$

The proposition below concerns the conditional probability $P_x[\sigma_R < T \mid \Lambda_R]$, or what is actually the same, the ratio $P_x[\sigma_R < T]/P_x(\Lambda_R)$. There we assume $\alpha < 2$, for we know that under (AS) uniformly for $0 \leq x \leq R$,

$$P_x[\sigma_R < T] \sim P_x(\Lambda_R) \sim V_d(x)/V_d(R) \quad \text{if} \quad \alpha = 2. \tag{8.20}$$

Proposition 8.3.1 *Suppose (AS) holds with* $0 < \alpha < 2$ *and let* $1/2 < \delta < 1$.

(i) *For* $1 < \alpha < 2$, *the following equivalences hold:*

$$p = 0 \iff \lim_{R \to \infty} \frac{P_x(\Lambda_R)}{V_d(x)/V_d(R)} = 1 \quad \text{for some/all } x \geq 0, \tag{8.21}$$

$$pq = 0 \iff \lim_{R \to \infty} P_x[\sigma_R < T \mid \Lambda_R] = 1 \quad \text{for some/all } x \in \mathbb{Z}, \tag{8.22}$$

the convergence of the limits above is uniform for $0 \leq x \leq R$, *and if* $pq \neq 0$, *then*

$$P_x[\sigma_R < T] \sim f\left(\frac{x}{R}\right)\frac{V_d(x)}{V_d(R)} \quad \text{uniformly for } 0 \leq x \leq R \text{ as } R \to \infty \tag{8.23}$$

for some (strictly) increasing and continuous function f such that $f(1) = 1$ and $f(0) = (\alpha - 1)/\alpha\hat\rho$ (an explicit expression of f will be given in (8.25)).

(ii) *If $\alpha = 1$, $\rho > 0$ and F is recurrent (necessarily $p \le q$), then*

$$\frac{P_x[\sigma_y < T]}{P_x(\Lambda_y)} \longrightarrow \begin{cases} (q-p)/q & \text{as } y \to \infty \text{ uniformly for } 0 \le x < \delta y, \\ 0 & \text{as } x \to \infty \text{ uniformly for } 0 \le y < \delta x. \end{cases}$$

(iii) *If $\alpha = 1$, $EX = 0$ and $p > q$ (entailing $\rho = 0$), then as $y \to \infty$*

$$\frac{P_x[\sigma_y < T]}{P_x(\Lambda_y)} \sim \begin{cases} (p-q)/p & \text{uniformly for } 0 \le x \le \delta y, \\[2mm] \dfrac{p-q}{p} \cdot \dfrac{U_a(y)/a(-y)}{U_a(x)/a(-x)} & \text{uniformly for } x \ge y/\delta. \end{cases}$$

Here the second case is true also for $p = q$ (because of (7.4) and (ii)).
[Some asymptotics of $U_a(x)/a(-x)$ are found in Remark 6.4.6(b) (given in the dual setting).]

(iv) *If $\rho > 0$ and F is transient (necessarily $\alpha \le 1$), then for $x \ge 0$, $y > 1$,*

$$P_x[\sigma_y < T]/P_x(\Lambda_y) \to 0 \quad \text{as} \quad |x - y| \to \infty. \tag{8.24}$$

Remark 8.3.2 The first case of (iii) is obtained as a special case of Theorem 7.1.4 and does not follow by duality from (ii); the proof of (iii) is much more involved than that of (ii). The crux of the issue lies in the verification of $P_x[Z(y) > \varepsilon y \mid \Lambda_y] \to 1$ as $x \to \infty$, $\varepsilon \downarrow 0$ in this order. When $\alpha = \hat\rho = 2p = 1$, if this convergence were true, then we could say $P_x[\sigma_y < T]/P_x(\Lambda_y)$ converges to 0 (as $y \to \infty$) for $x < \delta y$. A similar remark applies to the transient walks: (8.24) would be true if we could show $P_x[Z(y) > \varepsilon^{-1} \mid \Lambda_y] \to 1$ in the same successive limits as above.

For the proof of Proposition 8.3.1 we show the following

Lemma 8.3.3 (i) *If $1 < \alpha \le 2$, then as $R \to \infty$,*

$$P_x[\sigma_R < T] \sim \begin{cases} \left[\dfrac{(R/x)^{1-\alpha\rho}\mathfrak{h}_{\alpha\hat\rho}(R/x)}{\mathfrak{h}_{\alpha\rho}(1)}\right]\dfrac{V_d(x)}{V_d(R)} & \text{uniformly for } 1 \le x \le R, \\[4mm] \left[\dfrac{(R/x)^{\alpha\hat\rho}\mathfrak{h}_{\alpha\rho}(x/R)}{\mathfrak{h}_{\alpha\rho}(1)}\right]\dfrac{V_d(x)}{V_d(R)} & \text{uniformly for } x \ge R, \end{cases}$$

$$\tag{8.25}$$

in particular,

$$P_x[\sigma_R < T] \begin{cases} \to \begin{cases} (\alpha-1)\xi^{\alpha-1}\mathfrak{h}_{\alpha\hat\rho}(1/\xi) & \xi \le 1, \\ (\alpha-1)\mathfrak{h}_{\alpha\rho}(\xi) & \xi > 1, \end{cases} & (x/R \to \xi > 0), \\[4mm] \sim [(\alpha-1)/\alpha\rho](R/x)^{1-\alpha\hat\rho}\hat\ell(R)/\hat\ell(x) & (x/R \to \infty), \\[2mm] \sim [(\alpha-1)/\alpha\hat\rho]\,V_d(x)/V_d(R) & ([x \vee 1]/R \to 0); \end{cases}$$

$$\tag{8.26}$$

and for each $y = 0, 1, 2, \ldots,$

$$P_x[\sigma_y < T]x^{1-\alpha\hat{\rho}}\ell(x) \longrightarrow \frac{\alpha\hat{\rho}\, U_a(y)}{g_\Omega(y, y)} \quad \text{as } x \to \infty.$$

(ii) *Let* $\alpha = 1$ *and* $1 < \rho < 1$. *Then* $\sup_{x:|x-y|>\varepsilon y} P_x[\sigma_y < T] \to 0$ $(y \to \infty)$; *moreover, if (8.6) holds, as* $(x/R, y/R) \to (\xi, \eta) \in (0, 1)^2$ *with* $\xi \neq \eta$,

$$P_x[\sigma_y < T] \sim \frac{\kappa\rho\hat{\rho}}{a(x)L(x)}H(\xi, \eta) \sim \frac{1/L(x)}{[1/L]_*(x)}H(\xi, \eta). \tag{8.27}$$

Proof Because of the identity $P_x[\sigma_R < T] = g_\Omega(x, R)/g_\Omega(R, R)$, the first formula of (8.25) follows from Theorem 8.2.1. The derivation of the second one is similar. By $\lim_{\zeta \to \infty} \zeta^{\alpha\hat{\rho}-1}\mathfrak{h}_{\alpha\hat{\rho}}(\zeta) = 1$ (8.26) follows from (8.25) with the help of $\mathfrak{h}_\lambda(1) = 1/(\alpha - 1)$. Thus (i) is verified. (ii) follows immediately from Theorem 8.2.1 (ii). $\square$

The function $\zeta^{1-\alpha\rho}\mathfrak{h}_{\alpha\hat{\rho}}(\zeta) = \int_0^1 t^{\alpha\hat{\rho}-1}\left(1 - \zeta^{-1}(1 - t)\right)^{\alpha\rho-1} dt$ decreasingly approaches $1/\alpha\hat{\rho}$ as $\zeta \to \infty$ if $\alpha\rho < 1$ and $\mathfrak{h}_{\alpha\hat{\rho}}(\zeta) \equiv 1/(\alpha-1)$ if $\alpha\rho = 1$. This together with (6.12) and Lemma 8.3.3 yields that if $1 < \alpha < 2$, for $0 \leq x \leq R$,

$$\frac{V_d(x)}{V_d(R)} \geq P_x(\Lambda_R) \geq \frac{P_x[\sigma_R < T]}{1 + o(1)} \geq \frac{\alpha - 1}{\alpha\hat{\rho}} \cdot \frac{V_d(x)}{V_d(R)}\{1 + o(1)\}. \tag{8.28}$$

Proof (of Proposition 8.3.1) Let $\alpha > 1$. If $q > 0$ we have $\alpha\hat{\rho} < 1$ so that $\mathfrak{h}_{\alpha\rho}(\zeta) \sim (\alpha\rho)^{-1}\zeta^{-\gamma}$ $(\zeta \to \infty)$ with $\gamma = 1 - \alpha\hat{\rho} > 0$, and accordingly the second case of (8.25) implies that for each $\varepsilon > 0$,

$$\liminf_{R \to \infty} \inf_{z \geq (1+\varepsilon)R} P_z[T < \sigma_R] > 0,$$

while if $p > 0$ it follows that for a small $\varepsilon > 0$ and for all sufficiently large R

$$P_x\left[Z(R) > \varepsilon R \,\middle|\, \Lambda_R\right] \geq 1/2 \qquad (x < \delta R)$$

owing to Lemma 6.6.3. These together show that the condition $pq > 0$ implies $P_x[\sigma_R < T] \leq \theta V_d(x)/V_d(R)$ with $\theta < 1$. Now the equivalence (8.21) in (i) follows from Proposition 5.3.2, which also shows that if $p = 0$, $P_x[\sigma_R < T] \sim P_x(\Lambda_R)$ uniformly for $0 \leq x \leq R$ (this may be verified directly because of the first case of (8.26)). If $q = 0$, then $\alpha\hat{\rho} = 1$, so that $\mathfrak{h}_{\alpha\rho} = \mathfrak{h}_{\alpha-1} \equiv 1/(\alpha - 1)$ and (8.25) implies that $P_z[\sigma_R < T] \to 1$ uniformly for $R < z < MR$ for any $M > 1$ and we conclude that $P_x[\sigma_R < T \mid \Lambda_R] \to 1$ uniformly for $0 \leq x \leq R$ since for such x, $P_x[Z(R) < MR \mid \Lambda_R] \sim P_x(\Lambda_R)$ as $R \to \infty$ and $M \to \infty$. The asymptotic equivalence stated last in (i) is a reduced form of the first formula in (8.25).

(ii) follows from (7.2) of Theorem 7.1.1 ($\rho = 1$) and Corollary 7.3.3 ($0 < \rho < 1$) with the help of (6.1) and the asymptotic forms of $a(x)$ and $a(-x)$ given in (7.36).

The case $y \to \infty$ of (iii) follows from (7.15) of Theorem 7.1.4. The other case is cheaper and immediate from (7.49).

For the proof of (iv), suppose that F is transient. Then, using the oscillation of the r.w, we have

$$\lim_{x \to \infty} g_\Omega(x, x) = 1/q_\infty \quad \text{and} \quad \lim_{|x-y| \to \infty} g_\Omega(x, y) \to 0 \tag{8.29}$$

as a standard fact for a general transient r.w. (cf. Section A.5.3). If $\rho > 0$ and $R/4 \le x < R$, then $P_x(\Lambda_R)$ is bounded away from zero (cf., e.g., (6.8)), so that $P_x[\sigma_R < T \mid \Lambda_R] \le C P_x[\sigma_R < T]$, hence by (8.29)

$$P_x[\sigma_R < T]/P_x(\Lambda_R) \to 0 \quad \text{as} \quad |R - x| \to \infty. \tag{8.30}$$

Let $0 \le x < R/4$. Then if $0 < \rho < 1$, Lemma 6.6.3 entails

$$P_x[Z(R) \le \varepsilon R \mid \Lambda_R] \to 0 \quad \text{as} \quad R \to \infty \text{ and } \varepsilon \downarrow 0 \tag{8.31}$$

(uniformly in x), and (8.30) follows from (8.29) again. If $\rho = 1$ (when either (C3) or (C4) holds), using **L(2.1)** and **L(3.4)** in turn we see that

$$\frac{P_x[S_{B(\frac{1}{4}R)} > \frac{1}{2}R]}{P_x(\Lambda_R)} \le \frac{V_d(R)}{V_d(x)} \sum_{y=0}^{R/4} g_\Omega(x, y)\mu_+(R/4)$$

$$\le V_d(R)U_a(R/4)\mu_+(R/4) \to 0,$$

which leads to (8.30). $\qquad\qquad\qquad\qquad\qquad\qquad\qquad\qquad\qquad\qquad$ □

Remark 8.3.4 Let (C3) hold. Then, $u_a(x) \sim 1/\ell^*(x)$ and $V_d(x) \sim 1/\ell_\sharp(x)$, and uniformly for $0 \le x < \delta R$, $g_\Omega(x, R) \sim V_d(x)/\ell^*(R)$, so that

$$P_x[\sigma_R < T] \sim \frac{1}{\ell_\sharp(x)\ell^*(R)g_\Omega(R, R)}$$

(whether F is recurrent or transient and without assuming (AS)). Hence if F is transient, on taking $x = R/2$, (8.30) shows

$$\ell_\sharp(x)\ell^*(x) \to \infty.$$

This can be directly deduced from **L(3.4)** if $\liminf x\mu(x) > 0$ or from Lemma 6.4.3 if (AS) holds with $p < q$, but not otherwise.

Remark 8.3.5 By the expression of $\mathfrak{h}_\lambda(\zeta)$ given in (8.7), the formula (8.25) (valid for $1 < \alpha \le 2$) can be rewritten as

$$P_x[\sigma_R < T] \sim \begin{cases} \left[1 - \dfrac{x}{R}\right]^{\alpha-1} \displaystyle\int_{1-\frac{x}{R}}^1 \dfrac{(\alpha - 1)\,dw}{w^\alpha(1-w)^{1-\alpha\hat{\rho}}} & (1 \le x < R), \\[4mm] \left[\dfrac{x}{R} - 1\right]^{\alpha-1} \displaystyle\int_{1-\frac{R}{x}}^1 \dfrac{(\alpha - 1)\,dw}{w^\alpha(1-w)^{1-\alpha\rho}} & (x > R), \end{cases} \tag{8.32}$$

as $R \to \infty$ (uniformly in x). For a sub-class of F satisfying (AS), Kesten derived the above formula when $R/x \to \zeta > 0$, $\neq 1$ in a way different from ours (cf. Lemma 9 of [39] for the case $\zeta > 1$; the formula for the case $\zeta < 1$ is implicitly given in the proof of Lemma 2 of [40]). The above expression will be helpful in our computation of $g_{B(R)}(x, y)$ made in Section 8.5.

Let P_x^{Ω}, $x \geq 0$, be the law of the Markov chain determined by

$$P_x^{\Omega}[S_1 = x_1, \ldots, S_n = x_n] = P_x[S_1 = x_1, \ldots, S_n = x_n, n > T]\frac{V_d(x_n)}{V_d(x)} \qquad (8.33)$$

$(x, x_1, \ldots, x_n \geq 0)$, in other words, P_x^{Ω} is the h-transform with $h = V_d$ of the law of S killed as it enters Ω; P_x^{Ω} may be considered to be the law of S started at $x \geq 0$ conditioned never to enter Ω. From the defining expression (8.33) one deduces that

$$P_x^{\Omega}[\sigma_y < \infty] = \frac{V_d(y)}{V_d(x)}P_x[\sigma_y < T] \qquad (x \geq 0, y \geq 0). \qquad (8.34)$$

Because of this identity, the results for $P_x[\sigma_R < T]/P_x(\Lambda_y)$ given above together with those obtained in Chapters 6 and 7 and in Lemma 8.3.3 lead to the following

Corollary 8.3.6 *Suppose (AS) holds.*

(i) *If $1 < \alpha \leq 2$, then uniformly in $x \geq 0$,*

$$P_x^{\Omega}[\sigma_R < \infty] \sim f(x/R),$$

where $f(\xi)$ is a continuous function of $\xi \geq 0$ such that

$$for\ \xi \leq 1, \quad \begin{cases} f\ is\ identical\ to\ 1\ if\ \alpha = 2, \\ f\ equals\ the\ function\ appearing\ in\ Proposition\ 8.3.1\ if\ \ \alpha < 2, \end{cases}$$

for $\xi > 1$, $\ f(\xi) = (\alpha - 1)\xi^{-\alpha\hat{\rho}}\int_0^1 t^{\alpha\rho - 1}(\xi - 1 + t)^{\alpha\hat{\rho} - 1}\,dt$;

in particular, $P_x^{\Omega}[\sigma_R < \infty] \sim [(\alpha - 1)/\alpha\rho]R/x$ as $x/R \to \infty$.
(ii) *If $\alpha = 1$ and F is recurrent, then for each $0 < \delta < 1$, uniformly in $x \geq 0$ as $R \to \infty$,*

$$P_x^{\Omega}[\sigma_R < \infty]\begin{cases} \to (q - p)/q & p \leq q, \\ \sim \dfrac{p - q}{p} \cdot \dfrac{R\int_{R-x-1}^{R}\mu_+(t)\,dt}{-A(R)(x + 1)} \to 0 & p > q, \end{cases} \qquad for\ x < \delta R,$$

$$P_x^{\Omega}[\sigma_R < \infty]\begin{cases} \to 0 & p \leq q, \\ \sim \dfrac{p - q}{p} \cdot \dfrac{R/A(R)}{x/A(x)} & p > q, \end{cases} \qquad for\ x > R/\delta.$$

(iii) *If F is transient, then $P_x^{\Omega}[\sigma_R < \infty] \to 0$ whenever $|x - R| \to \infty$.*

Proof In view of (8.34), (i) follows from Proposition 8.3.1(i) when $x \leq R$ and from Lemma 8.3.3 when $x > R$. As for (ii), use (ii) and (iii) of Proposition 8.3.1 together with the estimate of $P_x(\Lambda_R)$ in Corollary 7.1.8 for the case $q < p, x < \delta R$. (iii) follows from Proposition 8.3.1(iv) when $\rho > 0$. If $\rho = 0$, one has only to recall $P_x(\Lambda_R)V_d(R)/V_d(x) \to 0$ for $x < \delta R$ (Lemma 6.4.7, see also (6.42)). □

Remark 8.3.7 Bertoin and Doney [5, Lemma 1] (where the random walks may be non-lattice and assumed only to be oscillating) derived

$$P_x(\Lambda_R)/P_0(\Lambda_R) \longrightarrow V_d(x)$$

(for any $x \geq 0$ fixed) and, using this, verified that

$$P_x^{\Omega}[S_1 = x_1, \ldots, S_n = x_n] = \lim_{R \to \infty} P_x[S_1 = x_1, \ldots, S_n = x_n \mid \Lambda_R].$$

8.4 Asymptotic Form of the Green Function $g_{B(R)}(x, y)$

Let $1 \leq \alpha \leq 2$. If $\alpha = 1$, suppose $0 < \rho < 1$, the case $\rho\hat{\rho} = 0$ being a special case of a relatively stable r.w. for which we have obtained rather fine results in Theorem 7.1.10. To state our results that admit the case $x/R \downarrow 0$ and/or $y/R \uparrow 1$ it is convenient to define ℓ and $\hat{\ell}$ by the identities $\ell(x) = x^{\alpha\rho}/U_a(x)$ and $\hat{\ell}(x) = x^{\alpha\hat{\rho}}/V_d(x)$ $(x \geq 1)$ rather than their specification by the asymptotic relations in (6.49). For $x = 0$, put $\ell(0) = 1$ and $\hat{\ell}(0) = 1/v^{\circ}$, so that for all $x = 0, 1, 2, \ldots$,

$$U_a(x) = (x \vee 1)^{\alpha\rho}/\ell(x) \quad \text{and} \quad V_a(x) = (x \vee 1)^{\alpha\hat{\rho}}/\hat{\ell}(x).$$

[In some cases, the results are stated only for $x, y \notin \{0, R\}$ for simplicity, although, because of the above identity, they are valid for $x = 0$ $(y = R)$ if x (and/or y) is replaced by $x \vee 1$ $((R - y) \vee 1)$ when $x < y$, and analogously when $y < x$.]

Define the constant $\kappa_{\alpha,\rho}$ by

$$\kappa_{\alpha,\rho} = \begin{cases} 1 & \text{if } \alpha = 2, \\ \dfrac{[\sin\alpha\rho\pi + \sin\alpha\hat{\rho}\pi]\{1 - 4pq\sin^2\frac{1}{2}\alpha\pi\}}{(\alpha - 1)B(\alpha\rho, \alpha\hat{\rho})|\sin\alpha\pi|} & \text{if } 1 < \alpha < 2, [5] \\ 1/2 & \text{if } \alpha = 1. \end{cases} \quad (8.35)$$

---

[5] If $p = q$, $\kappa_{\alpha,\rho} \to 1$ as $\alpha \uparrow 2$ and $\kappa_{\alpha,\rho} \to \frac{1}{2}$ as $\alpha \downarrow 1$ and $\rho \to \frac{1}{2}$.

Theorem 8.4.1 *Let* $1 \le \alpha \le 2$ *and* $0 < \rho < 1$. *Then for any* $\varepsilon > 0$ *and* $\delta < 1$,

$$g_{B(R)}(x,x) \sim \frac{2\kappa_{\alpha,\rho} \bar{a}(x)\bar{a}(R-x)}{\bar{a}(R)} \qquad as \ x \wedge (R-x) \to \infty$$

$$\sim \begin{cases} 2\kappa_{\alpha,\rho}\bar{a}(x)\left(1 - \frac{x}{R}\right)^{\alpha-1} & as \ x \to \infty \qquad for \ 0 \le x < \delta R, \\ 2\kappa_{\alpha,\rho}\bar{a}(R-x)\left(\frac{x}{R}\right)^{\alpha-1} & as \ R-x \to \infty \ for \ \varepsilon R < x \le R. \end{cases}$$

[Some explicit asymptotic forms of $2\kappa_{\alpha,\rho}\bar{a}(x)$ *are given in (8.41) below.]*

Remark 8.4.2 Let $\alpha = 1$ (without the restriction on ρ). Then for each $\delta < 1$,

$$g_{B(R)}(x,x) \sim g_{\Omega}(x,x) \quad as \ R \to \infty \text{ uniformly for } x < \delta R. \qquad (8.36)$$

If $\rho > 0$, this is readily verified by using Proposition 8.3.1(ii, iv) (with the help of (7.54)). If $\rho\hat{\rho} = 0$, (8.36) is a special case of Lemma 5.7.3.

For $1 \le \alpha \le 2$, $0 < \rho < 1$, define the function $K_{\rho}^{(\alpha)}(\xi,\eta)$ of $(\xi,\eta) \in [0,1]^2$ by

$$K_{\rho}^{(\alpha)}(\xi,\xi) = \begin{cases} \dfrac{1}{\alpha-1}[\xi(1-\xi)]^{\alpha-1} & for \ \alpha > 1, \\ \infty & for \ \alpha = 1, \end{cases}$$

$$K_{\rho}^{(2)}(\xi,\eta) = \xi \wedge \eta - \xi\eta,$$

and for $1 \le \alpha < 2$ and $\xi \ne \eta$,

$$K_{\rho}^{(\alpha)}(\xi,\eta) = \begin{cases} |\xi-\eta|^{\alpha-1} \displaystyle\int_0^{\frac{\xi(1-\eta)}{\eta(1-\xi)}} (1-w)^{-\alpha} w^{\alpha\rho-1} \, dw & \xi < \eta, \\ |\xi-\eta|^{\alpha-1} \displaystyle\int_0^{\frac{\eta(1-\xi)}{\xi(1-\eta)}} (1-w)^{-\alpha} w^{\alpha\rho-1} \, dw & \xi > \eta. \end{cases} \qquad (8.37)$$

[For $\alpha = 2$ with $\rho = \hat{\rho} = 1/2$ the expression on the RHS of (8.37) reduces to $K_{\rho}^{(2)}$.] If $\alpha > 1$, $K_{\rho}^{(\alpha)}$ is continuous on $[0,1]^2$. In a few particular cases we have explicit forms of $K_{\rho}^{(\alpha)}(\xi,\eta)$: if $q = 1 < \alpha$ (when $\alpha\rho = 1$),

$$K_{1/\alpha}^{(\alpha)}(\xi,\eta) = \begin{cases} (\alpha-1)^{-1}[\xi(1-\eta)]^{\alpha-1} & \xi \le \eta, \\ (\alpha-1)^{-1}\left([\xi(1-\eta)]^{\alpha-1} - [\xi-\eta]^{\alpha-1}\right) & \xi > \eta; \end{cases} \qquad (8.38)$$

also, for $1 \le \alpha \le 2$,

$$K_{\rho}^{(\alpha)}(\xi,\eta) \sim (\alpha\hat{\rho})^{-1}|\xi-\eta|^{\alpha-1}\left[\frac{\xi(1-\eta)}{\eta(1-\xi)}\right]^{\alpha\hat{\rho}} \qquad as \ \xi \wedge (1-\eta) \to 0; \qquad (8.39)$$

and similarly for the case $p = 1 < \alpha$ and the case $(1-\xi) \wedge \eta \to 0$, respectively.

Moreover, if $\alpha = 1 = 2\rho$,

$$K^{(1)}_{1/2}(\xi, \eta) = \log \frac{\xi \vee \eta - \xi\eta + \sqrt{(\xi - \xi^2)(\eta - \eta^2)}}{\xi \vee \eta - \xi\eta - \sqrt{(\xi - \xi^2)(\eta - \eta^2)}} \qquad (\xi \neq \eta).$$

In the following theorem (and its corollary) we write

$$\xi = x/R \quad \text{and} \quad \eta = y/R.$$

Theorem 8.4.3 (i) *Let* $1 < \alpha \leq 2$. *Then for each* $\varepsilon > 0$, *uniformly for* $(x, y) \in (0, R)^2$, *as* $x \vee y \to \infty$ *and* $(R - x) \vee (R - y) \to \infty$,

$$g_{B(R)}(x, y) \sim \frac{\alpha^2 \rho\hat\rho\, \ell(R)\hat\ell(R)R^{\alpha-1}}{\hat\ell(x)\ell(y)\ell(R - x)\hat\ell(R - y)} K^{(\alpha)}_\rho(\xi, \eta). \tag{8.40}$$

(ii) *Let* $\alpha = 1$ *and suppose that condition* (8.6) *holds, namely* $0 < \rho < 1$ *and if* $\rho = 1/2$, *either of the following conditions holds:*

(a) *the law of* X *is symmetric and* $L(x)$ *is almost decreasing,*
(b) *either* $p(x) = O(\mu_+(x)/x)$ *or* $p(-x) = O(\mu_-(x)/x)$ $(x \to \infty)$.

Then (8.40) *holds under the additional restriction* $|\xi - \eta| > \varepsilon$ *(for each* $\varepsilon > 0$*).*
[Note that $\ell(x)\hat\ell(x) \sim C_{\alpha,\rho}L(x)$ *with* $C_{\alpha,\rho}$ *equal to* $\frac{1}{2}$ *if* $\alpha = 2$ *and* $1/\kappa$ *if* $\alpha < 2$.]

Remark 8.4.4 Since $U_a(x)V_d(x) \sim x^\alpha/\ell(x)\hat\ell(x)$, it follows from (8.40) that

$$g_{B(R)}(x, y) \sim \frac{\alpha^2 \rho\hat\rho\, U_a(R)V_d(R)}{R} K^{(\alpha)}_\rho(\xi, \eta) \quad \text{uniformly for } (\xi, \eta) \in (\varepsilon, 1 - \varepsilon)^2.$$

Since $U_a(x)V_d(x)/x^2 \to 2/\sigma^2$, this yields the classical result given in [71] for the case $\sigma^2 < \infty$. For $1 \leq \alpha < 2$, $U_a(x)V_d(x) \sim \kappa/\mu(x)$ according to Lemma 6.5.1. If $1 - \psi(\theta) \sim Q|\theta|^\alpha$, then $x^\alpha\mu(x) \to 2Q\pi^{-1}\Gamma(\alpha) \sin \frac{1}{2}\alpha\pi$, and we obtain, as a special case of (8.40), the equivalence $g_{B(R)}(x, y)/R^{\alpha-1} \sim \left[Q(\Gamma(\frac{1}{2}\alpha))^2\right]^{-1} K^\alpha_\rho(\xi, \eta)$ $(0 < \xi, \eta < 1)$, which is the same as Kesten's result (in the case $\alpha > 1$) mentioned near the end of the introduction.

Dividing the relation (8.40) by $g_{B(R)}(y, y)$ we can obtain the asymptotic form of $P_x[\sigma_y < \sigma_{B(R)}]$, the hitting probability of y by the walk killed as it exits the interval $[0, R]$. Recall $\kappa_{\alpha,\rho} = \alpha/2$ for $\alpha \in \{1, 2\}$ and observe that

$$2\kappa_{\alpha,\rho} = \alpha^2 \rho\hat\rho\, \kappa\kappa_\alpha / \left[(\alpha - 1)^2(2 - \alpha)\right] \quad \text{for } 1 < \alpha < 2.$$

Then from Proposition 4.2.1(i) it holds that, as $x \to \infty$,

$$2\kappa_{\alpha,\rho}\bar{a}(x) \sim \begin{cases} x/\int_0^x t\mu(t)\,dt \sim 2x/L(x) & \alpha = 2, \\ (\alpha - 1)^{-1}\kappa\alpha^2\rho\hat\rho/[x\mu(x)] & 1 < \alpha < 2, \\ \kappa\rho\hat\rho\,[1/L]_*(x) & \alpha = 1\ (0 < \rho < 1). \end{cases} \tag{8.41}$$

(Here $[1/L]_*(x) = \int_0^x [tL(t)]^{-1}\,dt$.) In the following corollary, we state the result only in the case when $R - y \to \infty$ and either x/R is close to zero or y/R is close to unity. [For the other cases, see Remark 8.4.6, Proposition 8.5.3 ($\alpha > 1$) and Corollary 8.5.7 ($\alpha = 1$).] To include the extremes $x = 0$ and $y = R$ we write

$$\overline{\xi} = (x \vee 1)/R \quad \text{and} \quad \overline{1-\eta} = ([R - y] \vee 1)/R.$$

Corollary 8.4.5 (i) *Let* $1 < \alpha \leq 2$. *Then for any* $\varepsilon > 0$, *uniformly for* $x, y \in [0, R]$, *as* $\overline{\xi} \wedge \overline{1-\eta} \to 0$ *under* $\varepsilon < \xi + \eta < 2 - \varepsilon$ *and* $R - y \to \infty$

$$P_x[\sigma_y < \sigma_{B(R)}] \sim \frac{(\alpha - 1)\hat{\ell}(R)\ell(R - y)}{\alpha\hat{\rho}\,\hat{\ell}(x)\ell(R)} \cdot \frac{\overline{\xi}^{\alpha\hat{\rho}}\,\overline{1-\eta}^{1-\alpha\rho}}{\eta^{\alpha\hat{\rho}}(1 - \xi)^{1-\alpha\rho}}. \tag{8.42}$$

(ii) *Let* $\alpha = 1$ *and condition* (8.6) *hold. Then, as* $\overline{\xi} \wedge \overline{1-\eta} \to 0$, *under the same restriction on* x, y *as above,*

$$P_x[\sigma_y < \sigma_{B(R)}] \sim \frac{\hat{\ell}(R)}{\kappa\hat{\rho}\,\hat{\ell}(x)\ell(R)\hat{\ell}(R - y)[1/L]_*(R - y)} \cdot \left[\frac{\overline{\xi}\,\overline{1-\eta}}{\eta(1 - \xi)}\right]^{\rho}.$$

Proof By Lemma 6.5.1 $\ell(x)\hat{\ell}(x)/L(x)$ tends to $1/2$ if $\alpha = 2$ and to $1/\kappa$ if $1 \leq \alpha < 2$. On using this, (8.41) and the asymptotic form of $K_\rho^{(\alpha)}$ given in (8.39), observe that for $1 < \alpha \leq 2$, as $\overline{\xi} \wedge \overline{1-\eta} \to 0$ under the restriction in (i),

$$\frac{\alpha^2\rho\hat{\rho}K_\rho^{(\alpha)}(\xi, \eta)R^{\alpha-1}}{[2\kappa_{\alpha,\rho}\bar{a}(R - y)\eta^{\alpha-1}]\,\hat{\ell}(R - y)} \sim \frac{(\alpha - 1)|\eta - \xi|^{\alpha-1}}{\alpha\hat{\rho}[\eta(1 - \eta)]^{\alpha-1}} \left[\frac{\overline{\xi}\,\overline{1-\eta}}{\eta(1 - \xi)}\right]^{\alpha\hat{\rho}} \ell(R - y)$$

$$\sim \frac{\alpha - 1}{\alpha\hat{\rho}} \frac{\overline{\xi}^{\alpha\hat{\rho}}\,\overline{1-\eta}^{1-\alpha\rho}}{\eta^{\alpha\hat{\rho}}(1 - \xi)^{1-\alpha\rho}} \ell(R - y)$$

(observe $|\xi - \eta|/\eta \sim 1 - \xi$ for the second equivalence). Then, recalling the remark about ℓ and $\hat{\ell}$ given at the beginning of this section and noting $(1 - \xi) \wedge \eta > \frac{1}{2}\varepsilon$ in the specified way of taking a limit, one readily deduces the asserted equivalence of (i) from Theorems 8.4.3 and 8.4.1. The proof of (ii) is similar. □

Remark 8.4.6 (a) The equivalences (i) and (ii) above hold in particular when $\eta \uparrow 1$, $\xi < 1 - \varepsilon$, and $R - y \to \infty$, in which the last condition must be imposed, being necessary to get an estimate of $g_{B(R)}(y, y)$ as far as one uses Theorem 8.4.1. However, when $w := R - y = o(R)$ – in particular when w is fixed – one can apply Lemma 8.2.4, which ensures $g_{B(R)}(y, y) = g_{B(R)}(w, w) \sim g_\Omega(w, w)$. If $\alpha > 1$, from (8.39) and (8.40) one can accordingly deduce

$$P_x[\sigma_{R-w} < \sigma_{B(R)}] \sim \frac{\alpha\rho V_{\mathrm{d}}(w)(1 - \xi)^{\alpha\rho-1}\overline{\xi}^{\alpha\hat{\rho}}}{g_\Omega(w, w)\hat{\ell}(x)/\hat{\ell}(R)} \cdot \frac{R^{\alpha\rho-1}}{\ell(R - x)} \qquad (w, x \geq 0) \tag{8.43}$$

as $R - x \to \infty$ and $w/R \to 0$. Noting $V_d(0) = g_\Omega(0,0) = v^\circ$, one sees that, in the particular case $y = R$ and for $x < \delta R$, the RHS above differs from the expression that results from substituting R for y (hence R^{-1} for $\overline{1 - \eta}$) in the formula of (8.42) by the factor $\alpha^2 \rho \hat{\rho}/(\alpha - 1)$.[6] The same remarks apply to the case $\alpha = 1$.

(b) By the identity $g_{B(R)}(x, y) = \hat{g}_{B(R)}(y, x) = \hat{g}_{B(R)}(R - x, R - y)$, we have $P_x[\sigma_y < \sigma_{B(R)}] = \hat{P}_{R-x}[\sigma_{R-y} < \sigma_{B(R)}]$, where $\hat{P}_y$ denotes the law of the dual walk started at y. On applying this, we infer from (8.42) that as $\overline{1 - \xi} \wedge \overline{\eta} \to 0$ under $\varepsilon < \xi + \eta < 2 - \varepsilon$ and $y \to \infty$,

$$P_x[\sigma_y < \sigma_{B(R)}] \sim \frac{(\alpha - 1)\ell(R)\hat{\ell}(y)}{\alpha \rho \, \ell(R - x)\hat{\ell}(R)} \cdot \frac{\overline{1 - \xi}^{\,\alpha\rho} \, \overline{\eta}^{\,1-\alpha\hat{\rho}}}{(1 - \eta)^{\alpha\rho} \xi^{1-\alpha\hat{\rho}}}. \qquad (8.44)$$

As in Remark 8.2.3, Theorem 8.4.3 yields the following

Corollary 8.4.7 *Let $1 \le \alpha \le 2$ and $Y(t)$ be the same limit process as in Remark 8.2.3. Suppose that condition (8.6) holds if $\alpha = 1$. Then, the Green kernel for Y killed upon exiting the interval $[0, r]$ ($r > 0$) has the version given by*

$$G^Y_{B(r)}(\xi, \eta) = \tilde{\kappa} \, r^{\alpha-1} K_\rho^{(\alpha)}\left(\frac{\xi}{r}, \frac{\eta}{r}\right) \qquad (0 \le \xi, \eta \le r),$$

and

$$\lim_{n \to \infty} \frac{c_n}{n} g_{B(c_n r)}(c_n \xi, c_n \eta) = G^Y_{B(r)}(\xi, \eta) \qquad (0 \le \xi, \eta \le r), \qquad (8.45)$$

where $\tilde{\kappa}$ is the constant given in (8.17) and the convergence in (8.45) is locally uniform on $(0, r)^2$ if $\alpha > 1$ and on $(0, r)^2 \cap \{\xi \ne \eta\}$ if $\alpha = 1$.

We shall prove Theorem 8.4.1 in Section 8.5. For the proof of Theorem 8.4.3, we shall deal with the case when both ξ and η are bounded away from 0 and 1 in Section 8.5 (see Proposition 8.5.3, Corollary 8.5.4 for $\alpha > 1$, and Proposition 8.5.6, Corollary 8.5.7 for $\alpha = 1$). The other case will be addressed in Section 8.6, where we shall complete proof of Theorem 8.4.3.

8.5 The Scaling Limit of $g_{B(R)}(x, y)$

In this section, we follow Kesten [40] to seek a scaling limit of the Green function $g_{B(R)}(x, y)$. It will turn out that for $\alpha > 1$, $g_{B(R)}(x, y)/\bar{a}(R)$ converges as $(x/R, y/R) \to (\xi, \eta)$ to a function that is a natural generalisation of that found by Kesten [40] for a similar scaling limit. By duality, the asymptotic form of $g_{B(R)}(x, y)$ for $x \le y$ implies that for $y \le x$, and vice versa.

---

[6] Of course (8.43) conforms to (8.42) if $w \to \infty$ (under $w = o(R)$) and $\xi < 1 - \varepsilon$.

Let ϑ_n be the usual shift operator of r.w. paths and consider

$$\sigma_x \circ \vartheta_{\sigma_{B(R)}} = \inf\{n > \sigma_{B(R)} : S_n = x\},$$

the epoch when S visits x for the first time after exiting the interval $[0, R]$. The proof of the following lemma, crucial for our verification of the main results of this section, is essentially an adaptation of proof of Lemma 2 of Kesten [40].

Lemma 8.5.1 *Let $1 < \alpha \leq 2$. Then uniformly for $0 \leq x \leq R$, as $R \to \infty$,*

$$P_x[\sigma_x \circ \vartheta_{\sigma_{B(R)}} < T, \Lambda_R] = 1 - (1 - x/R)^{\alpha-1} + o(1). \tag{8.46}$$

Proof If $x/R \to 0$, then $P_x[\Lambda_R] \to 0$, and the asserted equality follows. The case $x/R \to 1$ is disposed of analogously [e.g., deduce $P_z[\sigma_x < T] \to 1$ as $x/z \to 1$ from Proposition 8.3.1(i)]. Hence we may suppose that $\varepsilon < x/R < (1 - \varepsilon)$ for an $\varepsilon > 0$. If $p = 0$ or $\alpha = 2$, it follows that $P_x(\Lambda_R) \sim V_d(x)/V_d(R)$ and Z is r.s., and hence by (8.25) $P_x[\sigma_x \circ \vartheta_{\sigma_{B(R)}} < T, \Lambda_R] \sim (x/R)^{\alpha\hat{\rho}}\mathfrak{h}_1(R/x)/\mathfrak{h}_1(1)$, which yields (8.46) since $\alpha\hat{\rho} = \alpha - 1$ and $\mathfrak{h}_1(\zeta)/\mathfrak{h}_1(1) = \zeta^{\alpha-1} - (\zeta - 1)^{\alpha-1}$.

Let $p > 0$ so that $\alpha\rho < 1$. Owing to Rogozin's result (2.28) the method of derivation in [40] applies to our setting. By virtue of our results obtained previously, we can bypass the first part of the proof in [40]: making the decomposition

$$P_x[\sigma_x \circ \vartheta_{\sigma_{B(R)}} < T, \Lambda_R] = \sum_{y=1}^{\infty} P_x[Z(R) = y; \Lambda_R]P_{R+y}[\sigma_x < T]$$

and putting $\xi = x/R$, we apply the second formula of (8.32) and (2.28) to see that the probability on the LHS is asymptotically equivalent to

$$(\alpha - 1)(1 - \xi)^{\alpha\rho}\xi^{\alpha\hat{\rho}}\pi^{-1}\sin\alpha\rho\pi\times \tag{8.47}$$

$$\times \int_0^{\infty} \frac{z^{-\alpha\rho}(1+z)^{-\alpha\hat{\rho}}}{z + 1 - \xi}\, dz \cdot \left(\frac{1+z-\xi}{\xi}\right)^{\alpha-1} \int_{\frac{1+z-\xi}{1+z}}^{1} w^{-\alpha}(1-w)^{\alpha\rho-1}\, dw.$$

Below, we duplicate the latter half of the proof in [40] with appropriate modifications. First, interchange the order of integration to have the repeated integral in the form

$$\xi^{-\alpha+1}\int_{1-\xi}^{1} w^{-\alpha}(1-w)^{\alpha\rho-1}\, dw \int_0^{\frac{w-1+\xi}{1-w}} z^{-\alpha\rho}(1+z)^{-\alpha\hat{\rho}}(z+1-\xi)^{\alpha-2}\, dz.$$

Note that $\pi^{-1}\sin\alpha\rho\pi = 1/\Gamma(\alpha\rho)\Gamma(1 - \alpha\rho)$. Then, by changing the variable $v = (z + 1 - \xi)[(z + 1)(1 - \xi)w]^{-1}$ for the inner integral, (8.47) becomes

$$\frac{(\alpha - 1)(1 - \xi)^{\alpha-1}}{\Gamma(\alpha\rho)\Gamma(1 - \alpha\rho)} \int_{1-\xi}^{1} w^{-\alpha}(1 - w)^{\alpha\rho-1} \, dw \int_{1/w}^{\frac{1}{1-\xi}} v^{\alpha-2} w^{\alpha-1}(wv - 1)^{-\alpha\rho} \, dv$$

$$= \frac{(\alpha - 1)(1 - \xi)^{\alpha-1}}{\Gamma(\alpha\rho)\Gamma(1 - \alpha\rho)} \int_{1}^{\frac{1}{1-\xi}} v^{\alpha-2} \, dv \int_{1/v}^{1} (1 - w)^{\alpha\rho-1} w^{-1}(vw - 1)^{-\alpha\rho} \, dw$$

$$= 1 - (1 - \xi)^{\alpha-1},$$

where for the last equality, we make the change of variable $w \mapsto t = (1 - w)[w(v - 1)]^{-1}$, which transforms the inner integral of the middle expression into $\int_0^\infty t^{-\alpha\rho}(1 + t)^{-1} \, dt = \Gamma(\alpha\rho)\Gamma(1 - \alpha\rho)$. This finishes the proof of the lemma. □

Lemma 8.5.2 *For all* $1 \leq \alpha \leq 2$ *and each* $\delta < 1$, *as* $R \to \infty$,

$$g_{B(R)}(x, x) = \left[(1 - x/R)^{\alpha-1} + o(1)\right] g_\Omega(x, x) \quad \text{uniformly for } 0 \leq x \leq \delta R.$$

Proof If $\alpha > 1$, this follows by substitution from Lemma 8.5.1 into the identity

$$g_\Omega(x, x) = g_{B(R)}(x, x) + P_x[\sigma_x \circ \vartheta_{\sigma_{B(R)}} < T, \Lambda_R] g_\Omega(x, x). \tag{8.48}$$

For $\alpha = 1$, the result is true without restriction on ρ (see (8.36)). □

Proof (of Theorem 8.4.1) The case $x < \delta R$ is immediate from (8.14) and Lemma 8.5.2. The case $x > \varepsilon R$ is its dual. □

By virtue of Theorem 8.4.1 the evaluation of $g_{B(R)}(x, y)$ follows from that of $P_x[\sigma_y < \sigma_{B(R)}]$. Dividing the identity (7.54), namely

$$\text{Eq(7.54)} \quad g_{B(R)}(x, y) = g_\Omega(x, y) - \sum_{z=1}^{\infty} P_x[S_{\sigma_{B(R)}} = R + z] g_\Omega(R + z, y),$$

by $g_{B(R)}(y, y)$ and passing to the limit with the help of Lemma 8.5.2, (8.25) and (2.28), we see that if $\alpha > 1$, there exists

$$u(\xi, \eta) := \lim_{x/R \to \xi, y/R \to \eta} P_x[\sigma_y < \sigma_{B(R)}], \tag{8.49}$$

and for $0 < \eta \leq \xi < 1$, the function $u(\xi, \eta)$ is written as

$$u(\xi, \eta) = \frac{\mathfrak{h}_1(\xi/\eta) - \mathfrak{h}_1(1/\eta)}{\mathfrak{h}_1(1)(1 - \eta)^{\alpha-1}} \quad \text{if } \alpha\rho = 1; \text{ and} \tag{8.50}$$

$$u(\xi, \eta) = \frac{\mathfrak{h}_{\alpha\rho}(\xi/\eta)}{\mathfrak{h}_{\alpha\rho}(1)(1 - \eta)^{\alpha-1}}$$

$$- k_{\xi,\eta} \int_0^\infty \frac{t^{-\alpha\rho}(1 + t)^{-\alpha\hat\rho}}{1 + t - \xi} \cdot \frac{\mathfrak{h}_{\alpha\rho}((1 + t)/\eta)}{\mathfrak{h}_{\alpha\rho}(1)} \, dt \quad \text{if } \alpha\rho < 1, \tag{8.51}$$

where $k_{\xi,\eta} = (1 - \eta)^{1-\alpha}(1 - \xi)^{\alpha\rho}\xi^{\alpha\hat\rho}\pi^{-1}\sin\alpha\rho\pi$. Note that for $\alpha > 1$,

$$\alpha\rho = 1 \iff \text{ either } p = 0 \text{ or } \alpha = 2 \iff Z \text{ is r.s.}$$

It seems hard to directly obtain a neat natural expression of the integral on the RHS of (8.51) unless $q = 0$, and we use the above result only to ensure that the limit exists, the convergence is uniform in $x \in (\eta R, \delta R)$ for each $0 < \eta < 1$ and $\delta < 1$ and the limit is continuous in $\xi \in (\eta, 1)$ (unless $\alpha\rho = 1$), and that as $x/R \to \xi$ and $y/R \to \eta$,

$$g_{B(R)}(x, y) \sim g_{B(R)}(y, y)u(\xi, \eta).$$

The method employed to prove the following proposition is also due to Kesten [40].

Proposition 8.5.3 *Let* $1 < \alpha \leq 2$ *and* $u(\xi, \eta)$ *be defined by (8.49). Then for* $0 < \eta < \xi < 1$,

$$u(\xi, \eta) = (\alpha - 1)\left(\frac{\xi - \eta}{\eta(1 - \eta)}\right)^{\alpha-1}\int_0^{\frac{\eta(1-\xi)}{\xi(1-\eta)}}(1 - w)^{-\alpha}w^{\alpha\rho-1}\,dw. \qquad (8.52)$$

Proof Fix r to be any number in $(\eta, 1)$, then putting $R' = \lfloor rR \rfloor$, we decompose

$$g_{B(R)}(y, y) = g_{B(R')}(y, y) + \sum_{x=R'+1}^{R} P_y[Z(R') = x - R', \Lambda_{R'}]g_{B(R)}(x, y). \qquad (8.53)$$

Note that by Theorem 8.4.1, as $y/R \to \eta$,

$$\frac{g_{B(R')}(y, y)}{g_{B(R)}(y, y)} \sim \left[\frac{1 - y/R'}{1 - y/R}\right]^{\alpha-1}.$$

Let $\alpha\rho < 1$. Divide both sides of (8.53) by $g_{B(R)}(y, y)$ and pass to the limit as $R \to \infty$ with the comment given after (8.51) taken into account. Then, using (2.28) we see that

$$1 - \left(\frac{1 - \eta/r}{1 - \eta}\right)^{\alpha-1} = \frac{\sin\alpha\rho\pi}{\pi}(r - \eta)^{\alpha\rho}\eta^{\alpha\hat\rho}\int_r^1\frac{(\xi - r)^{-\alpha\rho}\xi^{-\alpha\hat\rho}}{\xi - \eta}u(\xi, \eta)\,d\xi, \qquad (8.54)$$

or what is the same thing,

$$\frac{\pi\eta^{-\alpha\hat\rho}}{\sin\alpha\rho\pi}\left(\frac{1}{(r - \eta)^{\alpha\rho}} - \frac{(r - \eta)^{\alpha\hat\rho-1}}{[(1 - \eta)r]^{\alpha-1}}\right) = \int_r^1\left[\frac{\xi^{-\alpha\hat\rho}}{\xi - \eta}u(\xi, \eta)\right](\xi - r)^{-\alpha\rho}\,d\xi.$$

This may be viewed as Abel's integral equation, to be solved for $\frac{\xi^{-\alpha\hat\rho}}{\xi-\eta}u(\xi, \eta)$. For the computation of the solution to be tractable, we introduce the new variables $t = \xi - \eta$ and $s = r - \eta$, which transforms the above equation into[7]

---

[7] This seemingly circuitous step simplifies the computation in the next step.

$$\frac{\pi}{\sin \alpha \rho \pi} \left(\frac{\eta^{-\alpha \hat{\rho}}}{s^{\alpha \rho}} - \frac{\eta^{-\alpha \hat{\rho}} s^{\alpha \hat{\rho}-1}}{[(1-\eta)(s+\eta)]^{\alpha-1}} \right) = \int_s^{1-\eta} \left[\frac{u(t+\eta, \eta)}{(t+\eta)^{\alpha \hat{\rho}} t} \right] \frac{dt}{(t-s)^{\alpha \rho}}.$$

The standard formula for the solution yields[8]

$$\frac{(t+\eta)^{-\alpha \hat{\rho}}}{t} u(t+\eta, \eta) = \frac{d}{dt} \int_t^{1-\eta} \left(\frac{\eta^{-\alpha \hat{\rho}}}{s^{\alpha \rho}} - \frac{\eta^{-\alpha \hat{\rho}} s^{\alpha \hat{\rho}-1}}{[(1-\eta)(s+\eta)]^{\alpha-1}} \right) (s-t)^{\alpha \rho-1} \, ds.$$

Changing the variable $s = vt$ transforms the derivative on the RHS into

$$\eta^{-\alpha \hat{\rho}} \frac{d}{dt} \int_1^{(1-\eta)/t} \left(\frac{1}{v^{\alpha \rho}} - \frac{v^{\alpha \hat{\rho}-1}}{[(1-\eta)(v+\eta/t)]^{\alpha-1}} \right) (v-1)^{\alpha \rho-1} \, dv$$

$$= -\frac{(\alpha-1)\eta^{1-\alpha \hat{\rho}}}{(1-\eta)^{\alpha-1} t^2} \int_1^{(1-\eta)/t} \frac{v^{\alpha \hat{\rho}-1}}{(v+\eta/t)^{\alpha}} (v-1)^{\alpha \rho-1} \, dv.$$

Hence, on going back to the variable $\xi = t + \eta$,

$$\frac{u(\xi, \eta)}{\xi^{\alpha \hat{\rho}}(\xi-\eta)} = \left[\frac{(\alpha-1)\eta^{1-\alpha \hat{\rho}}}{(1-\eta)^{\alpha-1} t^2} \int_1^{(1-\eta)/t} \frac{v^{\alpha \hat{\rho}-1}}{(v+\eta/t)^{\alpha}} (v-1)^{\alpha \rho-1} \, dv \right]_{t=\xi-\eta}.$$

Introduce $w = \eta(v-1)/\xi v$. Then $w|_{v=(1-\eta)/(\xi-\eta)} = \eta(1-\xi)/[\xi(1-\eta)]$ and for $t = \xi - \eta$, the last integral in the square brackets above becomes

$$\left(\frac{\xi-\eta}{\xi} \right)^{\alpha} \left(\frac{\xi}{\eta} \right)^{\alpha \rho} \int_0^{\frac{\eta(1-\xi)}{\xi(1-\eta)}} (1-w)^{-\alpha} w^{\alpha \rho-1} \, dw,$$

and substitution leads to

$$u(\xi, \eta) = (\alpha-1)(\xi-\eta)^{\alpha-1} [\eta(1-\eta)]^{-\alpha+1} \int_0^{\frac{\eta(1-\xi)}{\xi(1-\eta)}} (1-w)^{-\alpha} w^{\alpha \rho-1} \, dw.$$

Consequently, we have the formula of Proposition 8.5.3.

For the case $\alpha \rho = 1$, one may proceed as above with the arguments much simplified. However, by computing the RHS of (8.50), one can rather directly deduce that $u(\xi, \eta) = [\xi/\eta]^{\alpha-1} - [|\xi - \eta|/\eta(1-\eta)]^{\alpha-1}$, which coincides with (8.52). □

Note that as $y/R \to \eta \in (0, 1)$, $\bar{a}(y) \sim \bar{a}(R)\eta^{\alpha-1}$, hence

$$g_{B(R)}(y, y) \sim 2\kappa_{\alpha, \rho} \bar{a}(R)[\eta(1-\eta)]^{\alpha-1}.$$

---

[8] Let $c > b \geq 0$, $0 < \lambda < 1$ and $f(t)$ be a continuous function of $t \in (b, c]$. Then the transform $g(s) = \int_s^c f(t)(t-s)^{-\lambda} \, dt$ ($b < s \leq c$) is inverted by

$$f(t) = -[\pi^{-1} \sin \lambda \pi] \left(\frac{d}{dt} \right) \int_t^c g(s)(s-t)^{\lambda-1} \, ds,$$

as one sees by observing $\int_t^c (s-t)^{\lambda-1} \, ds \int_s^c f(r)(r-s)^{-\lambda} \, dr = B(\lambda, 1-\lambda) \int_t^c f(r) \, dr$.

Since, by considering the dual r.w., the explicit form of $u(\xi, \eta)$ for $\xi < \eta$ is obtained from that for $\eta < \xi$, we have the following corollary of Proposition 8.5.3.

Corollary 8.5.4 *Let* $1 < \alpha \le 2$. *For each* $\varepsilon > 0$, *as* $R \to \infty$,

$$g_{B(R)}(x, y) \sim 2(\alpha - 1)\kappa_{\alpha,\rho}\bar{a}(R)K_{\rho}^{(\alpha)}(x/R, y/R)$$

uniformly for $x, y \in [\varepsilon R, (1 - \varepsilon)R]$.

Now we turn to the case $\alpha = 1$, where we must find a result corresponding to (8.46), and define $u(\xi, \eta)$ differently from (8.49). Recall that the condition (8.6) implies $0 < \rho < 1$ and ensures that the strong renewal theorem holds for u_a and v_d.

Lemma 8.5.5 *Let* $\alpha = 1$ *and* (8.6) *hold. Then for each* $\varepsilon > 0$, *uniformly for* $\varepsilon R < x < (1 - \varepsilon)R$, *as* $R \to \infty$,

$$P_x[\sigma_x \circ \vartheta_{\sigma_{B(R)}} < T \circ \vartheta_{\sigma_{B(R)}}, \Lambda_R] \sim \frac{\kappa \rho \hat{\rho}}{a(x)L(x)} \log \frac{1}{1 - x/R}.$$

The above equivalence entails (in fact is equivalent to)[9]

$$g_{B(R)}(x, x) = g_{\Omega}(x, x)\left(1 - \frac{\kappa \rho \hat{\rho} + o(1)}{a(x)L(x)} \log \frac{1}{1 - x/R}\right). \tag{8.55}$$

Proof The proof proceeds in the same way as that of Lemma 8.5.1. Because of (8.27) and (8.7) we have in place of (8.47)

$$\frac{\kappa \rho \hat{\rho}(1 - \xi)^{\rho}\xi^{\hat{\rho}} \sin \rho\pi}{a(x)L(x)\pi} \int_0^{\infty} \frac{z^{-\rho}(1 + z)^{-\hat{\rho}}}{z + 1 - \xi} \, dz \int_{\frac{1+z-\xi}{1+z}}^1 w^{-1}(1 - w)^{\rho-1} \, dw. \tag{8.56}$$

The same procedure as before transforms the repeated integral above into

$$(1 - \xi)^{-\rho}\xi^{-\hat{\rho}} \int_1^{\frac{1}{1-\xi}} v^{-1} \, dv \int_{1/v}^1 (1 - w)^{\rho-1}w^{-1}(vw - 1)^{-\rho} \, dw,$$

of which the inner integral equals $\pi/\sin \rho\pi$. The rest is easy. $\square$

Proposition 8.5.6 *Let* $\alpha = 1$ *and* (8.6) *hold. Then for each* $0 < \eta < \xi < 1$, *there exists*

$$u(\xi, \eta) := \frac{1}{\kappa \rho \hat{\rho}} \lim_{(x/R, y/R) \to (\xi, \eta)} L(R)g_{B(R)}(x, y), \tag{8.57}$$

and

$$u(\xi, \eta) = \int_0^{\frac{\eta(1-\xi)}{\xi(1-\eta)}} (1 - w)^{-1}w^{\rho-1} \, dw = \mathfrak{h}_{\rho}^{(1)}\left(\frac{\xi(1 - \eta)}{\eta(1 - \xi)}\right). \tag{8.58}$$

---

[9] Note $E_x[g_{\Omega}(S_{\sigma(R,\infty)}, y); \Lambda_R] = P_x[\sigma_y \circ \vartheta_{\sigma_{B(R)}} < T \circ \vartheta_{\sigma_{B(R)}}, \Lambda_R]g_{\Omega}(y, y)$ in (7.54).

Proof The existence of the limit defining $u(\xi, \eta)$ follows from (8.10) (because of the identity (7.54) and (2.28)) as in case $\alpha > 1$. Let $\eta < r < 1$ and $R' = \lfloor rR \rfloor$ as in the proof of Proposition 8.5.3. Apply (8.55) and (8.11) in turn to deduce

$$g_{B(R)}(y, y) - g_{B(R')}(y, y) \sim g_\Omega(y, y) \frac{\kappa \rho \hat{\rho}}{a(y)L(y)} \log \frac{1 - y/R}{1 - y/R'}$$

$$\sim \frac{\kappa \rho \hat{\rho}}{L(R)} \log \frac{1 - \eta}{1 - \eta/r},$$

and by using (8.53), (8.57) and (2.28), rewrite the difference on the LHS by means of $u(\xi, \eta)/L(R)$. Then, dividing what ensues by $\kappa \rho \hat{\rho}/L(R)$, passing to the limit, and using the above equivalence we get

$$\log \frac{1 - \eta}{1 - \eta/r} = \frac{\sin \rho \pi}{\pi} (r - \eta)^\rho \eta^{\hat{\rho}} \int_r^1 \frac{(\xi - r)^{-\rho} \xi^{-\hat{\rho}}}{\xi - \eta} u(\xi, \eta) \, d\xi.$$

On substituting $r = s + \eta$ and $\xi = t + \eta$ this identity is rewritten as

$$\frac{\pi}{\sin \rho \pi} \cdot \eta^{-\hat{\rho}} s^{-\rho} \log \frac{1 - \eta}{1 - \eta/(s + \eta)} = \int_s^{1-\eta} \left[\frac{(t + \eta)^{-\hat{\rho}}}{t} u(t + \eta, \eta) \right] (t - s)^{-\rho} \, dt.$$

We regard this relation as Abel's integral equation and write down the standard formula for the solution as before, and substitute $\xi - \eta$ for t, which results in

$$\frac{u(\xi, \eta)}{\eta^{-\hat{\rho}} \xi^{\hat{\rho}} (\xi - \eta)} = - \left[\frac{d}{dt} \int_t^{1-\eta} \log \left[(1 - \eta)(1 + \eta s^{-1}) \right] s^{-\rho} (s - t)^{\rho-1} \, ds \right]_{t=\xi-\eta}.$$

Changing the variables, first by $v = s/t$ and then by $w = \eta(v - 1)/\xi v$, one infers that the RHS above is given as

$$- \left[\frac{d}{dt} \int_1^{(1-\eta)/t} \log \left[(1 - \eta) \left(1 + \frac{\eta}{vt} \right) \right] v^{-\rho} (v - 1)^{\rho-1} \, dv \right]_{t=\xi-\eta}$$

$$= \left[\frac{\eta}{t^2} \int_1^{(1-\eta)/t} \frac{v^{-\rho} (v - 1)^{\rho-1}}{v + \eta/t} \, dv \right]_{t=\xi-\eta}$$

$$= \frac{1}{\xi - \eta} \left(\frac{\xi}{\eta} \right)^{\rho-1} \int_0^{\frac{\eta(1-\xi)}{\xi(1-\eta)}} (1 - w)^{-1} w^{\rho-1} \, dw.$$

Finally, the identity $\rho - 1 = -\hat{\rho}$ concludes the proof of (8.58).   □

Observe

$$K_\rho^{(1)}(\xi, \eta) = \mathfrak{h}_\rho^{(1)} \left(\frac{\eta(1 - \xi)}{\xi(1 - \eta)} \right) \quad (\xi < \eta); \quad = \mathfrak{h}_\rho^{(1)} \left(\frac{\xi(1 - \eta)}{\eta(1 - \xi)} \right) \quad (\eta > \xi). \quad (8.59)$$

Then, because of $2\kappa_{\alpha,\rho} \bar{a}(x) \sim \kappa \rho \hat{\rho}[1/L]_*(x)$ ((8.41)) and $\kappa \rho \hat{\rho} = 2(\pi^{-1} \sin \rho \pi)^2$, $U_a(x)V_d(x) \sim \kappa x/L(x)$ (Lemma 6.5.1), we have the following corollary.

Corollary 8.5.7 *Let* $\alpha = 1$, $0 < \rho < 1$ *and assume* (8.6) *holds. Then for each* $\varepsilon > 0$, *as* $x/R \to \xi$ *and* $y/R \to \eta$,

$$g_{B(R)}(x,y) \sim \frac{2(\pi^{-1}\sin\rho\pi)^2}{[R\mu(R)]^{-1}}K_\rho^{(1)}(\xi,\eta) \sim \frac{\rho\hat\rho U_a(R)V_d(R)}{R}K_\rho^{(1)}(\xi,\eta) \quad and$$

$$P_x[\sigma_y < \sigma_{B(R)}] \sim \frac{\kappa\rho\hat\rho\, K_\rho^{(1)}(\xi,\eta)}{L(R)a(R)} \sim \frac{K_\rho^{(1)}(\xi,\eta)}{L(R)[1/L]_*(R)}$$

uniformly for $\xi, \eta \in [\varepsilon, (1-\varepsilon)]$ *such that* $|\eta - \xi| > \varepsilon$.

8.6 Asymptotics of $g_{B(R)}(x,y)$ Near the Boundary

Here we extend the result of the preceding section to the case $x/R \downarrow 0$ and/or $y/R \uparrow 1$. Recall that ℓ and $\hat\ell$ are specified so that $U_a(x) = (x \vee 1)^{\alpha\rho}/\ell(x)$ and $V_a(x) = (x \vee 1)^{\alpha\hat\rho}/\hat\ell(x)$ for all $x = 0, 1, 2, \ldots$ In the preceding section ξ and η have denoted the limits of x/R and y/R, whereas in this section – as in Section 8.2 – they denote the scaled variables themselves, namely

$$\xi = x/R \quad and \quad \eta = y/R$$

(with one exception in the argument leading to (8.70) below). We let $1 \le \alpha \le 2$ throughout. By duality, we have only to consider the case $x \le y$.

Lemma 8.6.1 *If* $\alpha = 2$, *then for each* $\delta < 1$, *as* $y \to \infty$,

$$g_{B(R)}(x,y) \sim V_d(x)\frac{1-\eta}{\ell(y)} \quad uniformly for\ 0 \le x \le y < \delta R. \tag{8.60}$$

Proof Let $0 \le x \le y < \delta R$. If $y/R \downarrow 0$, then $g_{B(R)}(x,y) \sim g_\Omega(x,y)$ and the result follows from Theorem 8.2.1 since for $\alpha = 2$, $\alpha\rho = \alpha\hat\rho = 1$ and $\mathfrak{h}_1 \equiv 1$. Let $y > \varepsilon_1 R$ with a constant $\varepsilon_1 > 0$. Then it follows that

$$\forall \varepsilon > 0, \quad \lim_{y\to\infty} P_x[Z(y) < \varepsilon y, \Lambda_y] = \frac{V_d(x)}{V_d(y)} \quad (uniformly for\ 0 \le x < y)$$

(Lemma 5.2.4) and $\inf_{z:y \le z < (1+\varepsilon)y} P_z[\sigma_y < \sigma_{B(R)}] \to 1$ as $R \to \infty$ and $\varepsilon \downarrow 0$ uniformly for $y \in (\varepsilon_1 R, \delta R)$. Therefore $P_x[\sigma_y < \sigma_{B(R)}] \sim V_d(x)/V_d(y)$, and by Theorem 8.4.1 (having been shown in the preceding section)

$$g_{B(R)}(x,y) \sim \frac{V_d(x)}{V_d(y)}g_{B(R)}(y,y) \sim \frac{V_d(x)\hat\ell(y)}{y}2\kappa_{2,\frac12}\bar a(y)(1-\eta).$$

Since $\kappa_{2,\frac12} = 1$ and $2\bar a(y) \sim y/m(y) \sim y^{-1}U_a(y)V_d(y) \sim y/\ell(y)\hat\ell(y)$, this yields the asserted equivalence. $\qquad\square$

Lemma 8.6.2 *If $\alpha = 2$, for $0 \leq x \leq y$,*

$$g_{B(R)}(x, y) \sim \frac{V_d(x)V_d(R - y)}{\ell(y)V_d(R)} \qquad as \; \xi \vee (1 - \eta) \to 0. \qquad (8.61)$$

Proof By the identity $g_{B(R)}(x, y) = g_{B(R)}(R - y, R - x)$, the equivalence in (8.60) is rephrased as

$$g_{B(R)}(x, y) \sim \frac{V_d(R - y)x}{\ell(R - x)R} \qquad as \; R - x \to \infty \qquad (8.62)$$
$$\text{uniformly for } \varepsilon R < x \leq y \leq R.$$

Put $R' = \lfloor R/2 \rfloor$. Then the lemma follows if we can show

$$g_{B(R)}(x, y) \sim P_x(\Lambda_{R'})g_{B(R)}(R', y) \qquad as \; \xi \vee (1 - \eta) \to 0. \qquad (8.63)$$

Indeed, recalling $P_x(\Lambda_y) \sim V_d(x)/V_d(y)$ for $\alpha = 2$, one infers from (8.62) that the RHS of (8.63) is written as $V_d(x)V_d(R - y)/[\ell(R)V_d(R)]\{1 + o(1)\}$. Thus the equivalence (8.61) follows from (8.63). To verify (8.63) we use the representation

$$g_{B(R)}(x, y) = \sum_{z=1}^{R-R'} P_x[Z(R') = z; \Lambda_{R'}]\, g_{B(R)}(R' + z, y).$$

For any $\varepsilon > 0$, $P_x[Z(R') > \varepsilon R \,|\, \Lambda_{R'}] \to 0$ and $g_{B(R)}(R' + z, y)$ has a nice regularity for $0 \leq z \leq \frac{1}{2}R'$ in view of Proposition 8.5.3, whereas the ratio $g_{B(R)}(y, y)/g_{B(R)}(R', y)$ tends to infinity. This means that we only have to show that

$$\sum_{z=\frac{1}{2}R'}^{R-R'} P_x[Z(R') = z; \Lambda_{R'}]\, g_{B(R)}(R' + z, y) = o\left(\frac{V_d(x)V_d(R - y)}{\ell(R)V_d(R)}\right). \qquad (8.64)$$

Writing down $P_x[Z(R') = z; \Lambda_{R'}]$ by means of $g_{B(R)}$ and $p(\cdot)$, one rewrites the above sum as

$$\sum_{z=\frac{1}{2}R'}^{R-R'} \sum_{w=0}^{R'} g_{B(R')}(x, w)p(R' - w + z)g_{B(R)}(R' + z, y). \qquad (8.65)$$

Introducing the variable $w' = R - R' - z$ one further rewrites this double sum as

$$\sum_{w'=0}^{\frac{1}{2}R'} \sum_{w=0}^{R'} g_{B(R')}(x, w)g_{B(R)}(R - y, w')p(R - w - w').$$

Noting that $\mathfrak{h}_1^{(2)}(\cdot) \equiv 1$, from Theorem 8.2.1(i) we infer

$$g_{\Omega}(x, w) \leq V_d(x)/\ell(w)\{1 + o(1)\} \qquad (x \geq 0, w \geq 1) \qquad (8.66)$$

$(o(1) \to 0$ as $x \vee w \to \infty)$, and after changing variables $w' = r - w$, we accordingly see that the double sum (8.65) is bounded above by a constant multiple of

$$V_d(x)V_d(R - y) \sum_{r=0}^{\frac{3}{2}R'} \sum_{w=0}^{r} \frac{p(R - r)}{\ell(w)\ell(r - w)}.$$

Observe that

$$\sum_{r=0}^{\frac{3}{2}R'} \sum_{w=0}^{r} \frac{p(R - r)}{\ell(w)\ell(r - w)} \leq C \sum_{r=0}^{\frac{3}{2}R'} \frac{p(R - r)r}{\ell^2(r)} \leq C' \frac{\mu_+(\frac{1}{4}R)R'}{\ell^2(R)} \leq C' \frac{\mu(\frac{1}{4}R)U_a(R)}{\ell(R)}.$$

Then, using $U_a(R)\mu(R) = o\,(1/V_d(R))$ (Lemma 6.5.1(i)) we can conclude (8.64), hence (8.63). This finishes the proof of the lemma.   □

Completion of the proof of Theorem 7.1.10(ii). Here we resume our unfinished proof. Let F be p.r.s. We verify (8.64), sufficient for the proof of (7.30), hence of Theorem 7.1.10(ii). To this end, we follow the proof above. Under the dual of (7.7), $A(x) \sim \ell^*(x)\ell_\sharp(x)$ and, according as F is recurrent or transient, $a(x)$ or $G(x) \asymp 1/A(x)$, hence

(*) $a(x) \leq C_0V_d(x)/\ell^*(x)$ or $G(x) \leq C_0V_d(x)/\ell^*(x)$.

Because $g_{B(R)}(x, y) \sim g_\Omega(x, y)$ for $0 \leq x \leq y < \frac{3}{4}R$ (Lemma 5.7.3), $P_x[Z(R') > \varepsilon R' \,|\, \Lambda_{R'}] \to 0$ and $P_x(\Lambda_{R'}) \sim V_a(x)/V_d(R)$, our task is reduced to showing (8.64) under (*) as well as under the condition $p(x) = O\,(x/\mu(x))$.

Let (*) hold. If F is recurrent it follows that $g_\Omega(x, y) \leq V_d(x)/\ell^*(y)\{C_0 + o(1)\}$ $(x \geq 0, y \geq 1)$. We also have $U_a(x)\mu(x) = o\,(1/V_d(x))$, F being r.s. Hence the proof above applies. If F is transient, by using $g_\Omega(x, y) \leq G(y - x)$ and condition (*) in turn observe that $\sum_{w=0}^{2x}[g_\Omega(x, w)]^2 \leq C_1xG^2(x) \leq C_1'x\,[V_d(x)/\ell^*(x)]^2$. Then from the inequality $g_\Omega(x, w) \leq C_2V_d(x)/\ell^*(w)$, $w > 2x$ we infer that

$$\sum_{w=0}^{r} [g_\Omega(x, w)]^2 \leq C_1' \frac{V_d^2(x)x}{[\ell^*(x)]^2} + C_2' \frac{V_d^2(x)r}{[\ell^*(r)]^2}$$

$$\leq C \frac{V_d^2(x)R}{[\ell^*(R)]^2} \qquad (0 < r \leq R').$$

The sum of $[g_\Omega(R - y, r - w)]^2$ admits a similar bound. Hence by the Schwarz inequality and $U_a(x) \sim x/\ell^*(x)$ we obtain

$$\sum_{w=0}^{r} g_\Omega(x, w)g_\Omega(R - y, r - w) \leq C \frac{V_d(x)V_d(R - y)U_a(R)}{\ell^*(R)},$$

which shows (8.64) in a similar way to the case $\alpha = 2$.

If $p(x) = O(x/\mu(x))$, then as in the last part of proof of Lemma 8.6.2 we infer that the double sum in (8.65) is at most a constant multiple of

$$\frac{\mu(\frac{1}{4}R)}{R} \sum_{z=\frac{1}{2}R'}^{R-R'} \sum_{w=0}^{R'} g_{B(R')}(x,w) g_{B(R)}(R'+z,y) = o\left(\frac{V_d(x)V_d(R-y)}{V_d(R)\ell^*(R)}\right).$$

Thus (8.64) is verified. The proof of Theorem 7.1.10(ii) is now complete. □

The following two lemmas and proposition, valid for all $1 \leq \alpha \leq 2$, are derived from the estimates of $g_\Omega(x,y)$ given in Section 8.2, not depending on the results of Section 8.4, and we provide proofs of them only for $1 < \alpha < 2$, the same proofs applying to the case $\alpha = 1$ with (8.6) satisfied.

Lemma 8.6.3 Let $1 \leq \alpha \leq 2$ and (8.6) hold if $\alpha = 1$. Then for each $\frac{1}{2} < \delta < 1$, uniformly for $1 - \delta < \xi < \delta$, as $\eta \uparrow 1$,

$$g_{B(R)}(x,y) \sim \alpha\rho(1-\xi)^{\alpha\rho-1}\xi^{\alpha\hat{\rho}} [U_a(R)/R] V_d(R-y);$$

and uniformly for $1 - \delta < \eta < \delta$, as $\xi \downarrow 0$,

$$g_{B(R)}(x,y) \sim \alpha\rho\eta^{\alpha\rho-1}(1-\eta)^{\alpha\hat{\rho}} [U_a(R)/R] V_d(x).$$

Proof If $\alpha = 2$, this follows from (8.60) (see (8.62)).

Let $1 < \alpha < 2$. By $g_{B(R)}(x,y) = g_{B(R)}(R-y,R-x)$, we have only to show the first half. Taking a positive constant $\varepsilon < (1-\delta)/2$, we put $R_1 = \lfloor (1-\varepsilon)R \rfloor$, so that

$$g_{B(R)}(x,y) = \sum_{z=1}^{R-R_1} P_x[Z(R_1) = z, \Lambda_{R_1}] g_{B(R)}(R_1+z,y) \quad (x < R_1). \quad (8.67)$$

Writing $w = R - y$ and $R_2 = R - R_1 (\sim \varepsilon R)$, by Lemma 8.2.4 we see that as $\eta \uparrow 1$,

$$g_{B(R)}(R_1+z,y) = g_{B(R)}(w, R_2-z) = g_\Omega(w, R_2-z)\{1-o_\varepsilon(1)\}, \quad (8.68)$$

where $o_\varepsilon(1) \to 0$ as $\varepsilon \downarrow 0$ (uniformly for $R_1 + z$).

First suppose $p > 0$, so that $\alpha\rho < 1$. Then for $1 - \delta < \xi < \delta$ and $y < R$, using (2.28) together with (8.8) of Theorem 8.2.1 and putting $r = R_1/R$ and $s = (w \vee 1)/R$, we have

$$\sum_{z=1}^{R_2} P_x[Z(R_1) = z, \Lambda_{R_1}] g_\Omega(w, R_2-z)$$

$$\sim \frac{(\sin\alpha\rho\pi)(r-\xi)^{\alpha\rho}\xi^{\alpha\hat{\rho}}}{\pi} \cdot \frac{\alpha^2\rho\hat{\rho}V_d(w)}{[w \vee 1]^{1-\alpha\rho}} \int_0^{\varepsilon-s} \frac{t^{-\alpha\rho}(t+r)^{-\alpha\hat{\rho}}}{t+r-\xi} \cdot \frac{\mathfrak{h}_{\alpha\hat{\rho}}([\varepsilon-t]/s)}{\ell(R(\varepsilon-t))} dt,$$

where we have used the fact that the contribution from $R_2 - w < z < R_2$ to the sum on the LHS is negligible (as is verified by an argument similar to that made below). Note $\mathfrak{h}_{\alpha\hat{\rho}}([\varepsilon-t]/s) \sim (\alpha\hat{\rho})^{-1}([\varepsilon-t]/s)^{\alpha\hat{\rho}-1}$ as $(\varepsilon-t)/s \to \infty$. Then we see that for $s < \varepsilon^2$, the above integral restricted to $[\varepsilon^2, \varepsilon-s]$, of which the remainder is negligible in comparison, is asymptotically equivalent to

$$\frac{1}{\alpha\hat\rho\,\ell(R)}\int_{\varepsilon^2}^{\varepsilon-s}\frac{t^{-\alpha\rho}(t+r)^{-\alpha\hat\rho}}{t+r-\xi}\left(\frac{\varepsilon-t}{s}\right)^{\alpha\rho-1}dt$$

$$=\frac{1}{\alpha\hat\rho\,\ell(R)}\int_{\varepsilon}^{1-s/\varepsilon}\frac{t^{-\alpha\rho}(\varepsilon t+r)^{-\alpha\hat\rho}}{\varepsilon t+r-\xi}\left(\frac{1-t}{s}\right)^{\alpha\rho-1}dt$$

$$=\frac{s^{1-\alpha\rho}\{\pi+o_\varepsilon(1)\}}{\ell(R)(1-\xi)\alpha\hat\rho\sin\alpha\rho\pi}$$

as $\varepsilon\to0$. Substituting the last expression, taking (8.68) into account and passing to the limit as $y/R\uparrow1$ and $\varepsilon\downarrow0$ in turn, one can easily deduce

$$g_{B(R)}(x,y)\sim\alpha\rho(1-\xi)^{\alpha\hat\rho-1}\xi^{\alpha\hat\rho}V_d(R-y)R^{\alpha\rho-1}/\ell(R),$$

which is the same as the formula of the lemma.

If $p=0$, then $P_x[Z(R_1)/R_1\in\cdot\,|\Lambda_{R_1}]$ ($R_1=\lfloor(1-\varepsilon)R\rfloor$) approaches the delta measure at zero, $P_x(\Lambda_{R_1})\sim V_d(x)/V_d(R_1)\sim[\xi/(1-\varepsilon)]^{\alpha\hat\rho}$ and $g_\Omega(w,R-R_1)\sim V_d(w)/\ell(R)$. By (8.67) and (8.68) it therefore follows that

$$g_{B(R)}(x,y)=\xi^{\alpha\hat\rho}\frac{V_d(R-y)}{\ell(R)}\{1+o(\varepsilon)\},$$

which leads to the asserted formula by letting $y/R\uparrow1$ and $\varepsilon\downarrow0$ in turn.   □

The following proposition refines Rogozin's result (2.28) in the case $x/R\to0$.

Proposition 8.6.4 *Let* $1\le\alpha\le2$.

(i) *If* $\alpha\rho=1$ *(equivalently, either* $p=0$, *or* $\alpha=\rho=1$ *or* $\alpha=2$), *then for each* $\varepsilon>0$, $P_x[Z(R)>\varepsilon R\,|\Lambda_R]\to0$ $(R\to\infty)$ *uniformly for* $0\le x\le R$.
(ii) *Suppose that* $\alpha\rho<1$ *and (8.6) holds if* $\alpha=1$. *Then, as* $x/R\to0$,

$$P_x[Z(R)\ge y,\Lambda_R]\sim\frac{\sin\alpha\rho\pi}{\pi}\int_\eta^\infty\frac{dt}{t^{\alpha\rho}(1+t)^{\alpha\hat\rho+1}}\cdot\frac{V_d(x)}{V_d(R)}\qquad(\eta\ge0);$$

in particular, as $x/R\to0$,

$$P_x(\Lambda_R)\sim[\alpha\hat\rho B(\alpha\rho,\alpha\hat\rho)]^{-1}V_d(x)/V_d(R).\qquad(8.69)$$

Proof We first note that (i) is a special case of Lemma 5.2.4.
Let $1<\alpha<2$ and $p>0$ ($\Rightarrow0<\alpha\rho<1$). Then, using the second formula of Lemma 8.6.3, one infers that as $x/R\downarrow0$,

$$P_x[Z(R)>y,\Lambda_R]=\sum_{z=0}^R g_{B(R)}(x,z)\mu_+(R-z+y)$$

$$\sim\alpha\rho V_d(x)\int_0^1(1-t)^{\alpha\hat\rho}t^{\alpha\rho-1}\frac{p\,U_a(R)L(R)}{[(1-t+\eta)R]^\alpha}dt.$$

By Lemma 6.5.1 $U_a(R)L(R)/R^\alpha \sim \kappa R^{-\alpha\hat\rho}\hat\ell(R) = \kappa/V_d(R)$. Thus

$$P_x[Z(R) > y, \Lambda_R] \sim [p\alpha\rho\kappa]\varphi(\eta)\frac{V_d(x)}{V_d(R)}, \quad \text{where} \quad \varphi(s) = \int_0^1 \frac{(1-t)^{\alpha\hat\rho}t^{\alpha\rho-1}}{(1+s-t)^\alpha}\,dt.$$

We have the corresponding result (2.28) for the limit as $x/R \to \xi > 0$, which must be the same as the one obtained from Corollary 8.5.4 as above and by comparing the asymptotic forms of the two integrals as $\xi \downarrow 0$ with the help of $p\alpha\rho\kappa = \pi^{-1}\sin\alpha\rho\pi \, [\alpha\hat\rho B(\alpha\rho, \alpha\hat\rho)]^{-1}$ we see that

$$\varphi(s) = \frac{\Gamma(\alpha\rho)\Gamma(1+\alpha\hat\rho)}{\Gamma(\alpha)} \int_s^\infty \frac{dt}{t^{\alpha\rho}(1+t)^{\alpha\hat\rho+1}}, \tag{8.70}$$

which yields the asserted equivalence of (i). Here we provide a direct verification of this identity. First, observe that the identities $\int_0^\infty t^{-\alpha\rho}(1+t)^{-\alpha\hat\rho-1}\,dt = B(\alpha, 1-\alpha\rho)$ and $\varphi(0) = \Gamma(1-\alpha\rho)\Gamma(\alpha\rho)$ together verify that the equality (8.70) holds for $s = 0$. It suffices to prove the equality for the derivatives. By the change of variable $w = [s/(s+1)]t/(1-t)$ we deduce

$$\varphi'(s) = -\alpha \int_0^1 \frac{(1-t)^{\alpha\hat\rho}t^{\alpha\rho-1}}{(1+s-t)^{\alpha+1}}\,dt$$

$$= \frac{-\alpha}{s^{\alpha\rho}(1+s)^{\alpha\hat\rho+1}} \int_0^\infty w^{\alpha\rho-1}(1+w)^{-\alpha-1}\,dw,$$

where the last integral equals $B(\alpha\rho, \alpha\hat\rho + 1)$, showing that $\varphi'(s)$ agrees with the derivative of the RHS of (8.70).

The proof in the case $\alpha = 1$ is omitted, as noted previously. $\square$

Lemma 8.6.5 *Let* $1 \le \alpha \le 2$ *and* (8.6) *hold if* $\alpha = 1$. *Then as* $x/R \downarrow 0$ *and* $y/R \uparrow 1$,

$$g_{B(R)}(x, y) \sim \frac{\alpha\rho V_d(x)V_d(R-y)}{V_d(R)\ell(y)}R^{\alpha\rho-1}$$

$$\sim \frac{\alpha\rho \, R^{\alpha-1}\hat\ell(R)}{\hat\ell(x)\ell(y)\hat\ell(R-y)}\left(\frac{(x \vee 1)([R-y] \vee 1)}{R^2}\right)^{\alpha\rho}.$$

Proof For $\alpha = 2$, the formula above is the same as that of Lemma 8.6.2. If $p = 0$ and $1 < \alpha < 2$, then $V_d(x) = x^{\alpha-1}/\hat\ell(x)$ and Z is r.s., so that for each $\varepsilon > 0$, $P_x[Z(R) > \varepsilon R \mid \Lambda_R] \to 0$, and the same proof for the case $\alpha = 2$ applies with an obvious modification (use $\mu_+/\mu \to 0$ to obtain $V_d(x)U_a(x)\mu_+(x) \to 0$).

Let $1 < \alpha < 2$ and $p > 0$ (so that $\alpha\rho < 1$). Then writing

$$g_{B(R)}(x, y) = \sum_{1 \le z \le R-R'} P_x[Z(R') = z, \Lambda_{R'}]g_{B(R)}(R' + z, y)$$

$(R' = \lfloor R/2 \rfloor)$ we apply Lemma 8.6.3 and Proposition 8.6.4(ii) to see that the sum above is asymptotically equivalent to

$$\frac{(\sin \alpha\rho\pi)V_{\mathrm{d}}(x)}{\pi V_{\mathrm{d}}(\frac{1}{2}R)} \int_0^1 \frac{\left(\frac{1}{2} - \frac{1}{2}t\right)^{\alpha\rho-1} \left(\frac{1}{2} + \frac{1}{2}t\right)^{\alpha\hat{\rho}}}{t^{\alpha\rho}(1+t)^{\alpha\hat{\rho}+1}} \, dt \cdot \frac{\alpha\rho \, U(R)V_{\mathrm{d}}(R-y)}{R}$$

(note that z corresponds to $R't$ in the integral approximation of the sum), and using the identity $\int_0^1 t^{-\alpha\rho}(1-t)^{\alpha\rho-1} \, dt/(1+t) = 2^{\alpha\rho-1}B(1-\alpha\rho, \alpha\rho)$, we obtain

$$g_{B(R)}(x,y) \sim \frac{\alpha\rho V_{\mathrm{d}}(x)U_{\mathrm{a}}(R)V_{\mathrm{d}}(R-y)}{V_{\mathrm{d}}(R)R}.$$

After replacing $U_{\mathrm{a}}(R)$ by $R^{\alpha\hat{\rho}}/\ell(y)$ and $V_{\mathrm{d}}(z)$ by $z^{\alpha\hat{\rho}}/\hat{\ell}(z)$ (for $z \in \{x, R\}$) we can rewrite the RHS as in the formulae of the lemma.   □

Proof (of Theorem 8.4.3) First, consider the case $\alpha = 2$. By duality, we have only to consider the case $x \le y < \delta R$ or $x \vee (1 - y) = o(R)$ when (8.60) and (8.61) are applicable, which can be written as the single relation

$$g_{B(R)}(x,y) \sim \frac{V_{\mathrm{d}}(x)V_{\mathrm{d}}(R-y)\ell(R)}{V_{\mathrm{d}}(R)\ell(y)\ell(R-x)} \qquad (y \to \infty)$$

$$= \frac{\ell(R)\hat{\ell}(R)}{\hat{\ell}(x)\ell(y)\ell(R-x)\hat{\ell}(R-y)} \cdot \frac{(x \vee 1)[(R-y) \vee 1]}{R}.$$

This equivalence conforms to that of Theorem 8.4.3.

Next let $1 \le \alpha < 2$. If either $\xi := x/R \downarrow 0$ and $\eta := y/R \uparrow 1$, then

$$K_\rho^{(\alpha)}(\xi, \eta) \sim (\alpha\hat{\rho})^{-1}|\xi - \eta|^{\alpha-1}[\xi(1-\eta)/\eta(1-\xi)]^{\alpha\hat{\rho}},$$

and the result follows from Lemma 8.6.5 in a similar way to the case $\alpha = 2$. Also for the cases $\eta \uparrow 1$ with $\xi < (1 - \varepsilon)$ and $\xi \downarrow 0$ with $\eta > \varepsilon$ one obtains the result from Lemma 8.6.5. If ξ and η are bounded away, respectively, from 0 and 1, the assertions (for $x \le y$) are the same as Corollaries 8.5.4 or 8.5.7 according as $1 < \alpha < 2$ or $\alpha = 1$. By duality, these together verify the formula of the theorem under the restriction $\varepsilon < \xi + \eta < 2(1 - \varepsilon)$ for each $\varepsilon > 0$.

It remains to consider the cases $\xi \vee \eta \downarrow 0$ and $\xi \wedge \eta \uparrow 1$. By Lemma 8.2.4 we have $g_{B(R)}(x,y) \sim g_{\Omega}(x,y)$ as $\xi \vee \eta \downarrow 0$, where the asymptotic form asserted in Theorem 8.4.3 conforms to that of $g_{\Omega}(x,y)$, which we have obtained in Theorem 8.2.1. Hence the required formula follows. Similarly we have $g_{B(R)}(x,y) \sim g_{\Omega}(R-y, R-x)$ as $\xi \wedge \eta \uparrow 1$, and hence the result. The proof of Theorem 8.4.3 is now complete.   □

8.7 Note on the Over- and Undershoot Distributions II

Here we present refined versions of some known results on the overshoot and undershoot distributions. Let $N^+(R)$, $N^-(R)$ and $\hat{Z}(R)$ be as in Section 6.7. ($N^+(R)$ and $N^-(R)$ are the exit times from the half lines $(-\infty, R]$ and $[-R, \infty)$, respectively, and $\hat{Z}(R) = S_{N^-(R)+1} + R$.)

The results of the following proposition, which are local versions of over- and undershoot estimates are essentially corollaries of Lemma 8.1.1 and Theorems 8.2.1 and 8.4.3, in view of the representations (6.65) and (6.66).

Proposition 8.7.1 *Let (AS) hold with $1 \leq \alpha < 2$. Suppose that $0 < \alpha\rho < 1$ and that $p(x) = O(\mu(x)/x)$ if $\alpha = 2\rho = 1$. Put $C_\lambda = \pi^{-1}\sin\lambda\pi$.*

(i) *If $1 < \alpha < 2$, then as $R \to \infty$ along with $y/R \to \eta > 0$,*[10]

(a) $RP_0[Z(R) = y] \longrightarrow C_{\alpha\rho}\eta^{-\alpha\rho}(1+\eta)^{-1}$,
(b) $RP_0[S_{N^+(R)} = R - x] \longrightarrow [C_{\alpha\rho}/B(\alpha\rho, \alpha\hat{\rho})]K(\eta)$, *where*

$$K(\eta) = |1 - \eta|^{\alpha-1} \times \begin{cases} \eta^{-\alpha}\int_0^\eta (1 - w)^{-\alpha}w^{\alpha\rho-1}\,dw & \eta < 1 \\ \int_0^{1/\eta}(1 - w)^{-\alpha}w^{\alpha\rho-1}\,dw & \eta > 1 \end{cases}$$

and $K(1) = (\alpha - 1)^{-1}$; the convergences are uniform on any compact set of $\eta > 0$.

(ii) *If $1 < \alpha < 2$, then as $R \to \infty$ along with $x/R \to \xi$ and $y/R \to \eta$,*

(a) $RP_x[Z(R) = y, \Lambda_R] \longrightarrow C_{\alpha\rho}(1 - \xi)^{\alpha\rho}\xi^{\alpha\hat{\rho}}\dfrac{\eta^{-\alpha\rho}(\eta + 1)^{-\alpha\hat{\rho}}}{\eta + 1 - \xi}$
$$\text{for } \xi \in (0, 1), \eta > 0,$$

(b) $RP_x[S_{N^+(R)} = y, \Lambda_R] \longrightarrow \dfrac{C_{\alpha\rho}}{B(\alpha\rho, \alpha\hat{\rho})}(1 - \eta)^{-\alpha}K_\rho^{(\alpha)}(\xi, \eta)$
$$\text{for } (\xi, \eta) \in (0, 1)^2,$$

where $K_\rho^{(\alpha)}(\xi, \eta)$ is given in (8.37) and the convergences are uniform on any compact set of the indicated range of ξ, η.

(iii) *If $\alpha = 1$, (b) of (i) (resp. (ii)) holds but with uniformity on any compact set of $\{\xi > 0 : \xi \neq 1\}$ (resp. $\{(\xi, \eta) : 0 < \xi, \eta < 1, \xi \neq \eta\}$) and (ia) and (iia) hold under the additional condition that $p(x) = O(\mu(x)/x)$ as $x \to \infty$.*

Proof PROOF OF (ia). According to Lemma 8.1.1 $u_a(x) \sim \alpha\rho x^{\alpha\rho-1}/\ell(x)$ under the assumption of the proposition. Recall (6.65) says that $P_0[Z(R) = x]$ is represented by the sum $\sum_{y=0}^{R} u_a(y)P[Z = x + R - y]$, which we break at $y \sim \varepsilon R$. We claim that

---

[10] Here (a) gives a local version of the well-known overshoot distribution (see (A.16)) when specialised to the renewal process generated by i.i.d. copies of Z, while the problem answered in (b) seems not to have been considered in the existing literature. In the renewal theory $Z(R)$ is usually considered to be the residual waiting time. The corresponding results for the spent waiting time also hold but are omitted, the proofs being immediate.

$$I_\varepsilon := R \sum_{y=0}^{\varepsilon R} u_a(y) P[Z = x + R - y] \to 0 \quad \text{as } R \to \infty \text{ and } \varepsilon \downarrow 0 \text{ in this order}$$

uniformly in $x \in [0, MR]$ for each $M > 1$. Noting that the sum above equals the probability that the ascending ladder height process visits some $y \le \varepsilon R$ and $Z(R) = x$ and that $g_{[R+1,\infty)}(0, y) = g_\Omega(R - y, R)$, one deduces

$$I_\varepsilon \le R P_0[\exists y \le \varepsilon R, \exists n < \sigma_{(R,\infty)}, S_n = y, X_{n+1} = R - y + x]$$

$$= R \sum_{y=-\infty}^{\varepsilon R} g_\Omega(R - y, R) p(R + x - y).$$

Since $g_\Omega(R - y, R) \le 2\bar{a}(R) \le O(R/m(R)) = O(1/R\mu(R))$ $(y \ge 0)$ and, by (8.8), $g_\Omega(R - y, R) = O\left(R^{\alpha-1}/\ell(R)\hat{\ell}(R)\right)$ $(y < 0)$, the above bound of I_ε yields

$$I_\varepsilon \le C \frac{\mu(R + x) - \mu(R + x - \varepsilon R)}{\mu(R + x)} + O\left(R^{-1} L(R)/\ell(R)\hat{\ell}(R)\right), \qquad (8.71)$$

showing the claim because of the regular variation of μ. Now it suffices to show

$$II_\varepsilon := \frac{R}{C_{\alpha\rho}} \sum_{y=\varepsilon R}^{R} u_a(y) P[Z = x + R - y] \longrightarrow \alpha\rho \int_\varepsilon^1 t^{\alpha\rho-1} (1 + \xi - t)^{-\alpha\rho-1} \, dt, \quad (8.72)$$

since the RHS tends to $1/[\xi^{\alpha\rho}(1 - \xi)]$ as $\varepsilon \downarrow 0$. By $R u_a(Rt) \sim \alpha\rho U_a(R) t^{\alpha\rho-1}$ we have

$$II_\varepsilon \sim \frac{\alpha\rho}{C_{\alpha\rho}} \int_\varepsilon^1 t^{\alpha\rho-1} U_a(R) P[Z \in R(1 + \xi - dt)],$$

while, since $P[Z > x] \sim (\alpha\rho)^{-1} C_{\alpha\rho}/U_a(x)$, so that

$$\frac{\alpha\rho}{C_{\alpha\rho}} U_a(R) P[Z > R(1 + \xi - t)] \longrightarrow (1 + \xi - t)^{-\alpha\rho},$$

the measure $[\alpha\rho/C_{\alpha\rho}] U_a(R) P[Z \in R(1 + \xi - dt)]$ converges weakly to $\alpha\rho(1 + \xi - t)^{-\alpha\rho-1} \, dt$. Thus we have (8.72), as required.

PROOF OF (iia). Because of (2.28), we have only to show the existence of the limit $\sum_{w=0}^R g_{B(R)}(x, w) p(y + R - w)$ as $R \to \infty$, which is verified by using Theorems 8.4.1 and 8.4.3 similarly to the proof of (ia) (much simpler).

PROOF OF (b) OF (i) TO (iii). The assertions follow from Theorems 8.2.1, 8.4.3 and Lemma 6.5.1 (entailing $p\alpha^2\rho\hat{\rho} L(x)/\ell(x)\hat{\ell}(x) \to C_{\alpha\rho}/B(\alpha\rho, \alpha\hat{\rho})$) because of the identity $P_0[S_{N^+(R)} = R - x] = g_\Omega(x, R)\mu_+(x)$.

PROOF OF (a) FOR $\alpha = 1$ STATED IN (iii). The diagonal entries $g_\Omega(x, x)$ and $g_{B(R)}(x, x)$ are of the same order of magnitude as $\bar{a}(x)$ for x bounded away from the boundaries of Ω and $B(R)$, respectively. For $\alpha = 1$, we have $\bar{a}(x) \gg 1/x\mu(x)$, so that the argument made above that leads to (8.71) breaks down, but under $p(x) = O(\mu(x)/x)$, it works well because of (6.17), and this verifies, in a similar way, that (ia) holds also for $\alpha = 1$. The same comment applies to (iia).

The proof of Proposition 8.7.1 is complete.   □

Chapter 9
Asymptotically Stable Random Walks Killed Upon Hitting a Finite Set

We suppose, as in the preceding chapter, that (AS) holds and $|1 - \alpha| \vee \rho\hat{\rho} \neq 0$ (i.e., $\rho\hat{\rho} > 0$ if $\alpha = 1$), unless otherwise stated explicitly, and study the r.w. S killed upon hitting *a non-empty finite subset* of $\mathbb{Z}$ that we shall denote by B throughout this chapter. When $1 \leq \alpha \leq 2$ and S is of the normal attraction if $\alpha = 1$, B. Belkin proved in [2] that conditional law of S_n/c_n given $\sigma_B > n$ converges weakly to a certain distribution. In [3], he also obtained the corresponding functional limit theorem. His proofs are based on harmonic analysis in [2] and detailed analytic estimates of relevant probabilities in [3]. We shall give a proof of his results using a different approach, which is primarily probabilistic and an adaptation of the method devised by Bolthausen [10] in the study of the corresponding problem for a half-line, and extend them – with some refinement – to all $0 < \alpha \leq 2$. The method applies to the case of conditioning the r.w. on various events; as a typical instance, we apply the method to derive the escape probability from a finite set (Section 9.5). We also review in Section 9.9 some related results obtained previously: precise asymptotic estimates of the transition probability of the killed walk and $P_x[\sigma_B = n]$ in the case $1 < \alpha < 2$; the corresponding ones for the r.w. killed on the negative half line, and the functional limit theorem for the r.w. conditioned to stay positive forever.

We use the following notation. Let $Y = (Y(t))_{t \geq 0}$ be the same limit stable process as described in Remark 8.2.3: $S_{\lfloor tn \rfloor}/c_n \Rightarrow Y(t)$ (with c_n chosen by (2.17)), and let P_ξ^Y denote the law of Y starting at $\xi \in \mathbb{R}$, as before. Denote by $\mathfrak{p}_t(x)$ and $\mathfrak{f}^\xi(t)$ the density of the distribution of $Y(t)$ and of σ_0^Y, the first hitting time to the origin of Y started at ξ, respectively:

$$\mathfrak{p}_t(\xi) = \frac{P_0^Y[Y(t) \in d\xi]}{d\xi}, \qquad \mathfrak{f}^\xi(t) = \frac{d}{dt} P_\xi^Y[\sigma_0^Y \leq t];$$

there exist jointly continuous versions of these densities (for $t > 0$), which we will always choose. Owing to Gnedenko's local limit theorem [36], as $n \to \infty$,

$$c_n p^n(x) = \mathfrak{p}_1(x/c_n) + o(1), \tag{9.1}$$

K. Uchiyama, *Potential Functions of Random Walks in $\mathbb{Z}$ with Infinite Variance*, Lecture Notes in Mathematics 2338, https://doi.org/10.1007/978-3-031-41020-8_9

provided F is strongly aperiodic, where $o(1)$ is uniform in $x \in \mathbb{Z}$. We shall be concerned with the process Y killed upon hitting zero, whose transition density $\mathfrak{p}_t^0(\xi, \eta) = P_\xi^Y[Y(t) \in d\eta, \sigma_0 > t]/d\eta$ is expressed as

$$\mathfrak{p}_t^0(\xi, \eta) = \mathfrak{p}_t(\eta - \xi) - \int_0^t \mathfrak{f}^\xi(s)\mathfrak{p}_{t-s}^0(\eta)\,ds.$$

It is known that for our stable process Y

$$\begin{cases} 0 \text{ is regular for both } (0, 1) \text{ and } (-1, 0) \text{ except when } \alpha < 1 \text{ and } pq = 0, \\ 0 \text{ is regular for itself if and only if either } 1 < \alpha \le 2 \text{ or } 1 - \alpha = \rho\hat{\rho} = 0, \end{cases} \quad (9.2)$$

and

$$P_\xi^Y[\sigma_0 < \infty] = \begin{cases} 1 & \text{if } 1 < \alpha \le 2, \\ 0 & \text{if } \alpha = 1 \text{ with } \rho\hat{\rho} > 0 \text{ or } \alpha < 1 \end{cases} \quad (9.3)$$

for all $\xi \in \mathbb{R}$ (cf. Chap. 8 of [67], in particular Example 43.22).

9.1 The Potential Function for a Finite Set

This section is concerned with the potential function for the r.w. S killed upon hitting a finite set. For its description, we do not need (AS). The Martin boundary of the killed r.w. comprises two (extremal) points, $+\infty$ and $-\infty$, say, if $\sigma^2 < \infty$ (with $EX = 0$), while it consists of a single point if $\sigma^2 = \infty$; this corresponds to the fact that $EZE|\hat{Z}|$ is finite or infinite according as σ^2 is finite or infinite; one may equivalently say that both $H_{[0,\infty)}^{-y}$ and $H_{(-\infty,0]}^y$ converges to a probability measure as $y \to +\infty$ in the former case, while at least one of them converges to the null measure in the latter case. This makes an essential distinction between the two cases. We first consider the case $\sigma^2 = \infty$.

9.1.1 Recurrent Walks with $\sigma^2 = \infty$

Recall $g(x, y) = a(x) + a(-y) - a(x - y)$, which equals $g_{\{0\}}(x, y)$ for $x \ne 0$ and vanishes for $x = 0$. Since $g_{\{x_0\}}(x, y) = g_{\{0\}}(x - x_0, y - x_0)$ $(x_0 \in \mathbb{Z})$, this shows that if $x_0 \in B$,

$$g_B(x, y) = g_{\{x_0\}}(x, y) - E_x[g(S_{\sigma_B} - x_0, y - x_0)]. \quad (9.4)$$

The RHS is written as $a^\dagger(x - x_0) - E_x[a(S_{\sigma_B} - x_0)] + E_x[a(S_{\sigma_B} - y) - a(x - y)]$. Thus, passing to the limit as $y \to \pm\infty$, the following function is well-defined

$$u_B(x) := \lim_{|y| \to \infty} g_B(x, y) = a^\dagger(x - x_0) - E_x[a(S_{\sigma_B} - x_0)] \quad (x \in \mathbb{Z}) \quad (9.5)$$

(cf. [71, Theorem 30.1], [84, Eq(3.30)]). One also infers that as $|y| \to \infty$,

$$g_B(x, y) = u_B(x) + a(-y) - a(x - y) + o(1) \quad \text{uniformly in } x. \tag{9.6}$$

Using $H_B^{x+y}(z) = H_{B-y}^x(z-y)$, a simple change of variable yields $E_{x+y}[a(S_{\sigma_B} - x_0)] = E_x[a(S_{\sigma_{B-y}} + y - x_0)]$. Hence by (9.5) one sees that

$$u_B(x + y) = u_{B-y}(x). \tag{9.7}$$

From (9.5), some elementary manipulations show that u_B satisfies the equation

$$u_B(x) = E_x[u_B(S_1); S_1 \notin B] = \sum_{y \notin B} p(y - x) u_B(y) \qquad x \in \mathbb{Z}; \tag{9.8}$$

so that u_B, if restricted to $\mathbb{Z} \setminus B$, is a non-negative harmonic function for S killed upon hitting B (cf. [82, Section 7]), which is unique apart from a positive multiple. Thus $u_B(x)$ has the fundamental characteristic of a potential function; it plays the same role in the theory of r.w. killed on its entering B as $a^\dagger(x)$ does for that of r.w. killed on hitting the origin – note $u_{\{0\}} = a^\dagger$. In addition to this, $u_B(x)$, like $a^\dagger(x)$, has a simple positivity property, as is exhibited in the following

Lemma 9.1.1 (i) *If $\mu_+(x)\mu_-(x) > 0$ for all $x > 0$, then $u_B(x) > 0$ for all $x \in \mathbb{Z}$.*
(ii) *Suppose $\mu_+(x) = 0$ for some $x \geq 1$. Then $u_B(x) > 0$ for $x \geq \max B$ while for $x < \max B$, $u_B(x) > 0$ if and only if $P_x[\sigma_{[1+\max B, \infty)} < \sigma_B] > 0$; and there are only two alternatives: either $u_B(-x) = 0$ for all $x < \min B$ or $u_B(x) > 0$ for all $x \in \mathbb{Z}$.*
(iii) *Suppose $u_B(x) > 0$ for all $x \in \mathbb{Z}$. Then, for any $r \leq \min B$*

$$\lim_{x \to \infty} u_B(x) = \sum_{y \in [r, \infty) \setminus B} \frac{P[Z > y - r]}{EZ} u_B(y) \in (0, \infty] \qquad \text{if} \quad EZ < \infty,$$

and similarly if $E|\hat{Z}| < \infty$; in particular $\inf_{x \in \mathbb{Z}} u_B(x) > 0$.[1]

Proof For the proof of (i) and (ii), we can assume $\max B = 0$ without loss of generality. Suppose that $\mu_+(x) = 0$ for $x > 0$. Then as $x \to \infty$, $a(x) \to \infty$ and $u_B(x) \sim a(x)$ because of (9.5), and by (9.8) we obtain $u_B(x) > 0$ for $x > 0$ since the r.w. restricted on the half-line $[1, \infty)$ is irreducible therein. The same reasoning shows that $u_B(x) > 0$ if $P_x[S_n > 0, n < \sigma_B] > 0$ for some n, or, what amounts to the same, $P_x[\sigma_{[1,\infty)} < \sigma_B] > 0$. To show the converse suppose $P_x[\sigma_{[1,\infty)} < \sigma_B] = 0$. Then $x < 0$ and $g_B(x, y) = 0$ for all $y \geq 1$. Hence $u_B(x) = \lim_{y \to \infty} g_B(x, y) = 0$. If $u_B(x)$ is positive for some $x < \min B$, then it is also positive for all $x \in (-\infty, 0)$ by the same reasoning as above. Since $a(x) \to \infty$ if $\mu_+(x) = 0$ for some $x > 0$, we conclude (i) and (ii).

For the proof of (iii), observe that $g_B(-x, z) = \sum_{y \notin B} H_{[0,\infty)}^{-x}(y) g_B(y, z)$ for $x, y > 0$ if $B \subset [0, \infty)$. Then letting $z \to \infty$ and taking (9.7) into account, one sees for any finite B and $r \leq \min B$,

---

[1] The same criterion as given in (2.13) for $\lim_{x \to \pm\infty} a(x) < \infty$ applies to $u_B(x)$; see also (3.3).

$$u_B(-x + r) = u_{B-r}(-x) = \sum_{y \notin B-r} H^{-x}_{[0,\infty)}(y) u_{B-r}(y) \qquad (x > 0). \qquad (9.9)$$

Now suppose $EZ < \infty$. Then, since $H^{-x}_{[0,\infty)}(y)$ is 'essentially' increasing in x and converges to $P[Z > y]/EZ$ as $x \to \infty$ (see (4.38)), it follows that

$$\lim_{x \to \infty} u_B(-x) = \sum_{y \in [0,\infty) \backslash (B-r)} \frac{P[Z > y]}{EZ} u_{B-r}(y), \qquad (9.10)$$

which is the same as the formula of (iii) because of $u_{B-r}(y) = u_B(y + r)$.   □

9.1.2 Case $\sigma^2 < \infty$

Here, any non-negative harmonic function for S killed upon hitting any non-empty finite set is represented as a convex combination of two linearly independent ones. Pick any $x_0 \in B$ and put

$$u_B(x) = a^\dagger(x - x_0) - E_x[a(S_{\sigma_B} - x_0)],^2 \quad w_B(x) = (x - E_x[S_{\sigma_B}])/\sigma^2,$$

$$g^\pm_B(x) = \lim_{y \to \pm \infty} g_B(x, y) \quad \text{and} \quad K_B(x,y) := \frac{g^+_B(x)g^-_{-B}(-y) + g^-_B(x)g^+_{-B}(-y)}{2}.$$

By (9.4) and (2.10) one observes that the limits above exist and

$$g^+_B(x) = u_B(x) + w_B(x) \quad \text{and} \quad g^-_B(x) = u_B(x) - w_B(x), \qquad (9.11)$$

and accordingly that $u_B(x) = \frac{1}{2}[g^+_B(x) + g^-_B(x)]$ and

$$K_B(x,y) = u_B(x)u_{-B}(-y) - w_B(x)w_{-B}(-y). \qquad (9.12)$$

As before, one can show that both g^+_B and g^-_B are non-negative harmonic functions for the killed walk in the sense that $g^\pm_B(x) = \sum_{z \in \mathbb{Z} \backslash B} p(z - x)g^\pm_B(z)$ for all $x \in \mathbb{Z}$, hence the functions u_B and w_B also are harmonic in the same sense, and $g^\pm_{-B}(-\cdot)$, $u_{-B}(-\cdot)$ and $w_{-B}(-\cdot)$ are dual-harmonic. The same arguments as made in the case $\sigma^2 = \infty$ show the following lemma. Write r_B and l_B for max B and min B, respectively.

Lemma 9.1.2 *For any non-empty finite $B \subset \mathbb{Z}$*

(a) $g^\pm_B(x) = E_x[g^\pm_B(S_1); S_1 \notin B]$, $\quad w^\pm_B(x) = E_x[w^\pm_B(S_1); S_1 \notin B]$;
(b) $g^+_B(x) > 0 \Longleftrightarrow P_x[\sigma_{[r_B+1,\infty)} < \sigma_B] > 0 \quad (\Leftarrow x \geq r_B)$;
(b') $g^-_B(x) > 0 \Longleftrightarrow P_x[\sigma_{(-\infty,l_B-1]} < \sigma_B] > 0 \quad (\Leftarrow x \leq l_B)$;
(c) *either $g^+_B(x) = 0$ for all $x < l_B$ or $\inf_{x \notin [l_B, r_B]} g^+_B(x) > 0$;*
(c') *either $g^-_B(x) = 0$ for all $x > r_B$ or $\inf_{x \notin [l_B, r_B]} g^-_B(x) > 0$.*

---

^{2} The RHS does not depend on the choice of x_0 (cf. [80]).

Let $x, y \notin [l_B, r_B]$ and write $x \rightsquigarrow y$ for $P_x[\sigma_y < \sigma_B] > 0$. Then

$$g_B^+(x)g_{-B}^-(-y) > 0 \iff x \rightsquigarrow y > r_B \text{ or } r_{-B} < -y \rightsquigarrow -x \iff x \rightsquigarrow y,$$

and the same condition for $g_B^-(x)g_{-B}^+(-y) = \hat{g}_{-B}^+(-x)\hat{g}_B^-(y) > 0$. It accordingly follows that for each pair of x and y in the complement of $[l_B, r_B]$,

$$K_B(x, y) > 0 \iff P_x[\sigma_y < \sigma_B] > 0.$$

9.1.3 Transient Walks

With the Green function $G(x) < \infty$, following [49] we define $u_B(x)$ by

$$u_B(x) = G(0)P_x[\sigma_B = \infty] = \frac{P_x[\sigma_B = \infty]}{P_0[\sigma_0 = \infty]} = \lim_{n \to \infty} \frac{P_x[\sigma_B > n]}{P_0[\sigma_0 > n]}; \qquad (9.13)$$

in particular

$$u_{\{0\}}(x) = G(0) - G(-x) + \delta_{0,x} = a^\dagger(x).$$

Evidently $u_B(x) > 0$ for all $x \in \mathbb{Z}$. The following analogue of (9.5) holds:

$$u_B(x) = a^\dagger(x - x_0) - E_x[a(S_{\sigma_B} - x_0); \sigma_B < \infty] \qquad (x \in \mathbb{Z}, x_0 \in B). \qquad (9.14)$$

Indeed, by $E_x[a(S_1)] = a^\dagger(x)$, $a(S_{\sigma_B \wedge n} - x_0)$ is a bounded martingale under P_x, $x \neq x_0 \in B$. Hence, we have

$$a^\dagger(x - x_0) = \lim E_x[a(S_{\sigma_B} - x_0); \sigma_B < n] + \lim E_x[a(S_n - x_0); \sigma_B > n]$$
$$= E_x[a(S_{\sigma_B} - x_0); \sigma_B < \infty] + G(0)P_x[\sigma_B = \infty],$$

showing (9.14) for $x \neq x_0$. Noting $u_B(x_0) = G(0) \sum_{y \notin B} P_y[\sigma_B = \infty] p(y - x_0) = E_{x_0}[u_B(S_1); S_1 \notin B]$ and applying (9.14) with $x \notin B$ lead to the result for $x = x_0$.

9.2 The r.w. Conditioned to Avoid a Finite Set; Statements of Results

Let $D_{[0,\infty)}$ be the collection of real-valued functions $f(t)$, $t \geq 0$, which are right continuous with left limits. Let $\mathcal{B}_t$ be the σ-field of $D_{[0,\infty)}$ generated by the cylinder sets composed by $f(s)$, $s \leq t$, and put $\mathcal{B}_\infty = \bigwedge_{t>0} \mathcal{B}_t$. The complement in $[0, \infty)$ of the closure of $\{t \geq 0 : f(t) = 0\}$, $f \in D_{[0,\infty)}$, is a disjoint union of open intervals. Denote by E_f the set of these intervals, each of which we call an excursion interval; the path restricted to an excursion interval is called the excursion on it. Denote by $I_a(f)$ the leftmost one among the excursion intervals of f whose length is larger than $a > 0$ and by $\mathfrak{t}_f$ (sometimes written as $\mathfrak{t}(f)$) the left endpoint of $I_1(f)$. It is easy to see that $\mathfrak{t}_f$ is $\mathcal{B}_\infty$-measurable and $\mathfrak{t}_Y < \infty$ a.s.(P_0^Y).

Denote by $\mathbf{P}^0$ the probability measure on $D_{[0,\infty)}$ induced from P_0^Y by the map $\omega \in \Omega \mapsto (Y_{\mathfrak{t}(Y)+t})_{t\geq 0} \in D_{[0,\infty)}$. Here and in the sequel, $D_{[0,\infty)}$ is considered to be the Polish space endowed with Skorokhod's topology given by the metric

$$d(f,g) := \inf_\lambda \sum_{m=1}^\infty [\|\lambda - \mathrm{id}\|_m + \|f \circ \lambda - g\|_m \wedge 1]\, 2^{-m}$$

$(f, g \in D_{[0,\infty)})$, where $\|f\|_m = \sup_{0\leq t\leq m} |f(t)|$, λ ranges over the space of homeomorphisms from $[0, \infty)$ onto itself, and $\mathrm{id}(t) = t$ (cf. [73], [30]). Write $Y_n(t)$ for $S_{\lfloor nt \rfloor}/c_n$ and denote by $\mathbf{P}_x^{n,B}$ the law on $D_{[0,\infty)}$ induced by Y_n from the conditional law of $Y_n(t)$ under P_x given $\sigma_B > n$: for $\Gamma \in \mathcal{B}_\infty$,

$$\mathbf{P}_x^{n,B}(\Gamma) = P_x[Y_n \in \Gamma \mid \sigma_B > n].$$

We designate by '$\Rightarrow$' the convergence of laws on the metric space $(D_{[0,\infty)}, d)$.

Theorem 9.2.1 *If $\sigma^2 = \infty$, for each x such that $\mathsf{u}_B(x) > 0$, as $n \to 0$,*

$$\mathbf{P}_x^{n,B} \Longrightarrow \mathbf{P}^0. \tag{9.15}$$

Below we state some refinements or extensions of the above convergence. Denote by $\mathfrak{y}_t$ the usual coordinate map defined on $D_{[0,\infty)}$: $f \in D_{[0,\infty)} \mapsto \mathfrak{y}_t(f) = f(t)$. From the definition of $\mathbf{P}^0$ we can represent the distribution function $\mathbf{P}^0[\mathfrak{y}_1 \leq \eta]$ as the repeated integral $\int_{-\infty}^\eta dr \int_{-\infty}^\infty \mathbf{P}^0[\mathfrak{y}_t \in d\xi] \mathfrak{p}_{1-t}^0(\xi, r)$ $(0 < t < 1)$, hence it has a continuous derivative, which we denote by $\mathfrak{q}$. By (9.15) it accordingly follows that

$$\int_{-\infty}^\eta \mathfrak{q}(r)\, dr = \mathbf{P}^0[\mathfrak{y}_1 \leq \eta] = \lim P_0[Y_n(1) \leq \eta \mid \sigma_B > n] \quad (\eta \in \mathbb{R}). \tag{9.16}$$

The next theorem gives a local version of (9.16). To make the formulae concise, we write x_n and y_n for x/c_n and y/c_n, respectively. Put

$$\kappa_\alpha^\sharp = \frac{\sin \pi/\alpha}{\mathfrak{p}_1(0)\pi}. \tag{9.17}$$

Below M will denote an arbitrarily chosen constant greater than unity.

Theorem 9.2.2 *Suppose F is strongly aperiodic and $\sigma^2 = \infty$. Then as $n \to \infty$,*

(i) *for each $x \in \mathbb{Z}$ with $\mathsf{u}_B(x) > 0$ and uniformly for $|y| \in [M^{-1}c_n, Mc_n]$,*

$$P_x[S_n = y, \sigma_B > n] \sim \kappa_\alpha^\sharp \mathsf{u}_B(x)\mathfrak{q}(y_n)/n; \tag{9.18}$$

(ii) *if $1 < \alpha \leq 2$, then $\mathfrak{f}^{-\eta}(1) = \kappa_\alpha^\sharp \mathfrak{q}(\eta)$ and*

$$P_y[\sigma_B = n] \sim \mathfrak{f}^{y_n}(1)/n \quad \text{uniformly for } |y| \in [M^{-1}c_n, Mc_n]. \tag{9.19}$$

Corollary 9.2.3 (i) *As $x_n \to \xi \neq 0$ and $k/n \to t > 0$,*

$$P_x[\sigma_B > k] \longrightarrow P_\xi^Y[\sigma_0 > t].^3 \qquad (9.20)$$

(ii) *Suppose F is strongly aperiodic and $\sigma^2 = \infty$. Then uniformly for $|x|, |y| \in [M^{-1}c_n, Mc_n]$,*

$$c_n P_x[S_n = y, \sigma_B > n] \sim \mathfrak{p}_1^0(x_n, y_n); \qquad (9.21)$$

and for each x with $\mathfrak{u}_B(x) > 0$, uniformly for $|\eta| \in [M^{-1}, M]$, as $n \to \infty$, $y_n \to \eta$ and $k/n \to t > 0$,

$$c_n P_x[S_k = y \mid \sigma_B > n] \longrightarrow \mathfrak{q}_t(\eta) P_\eta^Y[\sigma_0 > 1 - t], \qquad (9.22)$$

where

$$\mathfrak{q}_t(\eta) = t^{-1}\mathfrak{q}(t^{-1/\alpha}\eta).$$

This corollary is derived from Theorem 9.2.2 as follows. We may suppose $1 < \alpha \le 2$, the case $\alpha \le 1$ being directly ensured in view of (9.3). From the scaling relation of Y it follows that for every $\lambda > 0$,

$$\mathfrak{f}^{\lambda\xi}(t) = \lambda^{-\alpha}\mathfrak{f}^\xi(t\lambda^{-\alpha}) \qquad (\xi \in \mathbb{R}, t > 0). \qquad (9.23)$$

Let $k/n \to t \in (0, 1)$ and $x_n = x/c_n \to \xi \neq 0$ as $n \to \infty$. Then, by (9.19) it follows that for $j > k$,

$$P_x[\sigma_B = j] \sim \mathfrak{f}^{x_j}(1)/j = \mathfrak{f}^{x_n c_n/c_j}(1)/j \sim \mathfrak{f}^{x_n}(j/n)/n.$$

Hence $P_x[\sigma_B > k] \sim \int_t^\infty \mathfrak{f}^\xi(s)\,ds$, showing (9.20). Similarly if $1/M \le y_n \le M$, noting $P_x[S_k = y \mid \sigma_B > k] \sim \mathfrak{q}(y_k)/c_k$ one infers that

$$c_n P_x[S_k = y \mid \sigma_B > n] \sim \mathfrak{q}(y_n c_n/c_k) \frac{c_n P_x[\sigma_B > k]}{c_k P_x[\sigma_B > n]} P_y[\sigma_B > n - k]$$

$$\sim t^{-1}\mathfrak{q}(y_n t^{-1/\alpha}) P_{y_n}^Y[\sigma_0 > 1 - t],$$

where we have used the fact that the ratio in the middle expression tends to t^{-1} (see Lemma 9.3.3) and (9.46)). (9.21) is shown in the same way.

The conditioned r.w. is Markovian and so is its limit process. Once we obtain the convergence result of the law $P_0[S_{\lfloor tn \rfloor} \in \cdot \mid \sigma_0 > n]$, which determines the entrance law of $\mathbf{P}^0$ by virtue of (9.20), we can write down the transition probability function, which together with the entrance law determines the finite-dimensional distribution of the limit process.

---

[3] One can easily see the one-sided bound $\liminf P_x[\sigma_B > n] \ge P_\xi^Y[\sigma_0 > t]$ without resorting to (9.19), which we use in the proof of (9.20) below – to have (9.20) one needs some auxiliary information. Since $\sigma_B \le \sigma_0$ if $0 \in B$, (9.20) follows once we have it with $B = \{0\}$.

Corollary 9.2.4 *Let* $\sigma^2 = \infty$.

(i) *The excursion under* $\mathbf{P}^0$ *is a time-inhomogeneous Markov process; the transition function is given by*

$$\mathfrak{p}(s,\xi;t,\eta) = \frac{\mathfrak{p}^0_{t-s}(\xi,\eta)P^Y_\eta[\sigma_0 > 1 - t]}{P^Y_\xi[\sigma_0 > 1 - s]} \quad (\xi \neq 0, 0 \leq s < t \leq 1), \tag{9.24}$$

$$\mathfrak{p}(0,0;t,\eta) = \mathfrak{q}_t(\eta)P^Y_\eta[\sigma_0 > 1 - t] \quad (0 < t \leq 1, \eta \in \mathbb{R}), \tag{9.25}$$

$$\mathfrak{p}(s,\xi;t,\eta) = \int_{-\infty}^{\infty} \mathfrak{p}^0(\xi,s;\eta',1)\mathfrak{p}_{t-s}(\eta - \eta')\,d\eta' \quad (0 \leq s \leq t < 1, \xi \in \mathbb{R}),$$

$$\mathfrak{p}(s,\xi;t,\eta) = \mathfrak{p}_{t-s}(\eta - \xi) \quad (1 < s < t).$$

(ii) $\int_{-\infty}^{\infty} \mathfrak{q}_s(\xi)\mathfrak{p}^0_{t-s}(\xi,\eta)\,d\xi = \mathfrak{q}_t(\eta) \quad (0 < s < t, \eta \in \mathbb{R})$.

Proof (9.24) follows from the definition, and (9.25) from (9.22). The other formulae of (i) are immediate since $t_f + 1$ is a stopping time. If $t \leq 1$, (ii) follows from (9.24) and (9.25). One can easily dispose of the general case by using the scaling relations

$$\mathfrak{q}_t(\lambda\xi) = \lambda^{-\alpha}\mathfrak{q}_{t/\lambda^\alpha}(\xi) \quad \text{and} \quad \mathfrak{p}^0_t(\lambda\xi,\lambda\eta) = \lambda^{-1}\mathfrak{p}^0_{t/\lambda^\alpha}(\xi,\eta), \tag{9.26}$$

valid for any $\lambda > 0$, $\xi, \eta \in \mathbb{R}$ and $t > 0$. □

Corollary 9.2.5 (i) *If* X *is strongly aperiodic,* $1 \leq \alpha \leq 2$ *and* $\sigma^2 = \infty$, *then*

$$P_x[\sigma_B = n] \sim \begin{cases} (1 - \frac{1}{\alpha})\kappa^\sharp_\alpha u_B(x)c_n/n^2 & \text{for each } x \in \mathbb{Z} \text{ with } u_B(x) > 0, \\ \kappa^\sharp_\alpha \mathfrak{q}(-x_n)/n & \text{uniformly for } |x_n| \in [M^{-1}, M], \end{cases} \tag{9.27}$$

and for each $x \in \mathbb{Z}$ *with* $u_B(x) > 0$, *as* $n \to \infty$,

$$P_x[S_n = y \mid \sigma_B > n] \sim \begin{cases} (1 - \frac{1}{\alpha})u_{-B}(-y)/n & \text{for each } y \in \mathbb{Z}, \\ \mathfrak{q}(y_n)/c_n & \text{uniformly for } |y_n| \in [M^{-1}, M]. \end{cases} \tag{9.28}$$

(ii) $\mathfrak{f}^{-\eta}(t) = \kappa^\sharp_\alpha \mathfrak{q}_t(\eta)$ $(t > 0, \eta \in \mathbb{R})$ *if* $1 < \alpha \leq 2$ *and*

$$\mathfrak{q}(\xi) = \frac{|\xi|}{2c_\sharp}e^{-\xi^2/2c_\sharp} \quad \text{if } \alpha = 2; \quad \text{and} \quad \mathfrak{q}(\xi) = \frac{c/\pi}{c^2 + (\xi - b)^2} \quad \text{if } \alpha = 1,$$

where $c = c_\sharp\pi/2$ *and* $b = c \tan[\pi(\rho - \frac{1}{2})]$ *(so that* $\int_0^\infty \mathfrak{q}(\xi)\,d\xi = \rho$*).*

[For $1 < \alpha < 2$ *the asymptotic of* $\kappa^\sharp_\alpha\mathfrak{q}(\xi) = \mathfrak{f}^{\text{sgn}(\xi)}(1/|\xi|^\alpha)/|\xi|^\alpha$ *as* $\xi \to 0$ *and as* $|\xi| \to \infty$ *is known (cf. (9.96), (9.97). The formulae of (ii) for* $\alpha = 2$ *and* $\alpha = 1$ *with* $\rho = 1/2$ *(entailing* $b = 0$*) are derived from (9.29) in [2].]*

If $\alpha > 1$, the formulae of (ii) follow from Theorem 9.2.2, the scaling relation of $\mathfrak{f}^\xi(t)$, and the well-known formula for the hitting time distribution of Brownian motion. For $\alpha = 1$, by (9.3) Y starting at the origin never returns to it so that $t_f = 0$

a.s., entailing the formula of (ii) for $\alpha = 1$. The second case of (9.28) is a special case of (9.22). The first case of (9.27) follows from that of (9.28) whose proof we shall give after that of Theorem 9.2.2. The second case of (9.27) will be shown in Proposition 9.6.1.

In the case $u_B(x) = 0$ (entailing $\mu_+(x_0)\mu_-(x_0) = 0$ for some $x_0 > 0$), we know only $u_B(x) > 0$ either for all $x < \min B$ ($\mu_+(x_0) = 0$) or for all $x > \max B$ ($\mu_-(x_0) = 0$). We consider only the latter case; the other one can be treated by duality. Denote by $\mathbf{P}^{(-\infty,0]}$ the stable meander, associated with Y, of length 1 (cf. [4]). It is known that the law, under P_0, of $(S_{\lfloor nt \rfloor}/c_n)_{0 \le t \le 1}$ conditioned with $T > n$ converges weakly to $\mathbf{P}^{(-\infty,0]}$ [19], [25]. In the sequel, $\mathbf{P}^{(-\infty,0]}$ is understood to be extended naturally by Markov property to the probability measure on $D_{[0,\infty)}$.

Theorem 9.2.6 *For $x \in \mathbb{Z}$ with $u_B(x) = 0$, $\mathbf{P}_x^{n,B} \Rightarrow \mathbf{P}^{(-\infty,0]}$ as $n \to \infty$.*

The r.w. S never being to the left of $\min B$ under the setting of this theorem, the result for the case $B = (-\infty, -1]$ mentioned above convinces us of its truth. The proof, however, is not so trivial. We shall provide it in Section 9.7.

Remark 9.2.7 (a) In the case $\sigma^2 < \infty$ Belkin [3] obtains a result that asserts the convergence of $\mathbf{P}_x^{n,B}$ when x is fixed and the characterisation of the limit law that depends on B. In Section 9.9.3, we shall state a refined version obtained in [82].

(b) Let $1 \le \alpha \le 2$ and let $\Psi(\theta)$ stand for the characteristic function of $q(\xi)$: $\Psi(\theta) := \int_{-\infty}^{\infty} e^{i\theta\xi} q(\xi)\, d\xi = \lim_{n\to\infty} P_x\left[e^{i\theta S_n/c_n} \mid \sigma_B > n\right]$. Belkin [2] obtains that

$$\Psi(\theta) = 1 - \Phi(\theta) \int_0^1 (1-t)^{1/\alpha - 1} e^{-t\Phi(\theta)}\, dt. \tag{9.29}$$

(The derivation of this, which we shall not give, may be carried out from (9.45).)

Remark 9.2.8 We shall also consider a process conditioned to avoid B forever in Section 9.8. If $u_B(x) > 0$ for all x, this conditional process is defined as the Markov process on $\mathbb{Z}$ that is the Doob h-transform by u_B of the r.w. S restricted to $\mathbb{Z} \setminus B$.

The proofs of Theorems 9.2.1, 9.2.2 and 9.2.6 are given in Sections 9.3, 9.6, and 9.7, respectively. The other sections of this chapter address some related problems. These sections, other than Section 9.3, may be read independently.

Extensions to the case $x \to \infty$ for $\alpha = 2$

The results above are valid as $x \to \infty$ under $x = o(c_n)$ in most cases of $1 \le \alpha < 2$ (see Section 9.9.2 for $1 < \alpha < 2$ and Remark 9.2.13 at the end of this section for $\alpha = 1$). For $\alpha = 2$, on the other hand, it turns out (Proposition 9.2.9) that they may be valid under the restrictive condition of $L(x)/L(c_n) \to 0$, while if $L(x)/L(c_n)$ is bounded away from zero and unity, the limit distribution of $Y_n(1)$ conditioned to avoid a finite set becomes a mixture of $q(\eta)\, d\eta$ and the limit distribution of $P_0[Y_n(1) \in \cdot \mid T > n]$. Below we consider only the case $S_0 = x > 0$ under $\alpha = 2$.

Before stating the results, we present the basic identity due to Kesten and Spitzer [49], which the proof rests on. From $P_x[\sigma_B > n] = \sum_{k=0}^{\infty} P_x[\sigma_B = n + 1 + k]$, one infers that

$$P_x[\sigma_B > n] = \sum_{y \in \mathbb{Z}} g_B(x, y) P_y[S_{n+1} \in B, \sigma_B > n], \qquad (9.30)$$

a trivial extension of Eq(1) in p.161 of [71]. By the identity $P_y[S_{n+1} = z, \sigma_B > n] = P_{-z}[S_{n+1} = -y, \sigma_{-B} > n]$ ($y \notin B$), one obtains

$$P_y[S_{n+1} \in B, \sigma_B > n] = \sum_{z \in B} P_{-z}[S_{n+1} = -y, \sigma_{-B} > n]. \qquad (9.31)$$

If $x_0 \in B$, it in particular follows that

$$P_x[\sigma_B > n] \le P_x[\sigma_{x_0} > n] = \sum_{y \in \mathbb{Z}} g_{x_0}(x, y) P_{-x_0}[S_{n+1} = -y, \sigma_{-x_0} > n],$$

hence for all $x \in \mathbb{Z}$ and $n = 1, 2, \ldots,$

$$P_x[\sigma_B > n] \le 2\bar{a}^\dagger(x - x_0) P_0[\sigma_0 > n] \qquad (x_0 \in B) \qquad (9.32)$$

since $g_{x_0}(x, y) \le g_{x_0}(x, x) = 2\bar{a}^\dagger(x - x_0)$. ((9.32) is valid under our basic setting.)

Proposition 9.2.9 *Suppose* $\sigma^2 = \infty$ *and* $\alpha = 2$.

(i) *If* $\lim_{x \to \infty} m_-(x)/m(x) = q > 0$, *then as* $n \wedge x \to \infty$,

$$P_x[\sigma_B > n] \sim a(x) \left[1 + \left(\frac{1}{2q} - 1 \right) \frac{L(x)}{L(c_n)} \right] P_0[\sigma_0 > n] \quad for \ x = o(c_n),$$
$$(9.33)$$

and $\mathbf{P}_x^{B,n}[\mathfrak{y}_1 \in \cdot] \Rightarrow \mathbf{P}^0[\mathfrak{y}_1 \in \cdot]$ *under* $L(x)/L(c_n) \to 0$.
(ii) *If* $\lim_{x \to \infty} m_-(x)/m(x) = 0$ *(entailing* $a(x)/a(-x) \to 0$*), then*

$$P_x[\sigma_B > n] \le (P_x[T > n] + a(x) P_0[\sigma_0 > n]) \{1 + o(1)\} \qquad (9.34)$$

for $x > 0$, *and if* $\mu_-(x) \asymp x^{-\beta} L_-(x)$ ($x \to \infty$) *for some* $\beta > 2$ *and s.v.* L_-, *then as* $n \wedge x \to \infty$ *under* $L(x)/L(c_n) \to 0$,

$$P_x[\sigma_B > n] \sim P_x[T > n] + \mathsf{u}_B(x) P_0[\sigma_0 > n]. \qquad (9.35)$$

Remark 9.2.10 (a) It holds that $P_x[T > n] \sim \sqrt{2/\pi c_\sharp}\, V_{\mathrm{d}}(x)/V_{\mathrm{d}}(c_n)$ (cf. (9.68) and (i) of Remark A.3.3). If $m_-/m \to q$, by Proposition 4.2.1,

$$a(x) \sim 2qx/L(x) \quad and \quad a(-x) \sim 2(1 - q)x/L(x) \quad (x \to \infty). \qquad (9.36)$$

(b) In (i) above, let $x \wedge n \to \infty$ under $L(x)/L(c_n) = \lambda \in [0,1]$ and $x/c_n \to 0$. Then, using $P_0[\sigma_0 > n] \sim \kappa_\alpha^\# c_n/n$ (Lemma 9.3.3), one infers from (9.33) that

$$\frac{e^{i\theta x/c_n} - 1}{P_x[\sigma_B > n]} \to i \left[c_\# k_\lambda / \kappa_2^\# \right] \theta \quad \text{where } k_\lambda := \frac{\lambda}{2q + (1 - 2q)\lambda}. \tag{9.37}$$

Hence, on noting $\kappa_2^\# = \sqrt{2c_\#/\pi}$, from (9.45) it follows that for $\theta \in \mathbb{R}$,

$$\Psi_n^x(\theta) := E_x[e^{i\theta S_n/c_n} \mid \sigma_B > n] \longrightarrow \Psi(\theta) + ik_\lambda \sqrt{\frac{c_\# \pi}{2}} \, \theta e^{-c_\# \theta^2/2}. \tag{9.38}$$

(c) In (ii), by what is mentioned in (a), elementary computations yield that

$$P_x[T > n]/[\kappa^\# a(-x)c_n/n] \sim 2^{-1} \ell^*(x)/\ell^*(c_n) \quad \text{under } x/c_n \to 0$$

as $n \wedge x \to \infty$ (recall Lemma 6.5.1(ii)). Since $P_x[T > n] \le P_x[\sigma_B > n] \times \{1 + o(1)\}$, by the inequality (9.34) and $a(x)/a(-x) \to 0$ it follows that for any $\varepsilon > 0$, as $x \wedge n \to \infty$ under $\ell^*(x)/\ell^*(c_n) > \varepsilon$, $P_x[\sigma_B > n] \sim P_x[T > n]$, which entails (9.37) with $\left[c_\# k_\lambda / \kappa_2^\# \right]$ replaced by $\sqrt{c_\# \pi/2} \lim \hat{\ell}^*(x)/\hat{\ell}^*(c_n)$ (if this limit exists), and

$$\Psi_n^x(\theta) = \left[\Psi(\theta) + i \frac{\hat{\ell}^*(x)}{\hat{\ell}^*(c_n)} \sqrt{\frac{c_\# \pi}{2}} \, \theta e^{-c_\# \theta^2/2} \right] \{1 + o(1)\} \quad (\theta \in \mathbb{R}). \tag{9.39}$$

(d) The tightness of the sequence $(\mathbf{P}_x^{n,B})_{n=1}^\infty$ follows also for the case $x \wedge n \to \infty$ if the convergence of the law $\mathbf{P}_x^{n,B}[\eta_1 \in \cdot]$ is established (see the proof of Lemma 9.3.7).

Observing $(2\pi)^{-1}\sqrt{c_\# \pi/2} \int_{-\infty}^\infty \theta e^{-c_\# \theta^2/2} \sin \eta\theta \, d\theta = 2^{-1}\eta e^{-\eta^2/2c_\#}/c_\#$, from the convergence of Ψ_n^x given in (9.38) and (9.39) one obtains the following

Corollary 9.2.11 *Suppose* $\sigma^2 = \infty$ *and* $\alpha = 2$. *Let* $\lim_{x\to\infty} m_-(x)/m(x) = q$ *and* k_λ *be as given in (9.37) if* $q > 0$. *If* $q = 0$, *suppose that* $k_\lambda := \lim \hat{\ell}^*(x)/\hat{\ell}^*(c_n)$ *exists and* $u_B(x) > 0$ *for all* $x \in \mathbb{Z}$. *Then, as* $x \wedge n \to \infty$ *so that* $x = o(c_n)$ *and* $L(x)/L(c_n) \to \lambda \in [0,1]$ *if* $q > 0$ *and* $\liminf \ell^*(x)/\ell^*(c_n) > 0$ *if* $q = 0$,

$$P_x[Y_n(1) \in d\eta \mid \sigma_B > n] \Longrightarrow 2^{-1} [1 + \text{sgn}(\eta)k_\lambda] \, |\eta| e^{-\eta^2/2c_\#} \, d\eta/c_\#. \tag{9.40}$$

Remark 9.2.12 The local versions of (9.33), (9.35) and (9.40) hold (see the proof of Proposition 9.6.1). As in the proof of Proposition 9.6.3,[4] one can accordingly show that if $\lambda' = \lim L(|y|)/L(c_n)$ and $0 < q < 1$, then for $x \vee |y| = o(c_n)$ with $x > -M$,

$$P_x[S_n = y \mid \sigma_B > n] \sim \sqrt{\frac{\pi}{8c_\#}} \, [1 + \text{sgn}(y)k_\lambda k_{\lambda'}] \, \frac{P_{-y}[\sigma_B > n]}{c_n}.$$

---

[4] One needs to perform some additional algebraic manipulations.

Proof (of Proposition 9.2.9) The second relation of (i) follows from the first (see the proof of Lemma 9.3.5). For the verification of the latter, from (9.30) and (9.31) one infers that

$$\frac{P_x[\sigma_B > n]}{P_0[\sigma_0 > n]} = \sum_{y \in \mathbb{Z}} g_B(x, y) \sum_{z \in B} P_{-z}[S_{n+1} = -y \mid \sigma_{-B} > n] \frac{P_{-z}[\sigma_{-B} > n]}{P_0[\sigma_0 > n]}. \quad (9.41)$$

The last ratio approaches $u_B(-z)$ (see (9.46)) and $\sum_{z \in B} u_{-B}(-z) = 1$. Let $m_-/m \to q > 0$ and $u_B(x) > 0$. Then by (9.36), (9.6) and Proposition 4.3.3,

$$\frac{g_B(x, y)}{u_B(x)\{1 + o(1)\}} = \begin{cases} 1 - L(x)/L(c_n) & \text{for } y < -\varepsilon c_n, \\ 1 + pL(x)/[qL(c_n)] & \text{for } y > \varepsilon c_n, \end{cases} \quad (9.42)$$

provided $0 < x = o(c_n)$. With the help of the bound $g_B(x, y) \leq 2\bar{a}(x) \asymp a(x)$ together with (9.42) and the above expression of $P_x[\sigma_B > n]/P_0[\sigma_0 > n]$, an application of the weak convergence $\mathbf{P}_{-z}^{n,B} \Rightarrow \mathbf{P}^0$ deduces the first relation of (i).

For the proof of (ii) we may suppose $B \subset (-\infty, 0]$. Let $m_-/m \to 0$. Then

$$P_x[\sigma_B > n] = P_{x-1}[T > n] + \sum_{k=1}^{n} \sum_{y \notin B} P_x[S_k = y, \sigma_{(-\infty, 0]} = k]P_y[\sigma_B > n - k]$$

for $x > 0$. Observe that the above double sum restricted to $k > \varepsilon n$ is negligible in comparison with $P_x[T > n]$ as $n \to \infty$ for every $\varepsilon > 0$. Then, using (9.32) and $2\bar{a}(x) \sim a(-x)$ $(x \to \infty)$, valid under the present assumption, one sees that for any $M > 1$, the double sum may be written as

$$\sum_{k=1}^{\varepsilon n} \sum_{-Mx < y \leq 0, y \notin B} P_x[S_k = y, \sigma_{(-\infty, 0]} = k]P_y[\sigma_B > n] + \frac{r_{M,x}c_n}{n} + o(P_x[T > n]),$$

where $r_{M,x} \leq C \sum_{y < -Mx} H^x_{(-\infty, 0]}(y)a(y)$. If $\mu_-(x) \asymp x^{-\beta}L_-(x)$, then $r_{M,x}/a(x)$ approaches zero as $x \wedge M \to \infty$ (use (4.40)). By (i) and (9.46), $P_y[\sigma_B > n] \sim u_B(y)P_0[\sigma_0 > n]$ for $-Mx < y \leq 0$, provided $L(x)/L(c_n) \to 0$, while by the dual of (9.9), $\sum_{y \notin B} H^x_{(-\infty, 0]}(y)u_B(y) = u_B(x)$ for $x > 0$. These together verify the second half of (ii). The first half is similar (but much simpler). ☐

Remark 9.2.13 Let $\alpha = 1$ (with $\rho\hat{\rho} > 0$). If $a(y+1) - a(y) = O(a(y)/|y|)$ $(|y| \to \infty)$ (see Proposition 4.3.5 for a sufficient condition), then by (9.41) it follows that for $|x| = o(c_n)$

$$P_x[\sigma_B > n] \sim u_B(x)P_0[\sigma_0 > n]$$

and the formulae asserted above for x fixed are extended to $x = o(c_n)$ (observe $a(x)L(x) \to \infty$).

9.3 Proof of Theorem 9.2.1

The proof will be given by a series of lemmas. Let c_n be given as in (2.17) and write x_n for x/c_n and similarly for y_n. Let τ_n^B stand for the epoch when the first excursion of length larger than n starts:

$$\tau_n^B = \inf\{k : S_{k+j} \notin B \text{ for } j = 1, \ldots, n\}. \tag{9.43}$$

The following lemma, adapted from [10, Lemma 3.1], is central to the present approach.

Lemma 9.3.1 *For any positive integers n and m and real numbers* $a_1, \ldots, a_m$,

$$P_x[S_{\tau_n^B + k} \leq a_k, \ k = 1, \ldots, m]$$
$$= P_x[S_k \leq a_k, \ k = 1, \ldots, m \mid \sigma_B > n] P_x[\sigma_B > n]$$
$$+ \sum_{y \in B} P_x[S_{\tau_n^B} = y, \sigma_B \leq n] P_y[S_k \leq a_k, \ k = 1, \ldots, m \mid \sigma_B > n].$$

Proof Writing τ_n for τ_n^B, one has

$$P_x[S_{\tau_n + k} \leq a_k, \ k = 1, \ldots, m]$$
$$= \sum_{j=0}^{\infty} \sum_{y \in \mathbb{Z}} P_x[S_{j+k} \leq a_k, \ k = 1, \ldots, m \mid \tau_n = j, S_j = y] P_x[\tau_n = j, S_j = y].$$

For $j \geq 1$, observe that $\{\tau_n \geq j\} = \bigcap_{i=0}^{j-1}\{i + 1 \leq^\exists r \leq j \wedge (i+n), S_r \in B\}$, which depends only on $X_1, \ldots, X_j$, and that

$$\{\tau_n = j, S_j = y\} = \{\tau_n \geq j\} \cap \{S_j = y, S_{j+k} \notin B, \ k = 1, \ldots, n\} \quad (y \in B).$$

One then infers that the conditional probability above is expressed as

$$P_y[S_k \leq a_k, \ k = 1, \ldots, m \mid \sigma_B > n].$$

Since $\sum_{j=1}^{\infty} P_x[\tau_n = j, S_j = y] = P_x[S_{\tau_n} = y, \sigma_B \leq n]$, $P_x[S_j \in B \mid \tau_n = j] = 1$ ($y \in B$, $j > 0$) and $\{\tau_n = 0\} = \{\sigma_B > n\}$, we obtain the asserted identity. $\qquad\square$

Let $\mathbf{P}_x^{\mathrm{exc}(n,B)}$ stand for the probability measure on $D_{[0,\infty)}$ induced from P_x by the excursion process $Y_n^{\mathrm{exc}(n,B)}(t) := S_{\tau_n^B + \lfloor nt \rfloor}, t \geq 0$.

Lemma 9.3.2 *Let* $\sigma^2 = \infty$. *Then for each x with* $u_B(x) > 0$, $\mathbf{P}_x^{\mathrm{exc}(n,B)} \Rightarrow \mathbf{P}^0$. *In particular* $\mathbf{P}_0^{n,\{0\}} = \mathbf{P}_0^{\mathrm{exc}(n,\{0\})} \Rightarrow \mathbf{P}^0$.

Proof For $f \in D_{[0,\infty)}$, we have denoted by t_f the left end point of the first excursion interval (away from zero) of f of length > 1. Put $B(n) = \bigcup_{y \in B}[y - \frac{1}{2}, y + \frac{1}{2}]/c_n$ and denote by $t^{B(n)}(f)$ the left point of the first excursion away from $B(n)$ of length larger than 1. Note that $t^{B(n)}(Y_n^{\mathrm{exc}(B)}) = \tau_n^B$, the augmentation of B having no effect on the excursion intervals of $\mathbb{Z}$-valued paths. Define the map h and h_n from $D_{[0,\infty)}$ into itself, respectively, by

$$h(f) = f(\mathfrak{t}_f + \cdot) \quad \text{and} \quad h^{B(n)}(f) = f(\mathfrak{t}_f^{B(n)} + \cdot),$$

so that $\mathbf{P}^0 = P_0^Y \circ h(Y)^{-1}$ and $Y_n^{\mathrm{exc}(B)} = h^{B(n)}(Y_n)$. Since $P_x \circ Y_n^{-1} \Rightarrow P_0^Y$, by the continuity theorem (Theorem 5.5 of [7]) it suffices to show that

$$\text{if } f_n \to f, \text{ then } h^{B(n)}(f_n) \to h(f) \text{ for } \mathbf{P}^0\text{-almost every } f \in D_{[0,\infty)}$$

for any sequence $(f_n) \subset D_{[0,\infty)}$, where the convergence is, of course, in Skorokhod's topology. Below we apply some sample path properties of Y, whose proofs are postponed to the end of this section. Since Y can reach zero if and only if it continuously approaches zero (a.s.) (Lemmas 9.3.8) and since every excursion interval of f to the left of $\mathfrak{t}$ is of length less than 1 (Lemma 9.3.9), both $\mathfrak{t} : D_{[0,\infty)} \mapsto (0, \infty)$ and $h : D_{[0,\infty)} \mapsto D_{[0,\infty)}$ are continuous. One also sees that if $f_n \to f$, then for $P_x^Y \circ Y^{-1}$-almost all f,

$$\mathfrak{t}_{f_n}^{B(n)} - \mathfrak{t}_f^{B(n)} \to 0 \quad \text{and} \quad \mathfrak{t}_f^{B(n)} \to \mathfrak{t}_f.$$

Here it is effectively applied that $\mathfrak{t}_f^{B(n)}$ is defined with $B(n)$ instead of B/c_n. Since $h^{B(n)} \circ f_n(0) = O(1/c_n) \to 0 = h \circ f(0)$, our task reduces to showing that for sequences of $f_n \in D_{[0,\infty)}$ and positive real numbers s_n,

$$\text{if } f_n \to f, s_n \to s \text{ and } f_n(s_n) \to f(s), \text{ then } f_n(s_n + \cdot) \to f(s + \cdot).$$

This is an elementary matter concerning the convergence of a sequence in the space $(D_{[0,\infty)}, d)$, so we omit its proof. □

Lemma 9.3.3 *Suppose F is recurrent. Then*

$$P_0[\sigma_0 > n] \sim \begin{cases} \kappa_\alpha^\# c_n/n & \text{if } 1 < \alpha \le 2, \\ 1/[1/\tilde{L}]_*(n) & \text{if } \alpha = 1. \end{cases}$$

Here $\tilde{L}(k) = L(c_k)$ and $[1/\tilde{L}]_*(n) = \int_0^n [t\tilde{L}(t)]^{-1}$ and $\kappa_\alpha^\#$ is given in (9.17).

Proof Put $\varphi(s) = \sum_{n=0}^\infty P_0[S_n = 0]e^{-sn}$ and $\omega(s) = \sum_0^\infty P_0[\sigma_0 > n]e^{-sn}$. Then $\varphi(s)\omega(s) = 1/(1 - e^{-s})$. When $1 < \alpha \le 2$, from the local limit theorem (9.1), one obtains $\varphi(s) \sim \mathfrak{p}_1(0) \sum e^{-sn}/c_n \sim \mathfrak{p}_1(0)\Gamma(1 - 1/\alpha)s^{1/\alpha-1}[\tilde{L}(1/s)]^{-1/\alpha}$, and a standard Tauberian argument leads to the result for $1 < \alpha \le 2$.

Let $\alpha = 1$. Then by (9.1) again one has $P[S_n = 0] \sim \mathfrak{p}_1(0)/n\tilde{L}(n)$, so that

$$\varphi(s) \sim \mathfrak{p}_1(0)[1/\tilde{L}]_*(1/s), \quad \text{hence} \quad s\omega(s) \sim 1/[1/\tilde{L}]_*(1/s) \quad (s \downarrow 0).$$

To use the Tauberian argument, we need a Tauberian condition. To this end we apply Lemma 9.3.2 to see that the logarithm of $r(t) := P_0[\sigma_0 > t]$ is slowly decreasing, that is $\inf_{t'>t:t'-t\le\varepsilon t}\log[r(t')/r(t)]\to 0$ as $t\to\infty$ and $\varepsilon\downarrow 0$, which serves as the Tauberian condition (cf. [8, Theorem 1.7.5]). This condition is paraphrased as

$$\lim_{\varepsilon\downarrow 0}\lim_{t\to\infty}\sup P_0[t\le\sigma_0\le t+\varepsilon t\mid\sigma_0>t]=0,$$

whose validity one can easily ascertain by using Lemma 9.3.2.   □

Remark 9.3.4 When $\alpha=\rho=1$, $S_{\lfloor nt\rfloor}/B_n$ converges to t locally uniformly in probability (cf. Theorem A.4.1(d)), so that by Lemma 9.3.1 the conditional process $S_{\lfloor n\cdot\rfloor}/B_n$ under the law $P_0[\cdot\mid\sigma_0>n]$ also converges to t. The local limit theorem implies merely $P_0[S_n=0]=o(1/c_n)$. Even though this does not allow us to follow the proof above, we can still prove that $P_x[S_{\lfloor n\cdot\rfloor}/B_n\in\cdot\mid\sigma_B>n]$ converges to the law concentrating at the single path $f(t)\equiv t$. To see this we show that

(*) $P_0[\sigma_0>n]$ is s.v. as $n\to\infty$.

To this end, it suffices, in view of the proof of Lemma 9.3.3, to show $\varphi(s)$ is s.v. as $s\downarrow 0$. For simplicity, assume F is strongly aperiodic. Then

$$2\pi\varphi(s)=\sum_{n=0}^{\infty}e^{-sn}\int_{-\pi}^{\pi}\psi^n(\theta)\,d\theta\sim\Re\int_{-\delta}^{\delta}\frac{d\theta}{1-e^{-s}\psi(\theta)}\quad\text{for any }\delta>0.$$

By $1-\psi(\theta)=\frac{\pi}{2}\theta[L(1/\theta)+iA(1/\theta)]\{1+o(1)\}$ $(\theta\to 0)$, one infers

$$2\pi\varphi(s)\sim\zeta(s;\delta):=\int_{-\delta}^{\delta}\frac{s+\frac{\pi}{2}\theta L(1/\theta)}{[s+\frac{\pi}{2}\theta L(1/\theta)]^2+[\theta A(1/\theta)]^2}\,d\theta.$$

Since A is s.v., one has $\zeta(\lambda s;\delta)\sim\zeta(s;\delta/\lambda)\sim\zeta(s;\delta)$ for any $\lambda>0$. Thus $\varphi(s)$ is s.v., as desired. Now (*) is applicable. For the rest, one can proceed as in the sequel.

In order to obtain the convergence of the conditional process $\mathbf{P}_x^{n,B}$ to $\mathbf{P}^0$ we must verify the next lemma (due to Belkin [3]). For this purpose we employ the analytic approach based on the harmonic analysis as in [2] but with the help of Lemma 9.3.2.

Lemma 9.3.5 *For any $x\in\mathbb{Z}$ with $a^\dagger(x)>0$, as $n\to\infty$,*

$$\mathbf{P}_x^{n,\{0\}}[\mathfrak{y}_1\in\cdot]\Longrightarrow\mathbf{P}^0[\mathfrak{y}_1\in\cdot].$$

Proof Put $r_n^x=P_x[\sigma_0>n]$ and $f_n^x=P_x[\sigma_0=n]$. We first make the decomposition

$$P_x[S_n=y\mid\sigma_0>n]=\frac{1}{r_n^x}\left(P_x[S_n=y]-\sum_{k=1}^{n}f_k^x P_0[S_{n-k}=y]\right),\quad n\ge 1.$$

(Note both sides vanish for $y=0$.) Put $\phi_{n,x}(\theta)=E_x[e^{i\theta S_n}\mid\sigma_0>n]$. Then, from the above identity one obtains $\phi_{n,x}(\theta)=\left(\psi^n(\theta)e^{i\theta x}-\sum_{k=1}^{n}f_k^x\psi^{n-k}(\theta)\right)/r_n^x$, and after summing by parts one gets

$$\phi_{n,x}(\theta) = \frac{e^{i\theta x} - 1}{r_n^x} \psi^n(\theta) + 1 - \sum_{k=0}^{n-1} \frac{r_k^x}{r_n^x} \{\psi^{n-k-1}(\theta) - \psi^{n-k}(\theta)\}. \tag{9.44}$$

Now, putting $\Psi_n^x(\theta) = E_x[e^{i\theta Y_n(1)} \mid \sigma_0 > n] = \phi_{n,x}(\theta/c_n)$, this is rewritten as

$$\Psi_n^x(\theta) = 1 - \sum_{k=0}^{n-1} \frac{r_k^x}{r_n^x} \left[1 - \psi\left(\frac{\theta}{c_n}\right)\right] \psi^{n-k-1}\left(\frac{\theta}{c_n}\right) + \frac{e^{i\theta x/c_n} - 1}{r_n^x} \psi^n\left(\frac{\theta}{c_n}\right). \tag{9.45}$$

By Lemma 9.3.3 the last term tends to zero as $n \to \infty$ (note $n/c_n^2 \sim 1/L(c_n) \to 0$ for $\alpha = 2$). Let $a^\dagger(x) > 0$. Then, by (2.7), $r_k^x/r_n^x \sim r_n^0/r_n^x$ if $\varepsilon n < k < n$, and by (4.18, 4.19) (and (4.30) if $\alpha = 1$) $1 - \psi(\theta) \asymp \theta^\alpha L(1/\theta)$, which shows $(1 - \psi(\theta/c_n)) = O(1/n)$ (recall $c_n^\alpha/L(c_n) \sim n$). Noting $\sum_1^{\varepsilon n} r_k^0 \le C\varepsilon^{1/\alpha} n r_n^0$ we can therefore conclude $\Psi_n^x(\theta) = \Psi_n^0(\theta) + o(1)$. We know $\Psi_n^0(\theta)$ tends to the characteristic function of $\mathfrak{q}$ owing to Lemma 9.3.2. Thus we have for $\xi \in \mathbb{R}$,

$$\mathbf{P}_x^{n,\{0\}}[\mathfrak{y}_1 \le \xi] = P_x[S_n/c_n \le \xi \mid \sigma_0 > n] \longrightarrow \mathbf{P}^0[\mathfrak{y}_1 \le \xi]. \qquad \square$$

Our proof of the following lemma, which extends the result on $\mathbf{P}_x^{n,\{0\}}$ in Lemma 9.3.5 to $\mathbf{P}_x^{n,B}$ for any finite set B, rests on the formula due to Kesten and Spitzer [49] that states

$$\lim_{n\to\infty} \frac{P_x[\sigma_B > n]}{P_0[\sigma_0 > n]} = \mathsf{u}_B(x). \tag{9.46}$$

Lemma 9.3.6 *For any $x \in \mathbb{Z}$ with $\mathsf{u}_B(x) > 0$,*

$$\lim_{n\to\infty} \mathbf{P}_x^{n,B}[\mathfrak{y}_1 \le \xi] = \mathbf{P}^0[\mathfrak{y}_1 \le \xi] \qquad (\xi \in \mathbb{R}). \tag{9.47}$$

Proof By Lemma 9.3.5 we have $\mathbf{P}_x^{n,\{w\}} \Rightarrow \mathbf{P}^0$ if $\mathsf{u}_{\{w\}}(x) = a^\dagger(x - w) > 0$. Let $\mathsf{u}_B(x) > 0$. Suppose $0 \in B$ for simplicity and observe that for any real number ξ, the probability $\mathbf{P}_x^{n,B}[\mathfrak{y}_1 \le \xi]$ is expressed as

$$\mathbf{P}_x^{n,\{0\}}[\mathfrak{y}_1 \le \xi] \frac{P_x[\sigma_0 > n]}{P_x[\sigma_B > n]}$$

$$- \sum_{w \in B, w \ne 0} \sum_{k=1}^{\infty} P_x[S_{\sigma_B} = w, \sigma_B = k] \mathbf{P}_w^{n-k,\{0\}} \left[\mathfrak{y}_1 \le \frac{\xi c_n}{c_{n-k}}\right] \frac{P_w[\sigma_0 > n - k]}{P_x[\sigma_B > n]}.$$

Noting that $P_x[\sigma_B > M] \to 0$ as $M \to \infty$ and the ratio in the summand above approaches $a(w)/\mathsf{u}_B(x)$ for $k \le M$ (for each M), we pass to the limit as $n \to \infty$ to find

$$\lim \mathbf{P}_x^{n,B}[\mathfrak{y}_1 \le \xi] = \mathbf{P}^0[\mathfrak{y}_1 \le \xi]\{a^\dagger(x) - E_x[a(S_{\sigma_B})]\}/\mathsf{u}_B(x).[5]$$

Since $a^\dagger(x) - E_x[a(S_{\sigma_B})] = \mathsf{u}_B(x)$ under the present supposition that $0 \in B$, we obtain $\mathbf{P}_x^{n,B}[\mathfrak{y}_1 \le \xi] \to \mathbf{P}^0[\mathfrak{y}_1 \le \xi]$. $\qquad \square$

---

[5] If S is transient, $a(S_{\sigma_B})$ is understood to vanish when $\sigma_B = \infty$.

Proof (of Theorem 9.2.1) To complete the proof we use Theorem 9.2.2,[6] which we shall deduce in Section 9.6 from relation (9.47) together with Lemma 9.3.3. As mentioned previously, Theorem 9.2.2 entails the convergence of finite-dimensional distributions of $\mathbf{P}_x^{n,B}$, and in particular

$$\lim_{n\to\infty} P_x[|S_{\lfloor tn \rfloor}| > \eta c_n \mid \sigma_B > n] = \mathbf{P}^0[\mathfrak{y}_t > \eta]. \tag{9.48}$$

We show the tightness of the sequence $(\mathbf{P}_x^{n,B})$ in the next lemma, and hence conclude the convergence $\mathbf{P}_x^{n,B} \Rightarrow \mathbf{P}^0$. □

Lemma 9.3.7 *The sequence* $(\mathbf{P}_x^{n,B})$ *is tight for every* $x \in \mathbb{Z}$.

Proof If S is transient, then $P_x[\sigma_B > n] \to u_B(x)/G(0)$ and the assertion follows from the result for the unconditional law. Suppose S is recurrent in the sequel.

Since the excursion paths are right-continuous and start at zero a.s. $(\mathbf{P}^0)$, by (9.48) it follows that for any $\varepsilon > 0$, as $n \to \infty$ and $\delta \downarrow 0$ in this order

$$P_x[|S_{\lfloor \delta n \rfloor}| > \varepsilon c_n \mid \sigma_B > n] \longrightarrow 0. \tag{9.49}$$

This intuitively obvious result is crucial to the proof. Put $\tau_{n,\varepsilon} = \inf\{k : |S_k| > \varepsilon c_n\}$. In view of the tightness criterion as given by [7, Theorem 14.4], for the proof of the lemma it suffices to show that

$$P_x[\tau_{n,\varepsilon} \le \delta n \mid \sigma_B > n] \longrightarrow 0 \quad \text{as } n \to 0 \text{ and } \delta \downarrow 0 \text{ in this order} \tag{9.50}$$

since $(\mathbf{P}_y^{n,B})$ satisfies the tightness criterion uniformly for $|y| > \varepsilon' c_n$ for each $\varepsilon' > 0$ and $P_x[|S_{\lfloor \delta n \rfloor}| > \varepsilon' c_n] \uparrow 1$ as $\varepsilon' \downarrow 0$ by Lemma 9.3.6. Make the decomposition

$$P_x[\tau_{n,\varepsilon} \le \delta n, \sigma_B > n]$$

$$= \sum_{|y| > \varepsilon c_n} \sum_{0 \le k \le \delta n} P_x[\tau_{n,\varepsilon} = k, S_k = y, \sigma_B > k] P_y[\sigma_B > n - k].$$

Using the functional limit theorem for the unconditioned walk, observe that there exists a constant $C = C(\varepsilon)$ such that for all (k, y) in the range of the double sum above

$$P_y[\sigma_B > n - k] < C P_y[S_{\lfloor \delta n \rfloor - k} > \varepsilon c_n, \sigma_B > n - k]$$

for all sufficiently large n. Hence $P_x[\tau_{n,\varepsilon} < \delta n, \sigma_B > n]$ is less than

$$C \sum_{|y| > \varepsilon c_n} \sum_{0 \le k < \delta n} P_x[\tau_{n,\varepsilon} = k, S_k = y, \sigma_B > k] P_y[S_{\lfloor \delta n \rfloor - k} > \varepsilon c_n, \sigma_B > n - k].$$

Noting this double sum equals $P_x[|S_{\lfloor \delta n \rfloor}| > \varepsilon c_n, \sigma_B > n]$, one finds that (9.50) follows from (9.49). □

---

[6] What we actually need is (9.20) rather than the local theorems such as (9.22) and (9.19).

Some sample path properties of the limit stable process

In our proof of Lemma 9.3.2 we have taken for granted the sample property of Y that Y can reach any given point if and only if it continuously approaches the point. Here we provide a proof where we use the fact that the bridge $P_\xi^Y[\cdot \,|\, Y(1) = \eta]$ has the version, denoted by $P_{\xi,\eta}^Y[\cdot]$, which is uniquely defined for every pair of $\xi, \eta \in \mathbb{R}$ by

$$(*)\quad P_{\xi,\eta}^Y[\Gamma] = \lim_{\varepsilon \downarrow 0} P_\xi^Y\left[\Gamma \,\big|\, |Y(1) - \eta| < \varepsilon\right]$$

valid for every event Γ depending on $(Y(t))_{t < \delta}$ for any $\delta < 1$ (cf. [4, §III.3]).

Lemma 9.3.8 (i) *For every ξ, $P_\xi^Y[\exists t > 0, Y(t-) \neq Y(t) = 0] = 0$.*

 (ii) *For all $\xi, \eta \in \mathbb{Z}$, $P_{\xi,\eta}^Y[\exists t \in (0, 1], Y(t-) \neq Y(t) = 0] = 0$.*

 (iii) *For every ξ, $P_\xi^Y[\exists t > 0, Y(t) \neq Y(t-) = 0] = 0$.*

Proof For the proof of (i) it is sufficient to show that

$$P_\xi^Y[\exists t > 0, Y(t-) < -\varepsilon \text{ and } Y(t) = 0] = 0 \quad \text{for any } \varepsilon > 0; [7] \qquad (9.51)$$

by symmetry this entails $P_\xi^Y[\exists t > 0, |Y(t-)| > \varepsilon \text{ and } Y(t) = 0] = 0$. The event under the symbol P_ξ^Y in (9.51) is contained in $\cup_{0 < r \in \mathbb{Q}} \{Y(r) < -\varepsilon, S_{\sigma[-\varepsilon,\infty)} \circ \vartheta_r = 0\}$, whose probability vanishes since $P_\eta^Y[S_{\sigma[-\varepsilon,\infty)} = 0] = 0$ for all $\eta < -\varepsilon$. (9.51) is verified

For the proof of (ii) first suppose $\rho\hat\rho > 0$ (automatically holds if $\alpha > 1$). It follows from (i) that $P_{\xi,\zeta}^Y[\exists t \in (0, 1), Y(t-) \neq Y(t) = 0] = 0$ for almost every ζ (w.r.t. the Lebesgue measure), since $P_\xi^Y[Y(1) \in d\zeta]$ has a positive continuous density. By virtue of $(*)$ we accordingly obtain $P_{\xi,\eta}^Y[\exists t \in (0, \delta), Y(t-) \neq Y(t) = 0] = 0$. Letting $\delta \uparrow 1$, we have the desired result. One can easily modify the above argument to deal with the case $\rho\hat\rho = 0$.

The proof of (iii) follows from (ii) by time reversal. Put for $0 \leq r < s \leq 1$,

$$\Gamma_{(r,s)}(Y) = \{\exists t \in (r, s), \ Y(t-) \neq Y(t) = 0\},$$

and

$$\Gamma_{(r,s)}^*(Y) = \{\exists t \in (r, s), \ Y(t) \neq Y(t-) = 0\}.$$

Then, $\Gamma_{(0,s)}^*(Y) = \{\exists t \in (1 - s, 1), Y(1 - t) \neq Y((1 - t)-) = 0\}$, so that, on writing $Y^*(t) := Y((1 - t)-)$,

$$\Gamma_{(0,s)}^*(Y) = \Gamma_{(1-s,1)}(Y^*).$$

Since, by a theorem concerning duality, the law of $(Y^*(t))_{0 \leq t \leq 1}$ under $P_{\xi,\eta}^Y$ coincides with that of $(-Y(t))_{0 \leq t \leq 1}$ under $P_{-\eta,-\xi}^Y$ [4, Corollary II.1.3] and since $\Gamma_{(r,s)}(Y) = \Gamma_{(r,s)}(-Y)$, we can conclude that

---

[7] This can be deduced from Proposition III.2.2(i) of [4] since 0 is regular for $(0, \infty)$ w.r.t. Y.

$$P^Y_{\xi,\eta}[\Gamma^*_{(0,s)}(Y)] = P^Y_{\xi,\eta}[\Gamma_{(1-s,1)}(Y^*)]$$
$$= P^Y_{-\eta,-\xi}[\Gamma_{(1-s,1)}(-Y)] = P^Y_{\xi,\eta}[\Gamma_{(1-s,1)}(Y)] = 0.$$

Thus (iii) has been verified, since s can be made arbitrarily close to 1. □

Lemma 9.3.9 *For each $\xi \in \mathbb{R}$ and $c > 0$, with P^Y_ξ-probability 1 there is no excursion interval, away from zero, of Y whose length equals c.*

Proof For any rational number $r > 0$, let (g, d) be the excursion interval containing r, which exists a.s. (P^Y_ξ) because of the trivial fact that $Y(r) \neq 0$ a.s. Since $r - g$ is measurable w.r.t. $\mathcal{F}_r := \sigma(Y(s) : 0 \le s \le r)$, it follows that $P^Y_\xi[d - g = c] = \int P^Y_\eta[Y(c - r + t) = 0]P^Y_\xi[Y(r) \in d\eta, g \in dt] = 0.$ □

9.4 The Distribution of the Starting Site of a Large Excursion

Though not used in this treatise, the following proposition is of independent interest.. Recall τ^B_n is the random time defined in (9.43).

Proposition 9.4.1 *If S is recurrent, $\sigma^2 = \infty$ and $u_B(x) > 0$ for all x, then $\lim_{n \to \infty} P_x[S_{\tau^B_n} = y] = u_B(y)$ for $y \in B$ and $x \in \mathbb{Z}$.*

To prove this result we need to use the following lemma. In the sequel suppose $u_B(x) > 0$ for all x.

Lemma 9.4.2 *If S is recurrent, for any $x, y \in \mathbb{Z}$ and $w, w' \in B$,*

$$\lim_{n \to \infty} \frac{\sum_{m=0}^\infty P_x[\tau^B_n \ge m, S_m = w]}{\sum_{m=0}^\infty P_y[\tau^B_n \ge m, S_m = w']} = 1.$$

Proof It suffices to show the asserted relation when $x \in B$. Note that $u_B(x) > 0$ implies $P_x[S_{\sigma_B} = w] > 0$ for all $w \in B$. Let $x \in B$ and $u_B(x) > 0$. If $m \le n$, then $P_x[\tau_n \ge m, S_m = w] = P_x[S_m = w] \sim 1/c_m$ as $m \to \infty$. Noting that $\sigma_B \le m \wedge n$ if $\tau_n \ge m$ and $S_m = w$, we see

$$P_x[\tau^B_n \ge m, S_m = w] = \sum_y \sum_{j=0}^{n \wedge m} P_x[\sigma_B = j, S_j = y]P_y[\tau_n \ge m - j, S_{m-j} = w].$$

Put $J_n(x, w) = \sum_{m=0}^\infty P_x[\tau^B_n \ge m, S_m = w]$. Then after interchanging the order of summation and making a change of variables, we obtain

$$J_n(x, w) = \sum_{y \in B} \sum_{j=0}^n P_x[\sigma_B = j, S_j = y]J_n(y, w)$$
$$= \sum_{y \in B} P_x[S_{\sigma_B} = y]\{1 + o(1)\}J_n(y, w).$$

We apply this last relation to $J_n(y, w)$ and iterate the same procedure. The trace of the r.w. on B constitutes a Markov chain with the one-step transition probability $p_{x,y} = P_x[S_{\sigma_B} = y]$, which is ergodic. If π stands for the unique invariant measure of the chain, then it follows that

$$J_n(x, w) = \sum_{y \in B} \pi(y)\{1 + o(1)\}J_n(y, w),$$

showing $J_n(x, w) \sim J_n(y, w)$ for every $x, y \in B$. This result applies to the dual walk so that $\hat{J}_n(x, w) \sim \hat{J}_n(y, w)$, which in turn yields $\hat{J}_n(w, x) \sim \hat{J}_n(w, y)$. Since w is arbitrary, we have the relation of the lemma.   □

Proof (of Proposition 9.4.1) We may suppose $0 \in B$. Put

$$\zeta_n = \inf\{k > \tau_n^B : S_k = 0\},$$

the epoch when the excursion that begins at τ_n^B terminates.

Put $h_n(y, w) = P_x[S_{\tau_n^B} = y, S_{\zeta_n} = w]$ ($w \in B$) so that $P_x[S_{\tau_n^B} = y] = \sum_{w \in B} h_n(y, w)$. We have

$$h_n(y, w) = \sum_{m=n+1}^{\infty} P_x[S_{\tau_n^B} = y, \zeta_n = m, S_m = w]$$

$$= \sum_{m=n+1}^{\infty} \sum_{k=n+1}^{m} \hat{P}_w[\sigma_B = k, S_{\sigma_B} = y]\hat{P}_y[\tau_n^B \geq m - k, S_{m-k} = x],$$

where the second equality is obtained by considering the dual walk. After changing the order of summation, the repeated sum above becomes

$$\hat{P}_w[\sigma_B > n, S_{\sigma_B} = y] \sum_{m=0}^{\infty} \hat{P}_y[\tau_n^B \geq m, S_m = x].$$

Write the first probability above as $\sum_z \hat{P}_w[S_n = z, S_{\sigma_B} = y \mid \sigma_B > n]\hat{P}_w[\sigma_B > n]$, which we further rewrite as $\hat{P}_w[\sigma_B > n] \sum_{z \notin B} \hat{P}_w[S_n = z \mid \sigma_B > n]\hat{P}_z[S_{\sigma_B} = y]$. Since $\hat{P}_w[\sigma_B > n] \asymp \hat{P}_w[\sigma_0 > n]$ and $\{\sigma_B > n\} \subset \{\sigma_0 > n\}$, from Lemma 9.3.5 we have $\hat{P}_w[|S_n| > \sqrt{c_n} \mid \sigma_B > n] \to 1$ as $n \to \infty$. Combining this with the fact that $P_z[S_{\sigma_B} = y] \to \mathsf{u}_B(y)$ (as $|z| \to \infty$ under $\sigma^2 = \infty$) shows that as $n \to \infty$,

$$\hat{P}_w[\sigma_B > n, S_{\sigma_B} = y] = \mathsf{u}_B(y)\hat{P}_w[\sigma_B > n]\{1 + o(1)\},$$

hence

$$h_n(y, w) = \mathsf{u}_B(y)\hat{P}_w[\sigma_B > n] \sum_{m=0}^{\infty} \hat{P}_y[\tau_n^B \geq m, S_m = x]\{1 + o(1)\}.$$

By Lemma 9.4.2 we infer that the probability under the summation sign can be replaced by $\hat{P}_x[\tau_n^B \geq m, S_m = w]$, so that

$$h_n(y, w) = u_B(y)\hat{P}_w[\sigma_B > n] \sum_{m=0}^{\infty} \hat{P}_x[\tau_n^B \geq m, S_m = w]\{1 + o(1)\}. \qquad (9.52)$$

Since $1 = \hat{P}_x[\tau_n^B < \infty] = \sum_{w \in B} \sum_{m=0}^{\infty} \hat{P}_x[\tau_n^B \geq m, S_m = w]\hat{P}_w[\sigma_B > n]$, we can accordingly conclude $\sum_{w \in B} h_n(y, w) \sim u_B(y)$, the relation of the proposition. □

9.5 An Application to the Escape Probabilities From a Finite Set

Shimura [68] (in the case $\sigma^2 < \infty$) and Doney [19] (under (AS) with some restriction) applied Bolthausen's idea [10] to the r.w. conditioned on various events that depend only on the path $(S_n)_{0 \leq n \leq \sigma_Q}$ and obtained the corresponding conditional limit theorems. Here we follow them to derive asymptotics of the escape probability from a finite set $B \subset \mathbb{Z}$ that are addressed in Section 5.6 in the special case $B = \{0\}$ under $m_+/m \to 0$. We continue to suppose that (AS) holds and $\rho\hat{\rho} \neq 0$ if $\alpha = 1$. We also suppose that F is recurrent, the present problem being trivial for transient F. As in Section 5.6, let Q, R be two positive integers and put $\Delta(Q, R) = (-\infty, -Q] \cup [R, \infty)$.

Proposition 9.5.1 *As $R \to \infty$ along with $Q/R \to \lambda \in (0, \infty)$,*

$$P_0[\sigma_{\Delta(Q,R)} < \sigma_0] \sim \begin{cases} C(\lambda)R^{1-\alpha}L(R) & \text{if } 1 < \alpha \leq 2, \\ 1/[1/\tilde{L}]_*(R) & \text{if } \alpha = 1 \text{ and } F \text{ is recurrent,} \end{cases} \qquad (9.53)$$

for a certain constant $C(\lambda) > 0$ (see Remark 9.5.3 for $C(\lambda)$).

The proof is given after Lemma 9.5.2 below. The result will be extended to a general finite set at the end of the section.

We consider the conditioning on the following two events:

$$\Gamma_{n,s}^{\circ} = \{\sigma_0 > sn\} \quad \text{and} \quad \Gamma_{Q,R}^{*} = \{\sigma_{\Delta(Q,R)} < \sigma_0\}.$$

For $f \in D_{[0,\infty)}$ denote by σ_K^f the first hitting time to $K \subset \mathbb{R}$ of the path f and put

$$A_s^{\circ} = \{f : \sigma_0^f > s\} \quad \text{and} \quad A_{q,r}^{*} = \{f ; \sigma_{\Delta(q,r)}^f < \sigma_0^f\} \qquad (s, q, t > 0)$$

so that $\Gamma_{n,s}^{\circ} = \{Y_n \in A_s^{\circ}\}$ and $\Gamma_{Q,R}^{*} = \{Y_n \in A_{Q/c_n, R/c_n}^{*}\}$. Corresponding to each of A_s° and $A_{q,r}^{*}$ we can consider excursions of f, away from 0, that belong to them and define

$(g_s^{\circ}(f), d_s^{\circ}(f))$: the first excursion interval (g, d) of f s.t. $f(g + \cdot) \in A_s^{\circ}$,

$(g_{q,r}^{*}(f), d_{q,r}^{*}(f))$: the first excursion interval (g, d) of f s.t. $f(g + \cdot) \in A_{q,r}^{*}$.

Below we shall often omit f from notation: e.g., $Y_n(g_r)$ is written for $Y_n(g_r^\circ(Y_n))$ as usual. Observe for $j = 1, 2, \ldots$

$$\{g_{Q,R}^*(S_{\lfloor n \cdot \rfloor}) = j, S_j = 0\} = \{\sigma_{K(Q,R)} > j, \sigma_{K(Q,R)} \circ \vartheta_j < \sigma_0 \circ \vartheta_j, S_j = 0\}.$$

With the help of this, as in the proof of Lemma 9.3.1, we obtain

$$P_0[Y_n \in \Gamma \mid \Gamma_{Q,R}^*] = P_0\left[Y_n(g_{Q/c_n, R/c_n}^* + \cdot) \in \Gamma\right] \tag{9.54}$$

for any $\Gamma \in \mathcal{B}_\infty$. From (9.54) we obtain, in the same way as before, the following analogue of Lemma 9.3.2. Throughout the rest of this subsection, we suppose $1 \le \alpha \le 2$ and F is recurrent if $\alpha = 1$.

Lemma 9.5.2 *For each* $q, r \in (0, \infty)$, *as* $Q/c_n \to q, R/c_n \to r$,

$$P_0[Y_n \in \cdot \mid \Gamma_{Q,R}^*] \Longrightarrow P_0^Y[Y(g_{q,r}^* + \cdot) \in \cdot]. \tag{9.55}$$

Proof Since 0 is a regular point of Y, Lemma 9.3.2 and (9.55) together imply that as $n \to \infty$,

$$P_0[\Gamma_{Q,R}^* \mid \Gamma_{n,s}^\circ] \longrightarrow P_0^Y[\sigma_{\Delta(q,r)} \circ \vartheta_{g_s^\circ} < d_s^\circ - g_s^\circ], \tag{9.56}$$

$$P_0[\Gamma_{n,s}^\circ \mid \Gamma_{Q,R}^*] \longrightarrow P_0^Y[d_{q,r}^* - g_{q,r}^* > s]. \tag{9.57}$$

Put $\lambda = q/r$. Dividing (9.56) by (9.57) we have

$$P_0(\Gamma_{Q,R}^*)/P_0(\Gamma_{n,s}^\circ) \to C_\lambda(r, s),$$

where

$$C_\lambda(r, s) = \frac{P_0^Y[\sigma_{\Delta(\lambda r, r)} \circ \vartheta_{g_s^\circ} < d_s^\circ - g_s^\circ]}{P_0^Y[d_{q,r}^* - g_{q,r}^* > s]}.$$

Now, on putting

$$C_\lambda(r) = C_\lambda(r, 1)$$

and recalling the definitions of $\Gamma_{Q,R}^*$ and $\Gamma_{n,1}^\circ$, substitution from Lemma 9.3.3 yields

$$P_0[\sigma_{\Delta(Q,R)} < \sigma_0] \sim C_\lambda(r) P_0[\sigma_0 > n] \sim \kappa_\alpha^\sharp C_\lambda(r) c_n/n.$$

By $c_n/n \sim c_n^{1-\alpha} L(c_n)/c_\sharp$ and $c_n \sim R/r$, it follows that $c_n/n \sim r^{\alpha-1} R^{1-\alpha} L(R)/c_\sharp$. Hence

$$P_0[\sigma_{\Delta(Q,R)} < \sigma_0] \sim [\kappa_\alpha^\sharp/c_\sharp] C_\lambda(r) r^{\alpha-1} R^{1-\alpha} L(R). \tag{9.58}$$

Since this equivalence does not depend on the choice of r, we conclude

$$P_0[\sigma_{\Delta(Q,R)} < \sigma_0] \sim [\kappa_\alpha^\sharp/c_\sharp] C_\lambda(1) R^{1-\alpha} L(R), \tag{9.59}$$

showing the formula of Proposition 9.5.1 with $C(\lambda) = [\kappa_\alpha^\sharp/c_\sharp]C_\lambda(1)$ for $\alpha > 1$. The case when $\alpha = 1$ and F is recurrent is treated in a similar (simpler) way. $\qquad\square$

Remark 9.5.3 (a) Let $\mathbf{P}^{(q,r)}$ stand for the law on $D_{[0,\infty)}$ induced from P_0^Y by the map $\omega \in \Omega \mapsto Y(g_{q,r}^* + \cdot) \in D_{[0,\infty)}$. Then $C_\lambda(1) = \mathbf{P}^0[\sigma_{\Delta(\lambda,1)} < \sigma_0]/\mathbf{P}^{(\lambda,1)}[\sigma_0 > 1]$ can be written as

$$
\frac{\mathbf{P}^0[\sigma_{\Delta(\lambda,1)} \le 1] + \int_{\mathbb{R}} \mathbf{P}^0[\sigma_{\Delta(\lambda,1)} > 1, \mathfrak{y}_1 \in d\eta] P_\eta^Y[\sigma_{\Delta(\lambda,1)} < \sigma_0]}{\mathbf{P}^{(\lambda,1)}[\sigma_{\Delta(\lambda,1)} \ge 1] + \int_{[0,1]\times\mathbb{R}} \mathbf{P}^{(\lambda,1)}[(\sigma_{\Delta(\lambda,1)}, \mathfrak{y}_t) \in dt d\eta] P_\eta^Y[\sigma_0 > 1 - t]}.
$$

Note that comparing (9.60) and (9.61) shows that $C_\lambda(r)r^{\alpha-1} = C_\lambda(1)$.

(b) Let $\alpha = 2$. It is easy to see $P_0[\sigma_{\Delta(Q,R)} < \sigma_0] \sim [2\bar{a}(Q)]^{-1} + [2\bar{a}(R)]^{-1}$. Since $2\bar{a}(R) \sim 2R/L(R)$, it follows that $P_0[\sigma_{\Delta(Q,R)} < \sigma_0] \sim 2^{-1}(1+\lambda^{-1})L(R)/R$. By Proposition 9.5.1 and since $\kappa_\alpha^\sharp = \sqrt{2c_\sharp/\pi}$, this yields

$$
C_\lambda(1) = (1+\lambda^{-1})\sqrt{c_\sharp \pi/8}.
$$

To extend Proposition 9.5.1 to σ_B in place of σ_0 we show the following

Lemma 9.5.4 *On the LHS of (9.55) P_0 may be replaced by P_x for any $x \in \mathbb{Z}$.*

Proof Let $0 = t_0 < t_1 < \cdots < t_m$ and φ be a bounded continuous function on $\mathbb{R}^m$ and put $H(f) = \varphi(\mathfrak{y}_{t_0}, \ldots, \mathfrak{y}_{t_m})$, $t > 0$ and $W_{n,t} = H(Y_n(t + \cdot))$ for $t > 0$. Let $R/c_n \to r > 0$ and $Q/R \to \lambda > 0$ and put

$$
J_\varepsilon(x, n, t) = E_x[W_{n,t}, \sigma_\Delta \wedge \sigma_0 > \varepsilon n \,|\, \Gamma_{Q,R}^*],
$$

where $\Delta = \Delta(Q, R)$. Then by (9.54) it follows that

$$
\lim_{\varepsilon \downarrow 0} \lim_{n \to \infty} J_\varepsilon(x, n, t) = E_x[W_{n,t} \,|\, \Gamma_{Q,R}^*]. \tag{9.60}
$$

Writing $P_x\left(\Gamma_{Q,R}^*\right) = \sum_y P_x[Y_n(\varepsilon) = y/c_n, \sigma_\Delta \wedge \sigma_0 > \varepsilon n] P_y\left(\Gamma_{Q,R}^*\right)\{1 + o_\varepsilon(1)\}$, where $o_\varepsilon(1) \to 0$ as $\varepsilon \downarrow 0$ (uniformly in n), one infers that if $\varepsilon < t$,

$$
J_\varepsilon(x, n, t) \tag{9.61}
$$
$$
= \frac{\sum_y P_x[Y_n(\varepsilon) = y/c_n, \sigma_\Delta > \varepsilon n \,|\, \sigma_0 > \varepsilon n] E_y[W_{n,t-\varepsilon}; \Gamma_{Q,R}^*]}{\sum_y P_x[Y_n(\varepsilon) = y/c_n, \sigma_\Delta > \varepsilon n \,|\, \sigma_0 > \varepsilon n] P_y(\Gamma_{Q,R}^*)}\{1 + o_\varepsilon(1)\}.
$$

For $\eta \in \mathbb{R}$ put $h_n(\eta) = E_{c_n\eta}[W_{n,t-\varepsilon}; \Gamma_{Q,R}^*]$. Let $q_n = Q/c_n, r_n = R/c_n$. Then

$$
h_n(y_n) = E_y[H(Y_n(t - \varepsilon + \cdot)); \sigma_{\Delta(q_n,r_n)}(Y_n) < \sigma_0],
$$

where $y_n = y/c_n$. Since any ξ is regular for (ξ, ∞) and $(-\infty, \xi)$, we have

$$
h_n(y_n) \longrightarrow E_\eta^Y[H(Y(t - \varepsilon + \cdot)); \sigma_{\Delta(q,r)} < \sigma_0]
$$

if $\eta \notin \{q, r, 0\}$. Since $E_x[\varphi(Y_n(\varepsilon)); \sigma_{\Delta(Q,R)} > \sigma_0 \mid \sigma_0 > \varepsilon n]$ converges weakly to $\mathbf{E}^0[\varphi(\varepsilon^{1/\alpha}\mathfrak{y}_1); \sigma^f_{\Delta(q,r)} > \sigma^f_0]$ for any bounded continuous function φ by the continuity theorem, from (9.61) one infers that as $n \to \infty$,

$$
J_\varepsilon(x, n, t) =
$$
$$
\frac{\int \mathbf{P}^0[\mathfrak{y}_1 \in d\eta, \sigma_{\Delta(q,r)} > \varepsilon] \mathbf{E}^0_{\varepsilon^{1/\alpha}\eta}[H(\mathfrak{y}_{t-\varepsilon+\cdot}); \sigma_{\Delta(q,r)} < \sigma_0]}{\int \mathbf{P}^0[\mathfrak{y}_1 \in d\eta, \sigma_{\Delta(q,r)} > \varepsilon] \mathbf{P}^0_{\varepsilon^{1/\alpha}\eta}[\sigma_{\Delta(q,r)} < \sigma_0]} \{1 + o_\varepsilon(1)\},
$$

with the self-evident notation of $\mathbf{E}^0_\eta$ and $\mathbf{P}^0_\eta$ and f suppressed from σ^f. As $\varepsilon \downarrow 0$, the RHS above converges to a limit since the LHS converges for $x = 0$. Since the limit does not depend on x, (9.60) shows the convergence of finite-dimensional distributions of the conditional law. The tightness of it is verified as before.   □

With this lemma, we can follow the arguments given after Lemma 9.5.2 to conclude the following extension of Proposition 9.5.1.

Theorem 9.5.5 *As $R \to \infty$ along with $Q/R \to \lambda \in (0, \infty)$, for each $x \in \mathbb{Z}$,*

$$
P_x[\sigma_{\Delta(Q,R)} < \sigma_B] \sim
\begin{cases}
C(\lambda)u_B(x)R^{1-\alpha}L(R) & \text{if } 1 < \alpha \le 2 \text{ and } u_B(x) > 0, \\
u_B(x)/[1/\check{L}]_*(R) & \text{if } \alpha = 1 \text{ and } F \text{ is recurrent,}
\end{cases}
$$

where $C(\lambda) = [\kappa^\sharp_\alpha/c_\sharp]C_\lambda(1)$. [$C_\lambda(1)$ is given in Remark 9.5.3. Compare the result with (5.66); see also Remark 5.6.2.]

9.6 Proof of Theorem 9.2.2

Let F be recurrent and $\sigma^2 = \infty$. Write $x_n = x/c_n$ for $x \in \mathbb{Z}$ and put

$$
Q^n_B(x, y) = P_x[S_n = y, \sigma_B \ge n].
$$

Note that for $y \in B$, $Q^n_B(x, y) = P_x[S_n = y, \sigma_B = n]$ by definition. In the sequel, M denotes an arbitrarily fixed constant larger than 1.

Proposition 9.6.1 *Suppose F is strongly aperiodic and $\sigma^2 = \infty$.*

(i) *Uniformly for $|y| \in [M^{-1}c_n, Mc_n]$, as $n \to \infty$,*

$$
Q^n_B(x, y) = \kappa^\sharp_\alpha u_B(x)\mathfrak{q}(y_n)/n\{1 + o(1)\} \qquad \text{for } x \in \mathbb{Z} \text{ with } u_B(x) > 0 \quad (9.62)
$$

and

$$
P_{-y}[\sigma_B = n] = \sum_{x \in -B} Q^n_{-B}(x, y) \sim \kappa^\sharp_\alpha \mathfrak{q}(y_n)/n. \tag{9.63}
$$

(ii) *For each $x \in \mathbb{Z}$, $\lim_{\varepsilon \downarrow 0} \lim_{n \to \infty} \sup \sup_{|y| < \varepsilon c_n} nQ^n_B(x, y) = 0$.*

Using Lemma 9.3.3 and Theorem 9.2.1 we obtain the following:

If φ is a continuous function on $\mathbb{R}$ and I is a finite interval of the real line, then uniformly for $|y| < Mc_n$, as $n \to \infty$

$$\sum_{w:w/c_n \in I} Q_B^n(x, y-w)\varphi(w/c_n) = \kappa_\alpha^\sharp u_B(x)\frac{c_n}{n}\int_I \mathfrak{q}(y_n+\xi)\varphi(\xi)\,d\xi\{1+o(1)\}. \quad (9.64)$$

The proof of (9.62) is based on this relation and Gnedenko's local limit theorem.

Proof (of Proposition 9.6.1) [8] Taking $m = \lfloor \varepsilon^{2\alpha} n \rfloor$ with a small $\varepsilon > 0$ we decompose

$$Q_B^n(x, y) = \sum_{z \in \mathbb{Z} \setminus B} Q_B^{n-m}(x, z)Q_B^m(z, y).$$

Note that $c_m/c_{n-m} \sim \varepsilon^2/(1 - \varepsilon^{2\alpha})^{1/\alpha}$. We apply (9.64) in the form

$$n\sum_{|w|<\varepsilon c_{n-m}} Q_B^{n-m}(x, y - w)\frac{\varphi(w/c_m)}{c_m} \sim \frac{\kappa_\alpha^\sharp u_B(x)}{1-\varepsilon^{2\alpha}}\int_{|\xi|<\varepsilon} \mathfrak{q}(y_{n-m} + \xi)\frac{\varphi(\xi/\tilde{\varepsilon}^2)}{\tilde{\varepsilon}^2}\,d\xi,$$
$$(9.65)$$

valid for each $\varepsilon > 0$ fixed, where $\tilde{\varepsilon} = \varepsilon/(1 - \varepsilon^{2\alpha})^{1/2\alpha} \sim \varepsilon$ $(\varepsilon \to 0)$.

Let $|y_n| \in [1/M, M]$. It is easy to see that

$$\sup_{z:|z-y|<\varepsilon c_n} |Q_B^m(z, y) - p^m(y - z)| = p^m(y - z) \times o_\varepsilon(1).$$

Since $\varepsilon c_n/c_m \sim 1/\varepsilon$ and $\mathfrak{p}_1(\pm 1/\varepsilon) = O(\varepsilon^{\alpha+1})$ according to [92] (cf. [67] Eq(14.34–35)), we also have for all sufficiently large n,

$$\sup_{z:|z-y|\geq\varepsilon c_{n-m}} Q_B^m(z, y) < C\varepsilon^{\alpha+1}/c_m,$$

which, combined with the preceding bound, yields

$$\left| Q_B^n(x, y) - \sum_{|z-y|<\varepsilon c_{n-m}} Q_B^{n-m}(x, z)p^m(y - z)\{1 + o_\varepsilon(1)\} \right|$$

$$\leq C'\frac{P_x[\sigma_B > n - m]\varepsilon^{\alpha+1}}{c_m}.$$

On the LHS, $p^m(y - z)$ may be replaced by $\mathfrak{p}_1((y-z)/c_m)/c_m$ (whenever ε is fixed), whereas the RHS is dominated by a constant multiple of $u_B(x)[c_n/nc_m]\varepsilon^{\alpha+1} \sim u_B(x)\varepsilon^{\alpha-1}/n$ (provided $u_B(x) > 0$). Hence, after a change of variable

$$nQ_B^n(x, y) = n\sum_{|w|<\varepsilon c_{n-m}} Q_B^{n-m}(x, y - w)\frac{\mathfrak{p}_1(w/c_m)}{c_m}\{1 + o_\varepsilon(1)\} + u_B(x) \times o_\varepsilon(1).$$

---

[8] This proof of (9.62) is borrowed from [82].

Now, on letting $n \to \infty$ and $\varepsilon \to 0$ in this order, (9.65) shows that the RHS can be written as

$$\kappa_\alpha^\sharp u_B(x) \left\{ \int_{|\xi|<\varepsilon} \mathfrak{q}(y_n - \xi) \frac{\mathfrak{p}_1(\xi/\varepsilon^2)}{\varepsilon^2} \, d\xi + o_\varepsilon(1) \right\} = \kappa_\alpha^\sharp u_B(x) \{\mathfrak{q}(y_n) + o_\varepsilon(1)\}.$$

(9.66)

This verifies the formula (9.62). Time reversal shows (9.63). Thus (i) is verified.

Using the bound $Q_B^n(x, y) = O(1/c_n)$ and taking $y = 0$, $I = [-\varepsilon, \varepsilon]$ and $\varphi \equiv 1/c_n$ in (9.64) we have $\sum_{|z| \le \varepsilon c_n} Q_B^n(x, z) Q_B^n(z, y) \le C a^\dagger(x) n^{-1} \int_{-\varepsilon}^{\varepsilon} \mathfrak{q}(\eta) \, d\eta$. Since $\sum_{|z|>\varepsilon c_n} Q_B^n(x, z) = O(1/n)$ and $\sum_{|z|>\varepsilon c_n} Q_B^n(z, y) \to 0$ as $n \to \infty$, $y/c_n \to 0$, the sum over $|z| > \varepsilon c_n$ is of a smaller order of magnitude than $1/n$, showing (ii).   □

Lemma 9.6.2 *If $1 < \alpha \le 2$, then $\kappa_\alpha^\sharp \mathfrak{q}(\eta) = \mathfrak{f}^{-\eta}(1)$.*

Proof According to [4, Lemma VIII.13],

$$\kappa_\alpha^\sharp \int_0^t \mathfrak{f}^{-\eta}(s) \, ds = \int_0^t s^{-1+1/\alpha} \mathfrak{p}_{t-s}(\eta).$$

After taking Fourier transforms in η of both sides, differentiate with respect to t. Then

$$\kappa_\alpha^\sharp \int_{-\infty}^{\infty} \mathfrak{f}^{-\eta}(t) e^{i\theta\eta} \, d\eta = t^{-1+1/\alpha} - \Phi(\theta) \int_0^t s^{-1+1/\alpha} e^{-(t-s)\Phi(\theta)} \, ds.$$

The RHS with $t = 1$ coincides with the expression of the characteristic function of $\mathfrak{q}$ in (9.29). Thus the identity of the lemma follows.

Alternatively, one can proceed as follows. For $\varepsilon > 0$, let $J_n(\varepsilon)$ stand for the expectation $E_0[\varphi(Y_n(1)) < \varepsilon, n \le \sigma_0 < n + n\varepsilon \mid \sigma_0 > n]$ and make the decomposition

$$J_n(\varepsilon) = \sum_y P_y[n \le \sigma_0 < n + \varepsilon n] P_0[S_n = y \mid \sigma_0 > n] \varphi(y/c_n).$$

As $n \to \infty$, by Theorem 9.2.1, (9.62) and the equality in (9.63) the sum above equals $\kappa_\alpha^\sharp \sum_y \sum_{k=n}^{n+\varepsilon n} \mathfrak{q}(-y_{n+k})(n+k)^{-1}\mathfrak{q}(y_n)\varphi(y_n)/c_n\{1 + o(1)\}$, which converges to

$$\kappa_\alpha^\sharp \int_{-\infty}^{\infty} \mathfrak{q}(\eta)\varphi(\eta) \, d\eta \int_1^{1+\varepsilon} \frac{\mathfrak{q}(-\eta(1+t)^{-1/\alpha})}{1+t} \, dt,$$

while $J_n(\varepsilon)$ approaches $\mathbf{E}^0[\varphi(\mathfrak{y}_1), 1 \le \sigma_0^f \circ \vartheta_1 < 1 + \varepsilon]$, which can be written as $\int_{-\infty}^{\infty} P_\eta^Y[1 \le \sigma_0 < 1 + \varepsilon]\mathfrak{q}(\eta)\varphi(\eta) \, d\eta$. Dividing these two expressions by ε and letting $\varepsilon \downarrow 0$ we find $\mathfrak{f}^\eta(1) = \kappa_\alpha^\sharp \mathfrak{q}(-\eta)$, the dual of the required identity.   □

Proposition 9.6.3 *If F is strongly aperiodic, for $x, y \in \mathbb{Z}$ with $u_B(x)u_B(-y) > 0$,*

$$Q_B^n(x, y) \sim u_B(x)P_0[\sigma_0 = n]u_{-B}(-y).$$

Proof Suppose n is even and decompose $Q_B^n(x, y) = \sum Q_B^m(x, z)Q_B^m(z, y)$, where $m = n/2$. For any $\varepsilon > 0$, the sum restricted to $\varepsilon c_n < |z| < c_n/\varepsilon$ is written as

$$\sum_{\varepsilon c_n < |z| < c_n/\varepsilon} \kappa_\alpha^{\sharp} u_B(x) q\left(\frac{z}{c_m}\right) q\left(\frac{-z}{c_m}\right) \kappa_\alpha^{\sharp} u_{-B}(-y) \frac{1 + o(1)}{m^2}$$

owing to (9.62). Passing to the limit as $n \to \infty$ and $\varepsilon \downarrow 0$, by Proposition 9.6.1 one obtains

$$\frac{m^2}{c_m} Q_B^n(x, y) \longrightarrow u_B(x)u_{-B}(-y)(\kappa_\alpha^{\sharp})^2 \int_{-\infty}^{\infty} q(\zeta)q(-\zeta)\, d\zeta. \qquad (9.67)$$

For $B = \{0\}$ and $x = y = 0$, this becomes

$$\frac{m^2}{c_m} P_0[\sigma_0 = n] \longrightarrow (\kappa_\alpha^{\sharp})^2 \int_{-\infty}^{\infty} q(\zeta)q(-\zeta)\, d\zeta.$$

By substitution into the preceding relation one finds the equivalence of the proposition. $\qquad \square$

Proof (of Corollary 9.2.5) By the comments made after the statement of Corollary 9.2.5 it suffices to show the first case of (9.28). (9.67) shows that $P_x[\sigma_B = n]n^2/c_n$ converges to a positive multiple of $u_B(x)$. In particular, it follows that $(n^2/c_n)P_0[\sigma_0 = n] \to (1 - \alpha^{-1})\kappa_\alpha^{\sharp}$ because of Lemma 9.3.3. Hence an application of Proposition 9.6.3 concludes the proof. $\qquad \square$

As a byproduct of the proof above we see that $\int_{-\infty}^{\infty} q(\eta)f^{\eta}(1)\, d\eta$ is the limit of $P_0[\sigma_0 = 2n]n^2/[\kappa_\alpha^{\sharp}c_n]$, which equals $(1 - \alpha^{-1})2^{-2+1/\alpha}$. In the same way, with the help of the scaling property of $f^{\eta}(t)$ one readily sees that if $u_B(x) > 0$, for $t > 0$,

$$\int_{-\infty}^{\infty} q(\eta)f^{\eta}(t)\, d\eta = \lim_{n/k \to t} P_x[\sigma_B = k + n \mid \sigma_B > k] = \frac{(1 - \alpha^{-1})}{(1 + t)^{2 - 1/\alpha}}.$$

Recalling that $I_1(f)$ denotes the first excursion interval of f whose length is larger than 1, one easily infers that the above equalities entail the following

Corollary 9.6.4 *If $1 < \alpha \leq 2$, $\mathbf{P}^0$[the length of I_1 is larger than t] $= t^{-1+1/\alpha}$ for $t \geq 1$.*

9.7 Proof of Theorem 9.2.6

Put $\langle B \rangle = (-\infty, \min B] \cup B$. Recall $T = \sigma_\Omega$ and $\Omega = (-\infty, -1]$.

Lemma 9.7.1 *There exists the limit* $V_B(x) := \lim_{k \to \infty} E_x \left[V_d(S_k); k < \sigma_{\langle B \rangle} \right]$, *and*

$$P_x[\sigma_{\langle B \rangle} > n]/P_0[T > n] \to V_B(x)/v^\circ.$$

If $x \geq 0$ *and* $\Omega \subset \langle B \rangle$, *then* $V_B(x) = V_d(x) - E_x[V_d(S_{\sigma_{\langle B \rangle}})]$.
[The assertion is valid for every r.w. satisfying the basic assumption of this treatise.]

Proof The proof rests on Theorem 10 of [44], a weak version of which reads

$$\frac{P_x[T > n]}{P_0[T > n]} \longrightarrow \frac{V_d(x)}{v^\circ} \quad and \quad \frac{P_0[T > n+1]}{P_0[T > n]} \longrightarrow 1 \,^9 \qquad (9.68)$$

as $n \to \infty$. For each $k < n$,

$$P_x[\sigma_{\langle B \rangle} > n] = \sum_y P_x[S_k = y, \sigma_{\langle B \rangle} > k] P_y[\sigma_{\langle B \rangle} > n - k]. \qquad (9.69)$$

For any $M > 1$, $P_x[S_k < M \mid \sigma_{\langle B \rangle} > k] \to 0$ as $k \to \infty$ and $V_d(y)/V_d(y + 1) \to 1$ as $y \to \infty$. It also holds that $P_{y-r_B-1}[T > n-k] \leq P_y[\sigma_{\langle B \rangle} > n - k]$ $\leq P_{y-l_B}[T > n-k]$ (where $l_B = \min B, r_B = \max B$). Substituting these inequalities into the sum in (9.69) we then apply (9.68) to see that if one defines $r(n, k)$ by

$$\frac{P_x[\sigma_{\langle B \rangle} > n]}{P_0[T > n]} = \frac{E_x[V_d(S_k), \sigma_{\langle B \rangle} > k]}{v^\circ} \{1 + r(n, k)\},$$

then for any $\varepsilon > 0$, one can choose k so that $\limsup_{n \to \infty} |r(n, k)| < \varepsilon$. This shows the convergence of the LHS as $n \to \infty$, which, in turn, shows the first half of the lemma. If $x \geq 0$ and $\Omega \subset \langle B \rangle$, then under P_x, $V_d(S_{n \wedge \sigma_{\langle B \rangle}})$ is equal to $V_d(S_{n \wedge T \wedge \sigma_{\langle B \rangle}})$, hence a martingale, and letting $n \to \infty$ in the identity $V_d(x) = E_x[V_d(S_{n \wedge \sigma_{\langle B \rangle}})] = E_x[V_d(S_{\sigma_{\langle B \rangle}}); \sigma_{\langle B \rangle} \leq n] + E_x[V_d(S_n); \sigma_{\langle B \rangle} > n]$ leads to the second half. $\quad\square$

Lemma 9.7.2 *For any* $x \in \mathbb{Z}$ *the conditional law* $P_x[Y_n \in \cdot \mid T > n]$ *converges weakly to the law of the stable meander.*

Proof[10] For $x = 0$ this is the first example given in Section 3 of [19]. Given a 'nice' set $\Gamma \in \mathcal{B}_\infty$ that depends only on paths after an arbitrarily chosen $t_0 \in (0, 1)$, decompose $P_1[Y_n \in \Gamma \mid T > n]$ as the sum

$$\sum_{k < t_0 n} P_1[\sigma_0 = k, T > k] P_0[Y_n \in \vartheta_{k/n}(\Gamma) \mid T > n - k] \frac{P_0[T > n - k]}{P_1[T > n]}$$

$$+ P_1[Y_n \in \Gamma \mid \sigma_{\Omega+1} > n] \frac{P_1[\sigma_{\Omega+1} > n]}{P_1[T > n]} + \eta_n,$$

---

[9] In [44] strong aperiodicity is assumed, which is not needed for this weak version.
[10] An alternative proof can be given by examining those in [19], [68].

with $0 \le \eta_n \le P_1[t_0 n \le \sigma_0 \le n \mid T > n] \to 0$. Then, noting $P_1[\sigma_{\Omega+1} > n] = P_0[T > n]$ and employing (9.68), one infers that

$$\lim_{n \to \infty} P_1[Y_n \in \Gamma \mid T > n] = \frac{P_1[\sigma_0 < T] + 1}{V_d(1)/v^\circ} \lim_{n \to \infty} P_0[Y_n \in \Gamma \mid T > n].$$

Since $P_1[\sigma_0 < T] = P_0[S_T = -1] = v_d(1)/v^\circ$, the ratio on the RHS equals unity, and the assertion of the lemma is verified for $x = 1$. Induction generalises this to any $x > 0$ instead of 1, which, in turn, implies the result for any $x < 0$. $\qquad\square$

Proof (of Theorem 9.2.6) Let $\Gamma \in \mathcal{B}_\infty$ be as in proof of Lemma 9.7.2. We may suppose $\min B = 0$. Let $u_B(x) = 0$. Then $x > 0$ and $\sigma_B \le T$ a.s.(P_x) and one sees that

$$P_x[Y_n \in \Gamma, \sigma_{\langle B]} > n]$$

$$= P_x[Y_n \in \Gamma, T > n] - \sum_{k=1}^{n} \sum_{y \in B} P_x[\sigma_B = k, S_k = y] P_y[Y_n \in \vartheta_{k/n}(\Gamma), T > n - k].$$

The second probability in the summand of the double sum is asymptotically equivalent to $P_y[Y_n \in \Gamma \mid T > n](V_d(y)/v^\circ)P_0[Y > n]$ for each k fixed. After dividing both sides of the identity above by $P_x[\sigma_B > n]$, let $n \to \infty$ and apply the above two lemmas to see $\lim P_x[Y_n \in \Gamma \mid \sigma_{\langle B]} > n] = \lim P_0[Y_n \in \Gamma \mid T > n]$. This implies the convergence of finite-dimensional distribution when x is fixed. The tightness can be shown similarly to Lemma 9.3.7. $\qquad\square$

9.8 Random Walks Conditioned to Avoid a Finite Set Forever

Here we suppose $u_B(x) > 0$ for all $x \in \mathbb{Z}$.[11] For any recurrent walk there exists the limit law

$$\lim_{n \to \infty} P_x[(S_n)_{n=0}^{\infty} \in \cdot \mid \sigma_B > n]$$

(in the sense of finite-dimensional distributions); indeed it is a harmonic transform of the r.w. killed upon hitting B, the n-step transition law being given by the Doob h-transform:

$$Q_B^n(x, y)u_B(y)/u_B(x) \quad (x, y \notin B), \tag{9.70}$$

where $Q_B^n(x, y) = P_x[S_n = y, \sigma_B \ge n]$ as in Section 9.6.[12] Let $P_x^{\infty, B}$, $x \in \mathbb{Z}$, stand for the above limit law, which may be regarded as the probability law of the conditional process S_n given that it never visits the origin.

---

[11] If $u_B(x) = 0$ for some x, the r.w. stays in a half-line forever once it enters this half-line, and the problem addressed in this subsection essentially becomes the problem for a half-line.

[12] Note that $Q_B^n(x, y) = P_x[S_n = y, \sigma_B > n]$ for $y \notin B$.

It is observed in [84, Section 7] that

the conditional law $P_x[\,\cdot\,|\,\sigma_{\Delta(R)} < \sigma_B]$ converges to $P_x^{\infty,B}$ as $R \to \infty$,   (9.71)

where $\Delta(R) = (-\infty, -R] \cup [R, \infty)$, and that if $\sigma^2 = \infty$, for every $x \in \mathbb{Z}$,

(a) $P_x^{\infty,B}[\lim S_n = +\infty] = 1$ if $EZ < \infty$,

(b) $P_x^{\infty,B}[\lim\sup S_n = +\infty, \ \liminf S_n = -\infty] = 1$ if $EZ = E|\hat{Z}| = \infty$;   (9.72)

and if $\sigma^2 < \infty$, either $\lim S_n = +\infty$ or $\lim S_n = -\infty$ with $P_x^{\infty,B}$-probability one and

$$P_x^{\infty,B}[\lim S_n = +\infty] = \frac{u_B(x) + w_B(x)}{2u_B(x)} \qquad (x \in \mathbb{Z}).   \qquad (9.73)$$

One notices that if $\sigma^2 = \infty$, there is only one harmonic function, hence a unique Martin boundary point: $\lim_{|y|\to\infty} g_{\{0\}}(\cdot, y)/g_{\{0\}}(x_0, y) = a(\cdot)/a(x_0)$, so that two geometric boundary points $+\infty$ and $-\infty$ are not distinguished in the Martin boundary whereas the walk itself discerns them provided that either EZ or $E\hat{Z}$ is finite.

In the sequel we suppose that (AS) holds with $1 < \alpha \le 2$, $\sigma^2 = \infty$, and $p := \lim \mu_+(x)/\mu(x)$ (even if $\alpha = 2$), and consider the scaling limit of S_n under $P_x^{\infty,B}$. By (9.70) we have for $\Gamma \in \mathcal{B}_1$,

$$\frac{P_x^{\infty,B}[Y_n \in \Gamma]}{P_x[\sigma_B > n]} = \frac{E_x[u_B(S_n); Y_n \in \Gamma \,|\, \sigma_B > n]}{u_B(x)}.   \qquad (9.74)$$

For simplicity we let

$$c_{\#} = 1.$$

[For general $c_{\#}$, one has only to replace $\mathfrak{f}^\xi(t)$ by $\mathfrak{f}^\xi(c_{\#}t)c_{\#}$, $\mathfrak{q}(\xi)$ by $\mathfrak{q}(c_{\#}^{-1/\alpha}\xi)c_{\#}^{-1/\alpha}$, $\mathfrak{p}_t^0$ by $\mathfrak{p}_{c_{\#}t}^0$, L by $L/c_{\#}$ and so on.]

As one may expect, we shall see that the limit is an h-transform of the process on $\mathbb{R} \setminus \{0\}$ whose transition probability is given by $\mathfrak{p}_t^0(\xi, \eta)$. Put

$$\mathfrak{a}(\xi) = \frac{1 - \mathrm{sgn}(\xi)(p-q)}{\kappa_\alpha^\circ} \kappa_\alpha^{\#}|\xi|^{\alpha-1} \qquad (1 < \alpha \le 2),   \qquad (9.75)$$

where

$$\kappa_\alpha^\circ = \begin{cases} 1 & \text{if } \alpha = 2, \\[2mm] \dfrac{\kappa_\alpha}{(2-\alpha)(\alpha-1)} = \dfrac{\pi\left(1 + (p-q)^2 \tan^2 \frac{1}{2}\alpha\pi\right)}{|\tan\frac{1}{2}\alpha\pi|} & \text{if } 1 < \alpha < 2. \end{cases}   \qquad (9.76)$$

It follows that $\mathfrak{a}(\xi) = 0$ if either $p = 1$ and $\xi > 0$ or $p = 0$ and $\xi < 0$. By (4.23), $\bar{a}(x) \sim [\kappa_\alpha^\circ L(x)]^{-1}|x|^{\alpha-1}$ ($|x| \uparrow \infty$), hence $u_B(c_n) \sim a(c_n) \sim [2q/\kappa_\alpha^\circ]n/c_n$ and similarly for $u_B(-c_n)$; one accordingly has

$$\mathfrak{a}(\xi) = \lim_{n \to \infty} u_B(\xi c_n) P_x[\sigma_B > n]/u_B(x) \,^{13} \quad \text{if} \quad u_B(x) > 0. \tag{9.77}$$

Theorem 9.8.1 *Let* $1 < \alpha \leq 2$, $\sigma^2 = \infty$ *and* $p = \lim_{x \to \infty} m_+(x)/m(x)$.

(i) $\mathfrak{a}$ *is harmonic w.r.t.* $\mathfrak{p}_t^0$:

$$\mathfrak{a}(\xi) = \int_{-\infty}^{\infty} \mathfrak{p}_t^0(\xi, \eta) \mathfrak{a}(\eta) \, d\eta \quad (\xi \in \mathbb{R}). \tag{9.78}$$

(ii) *The probability law* $P_x^{\infty,B} \circ Y_n^{-1}$ *on* $D_{[0,\infty)}$ *converges weakly to the law of a Markov process on* $\mathbb{R}$ *of which the transition function is given by the h-transform* $\mathfrak{p}_t^\mathfrak{a}(\xi, \eta) := \mathfrak{p}_t^0(\xi, \eta) \mathfrak{a}(\eta)/\mathfrak{a}(\xi)$ $(\xi\eta \neq 0)$ *and the entrance law by*

$$\lim P_x^{\infty,B}[Y_n(t) \leq \eta] = \mathbf{E}^0[\mathfrak{a}(\mathfrak{y}_{t \vee 1}); \mathfrak{y}_t \leq \eta] = \int_{-\infty}^{\eta} \mathfrak{q}_t(\xi) \mathfrak{a}(\xi) \, d\xi. \tag{9.79}$$

($\mathbf{E}^0$ *is the expectation with respect to* $\mathbf{P}^0$ *that is characterised by Corollary 9.2.4(i).)*

(iii) $\lim_{\xi' \to 0} \mathfrak{p}_t^0(\xi', \xi)/\mathfrak{a}(\xi') = \mathfrak{q}_t(\xi)$, *with the understanding that when* $\alpha \neq 2$, *the limit is taken along* $\xi' > 0$ *if* $p = 0$ *and* $\xi' < 0$ *if* $p = 1$ *(so that* $\mathfrak{a}(\xi') > 0$).

(9.78) (as well as Proposition 9.8.3 below) is a special case of Theorem 2(iii) of Pantí [56]. We shall derive it from the corresponding property of a.

Remark 9.8.2 (a) Let $p = 1$. Then $\mathfrak{a}(\xi)$ is positive for $\xi < 0$ and zero for $\xi \leq 0$, and (9.79) gives a proper probability law that concentrates on the negative half-line – a remarkable difference from the limit of $P_x^{n,B}$. Note that for $\xi > 0$, $\mathfrak{p}_t^0(\xi, \eta) = \mathfrak{p}_t^{(\infty,0]}(\xi, \eta)$, so that both sides of (9.78) vanish.

(b) If $1 < \alpha \neq 2$, $\mathfrak{a}$ is a unique non-negative harmonic function in the sense that any non-negative function satisfying (9.78) is a constant multiple of $\mathfrak{a}$ (Lemma 9.9.3).

First, we explain how (i) and (ii) of Theorem 9.8.1 are derived, setting aside any necessary technical results, which we shall provide later. We derive (i) from the identity

$$(*) \qquad \sum_y P_x[S_n = y, \sigma_0 > n] a(y) = a(x) \quad (x \neq 0).$$

Let $L \equiv 1$ for simplicity. Then $a(c_n\xi)/c_n^{\alpha-1} \to c\mathfrak{a}(\xi)$ as $n \to \infty$. Divide both sides of $(*)$ by $c_n^{\alpha-1}$ and then let $n \to \infty$ along with $x_n = x/c_n \to \xi$. If $\sup_n \sum_y p^n(x, y)|y/c_n|^\nu < \infty$ for some $\nu > \alpha - 1$ (which we shall show in Proposition 9.8.5), by the trivial bound $P_x[S_n = y, \sigma_0 > n] \leq p^n(x, y)$ an application of (9.21) (in Corollary 9.2.3) yields the identity of (i) for $t = 1$. For general $t > 0$, the result follows the second scaling relation in (9.26).

---

[13] For general $c_\sharp \neq 1$, this limit becomes $\mathfrak{a}(\xi)/c_\sharp$.

For the proof of (ii), one has only to show the convergence of $(Y_n(t))_{0 \le t \le M}$ to the process specified therein under P_x^B for each $M \ge 1$. To this end it suffices to show it for $M = 1$; the general case follows from this particular case. From (9.74) it follows that for any finite interval I and $\Gamma \in \mathcal{B}_1$,

$$P_x^{\infty,B}[Y_n \in \Gamma, Y_n(1) \in I] = \frac{E_x[r_n^{B,x} u_B(S_n), Y_n \in \Gamma, Y_n(1) \in I \mid \sigma_B > n]}{u_B(x)},$$

where $r_n^{B,x} = P_x[\sigma_B > n]$. Combined with (9.77) and Theorem 9.2.1 this shows

$$\lim P_x^{\infty,B}[Y_n \in \Gamma, Y_n(1) \in I] = E^0[\mathfrak{a}(\mathfrak{y}_1); \Gamma, \mathfrak{y}_1 \in I]. \tag{9.80}$$

For $\Gamma = \{\xi \le \mathfrak{y}_t \le \eta\}$ $(-\infty < \xi < \eta < \infty)$ with $t \le 1$, this becomes (9.79) – but with the range $(-\infty, \eta]$ replaced by $[\xi, \eta]$ – because of (9.24), (9.25) and (9.78) (for $t > 1$, see (ii) of Corollary 9.2.4). To complete the proof of (ii) of Theorem 9.8.1 it suffices to show the following proposition (taking it for granted that $\mathfrak{a}$ is harmonic), for it entails the uniform integrability of $r_n^{B,x} u_B(S_n)$ under $P_x^{n,B} := P_x[\cdot \mid \sigma_B > n]$.

Proposition 9.8.3 *For every* $1 < \alpha \le 2$, $E^0[\mathfrak{a}(\mathfrak{y}_1)] = \int_{-\infty}^{\infty} \mathfrak{a}(\eta) \mathfrak{q}(\eta) \, d\eta = 1$.

Since $r_n^{B,x} u_B(S_n)/u_B(x) \sim \mathfrak{a}(Y_n(1))$, taking $\Gamma = D_{[0,\infty)}$ in (9.74) gives

$$1 = [u_B(x)]^{-1} E_x[u_B(S_n); \sigma_B > n] \sim E_x[\mathfrak{a}(Y_n(1)) \mid \sigma_B > n].$$

This yields the result asserted in Proposition 9.8.3 if $E_x[|Y_n(1)|^{\alpha-1+\delta} \mid \sigma_B > n]$ is bounded for some $\delta > 0$, which we show at the end of this subsection in the case $x = 0$, $B = \{0\}$, which is enough. Alternatively, one may try to carry out a direct computation of

$$E^0[\mathfrak{a}(\mathfrak{y}_1)] = \frac{2\kappa_\alpha^\sharp}{\kappa_\alpha^\circ} \left[q \int_0^\infty |\eta|^{\alpha-1} \mathfrak{q}(\eta) \, d\eta + p \int_{-\infty}^0 |\eta|^{\alpha-1} \mathfrak{q}(\eta) \, d\eta \right]. \tag{9.81}$$

If $\alpha = 2$, then $\mathfrak{q}(\eta) = \mathfrak{q}(-\eta)$ and $\kappa_\alpha^\circ = 1$, so that the RHS of (9.81) becomes $\kappa_\alpha^\sharp \int_0^\infty \eta \mathfrak{q}(\eta) \, d\eta$. Observing $\kappa_\alpha^\sharp \mathfrak{q}(\eta) = \mathfrak{f}^\eta(1) = (2\pi)^{-1/2} \eta e^{-\eta^2/2}$ one can easily see that it equals unity, as desired. If $p = q$, one can compute the RHS of (9.81), without much difficulty, by using Belkin's expression (9.29) for $\Psi(\theta)$ and the Parseval relation. The details are omitted. Unfortunately, this approach runs into a serious difficulty in the case $p \ne q$.

The proof of (iii) is postponed to the next section (see Lemma 9.9.4), where we shall give alternative proofs of (i) of Theorem 9.8.1 and Proposition 9.8.3 (Lemma 9.9.3(ii)).

Lemma 9.8.4 *Let $0 < \alpha \leq 2$. If $0 < v < \alpha$, then as $n \to \infty$.*

$$\sum_{x=-\infty}^{\infty} p^n(x)|x|^v = \frac{C(v)}{\pi} \int_{-\infty}^{\infty} \frac{1 - \psi^n(\theta)}{|\theta|^{v+1}} \cdot \frac{\sin \theta/2}{\theta/2} \, d\theta \{1 + o(1)\}, \qquad (9.82)$$

where $C(v) = -\Gamma(v+1)\cos[(v+1)\pi/2]$.

Proof Let $\lambda > 0$ and put $\chi_\lambda(s) = |s|^v e^{-\lambda|s|}$ and $\hat{\chi}_\lambda(\theta) = \int_{-\infty}^{\infty} \chi_\lambda(s) e^{i\theta s} \, ds$. Writing $b := (\lambda - i\theta) = \sqrt{\lambda^2 + \theta^2} e^{-i \tan^{-1} \theta/\lambda}$, substitute $s = t/b$ (so that $-\lambda s + i\theta s = -t$) to obtain

$$\frac{\hat{\chi}_\lambda(\theta)}{2} = \Re \, b^{-v-1} \int_0^\infty e^{-t} t^v \, dt = \frac{\Gamma(v+1) \cos[(v+1) \tan^{-1} \theta/\lambda]}{(\lambda^2 + \theta^2)^{(v+1)/2}};$$

in particular $\hat{\chi}_\lambda(\theta)|\theta|^{v+1}$ is bounded and tends to $2C(v)$ as $\lambda \downarrow 0$. Let h be the $\mathbb{Z}$-valued function on $\mathbb{R}$ defined by $h(t) = x$ for $-\frac{1}{2} < t - x \leq \frac{1}{2}, x \in \mathbb{Z}$. Then

$$\int_{-\infty}^{\infty} p^n(h(t)) e^{i\theta t} \, dt = \sum_y \int_{y-\frac{1}{2}}^{y+\frac{1}{2}} p^n(y) e^{i\theta t} \, dt = \psi^n(\theta) \frac{\sin \theta/2}{\theta/2}.$$

Noting that $\chi_\lambda(\theta)$ is integrable, one can easily verify $\chi_\lambda(t) \int_{-\infty}^{\infty} \chi_\lambda(t) \, dt = 0$. By virtue of Parseval's identity we accordingly find

$$\int_{-\infty}^{\infty} p^n(h(t)) \chi_\lambda(t) \, dt = \frac{1}{2\pi} \int_{-\infty}^{\infty} (\psi^n(\theta) - 1) \frac{\sin \theta/2}{\theta/2} \hat{\chi}_\lambda(\theta) \, d\theta.$$

It is easy to see $\sum_{x=-\infty}^{\infty} p^n(x)|x|^v = \int_{-\infty}^{\infty} p^n(h(t))|t|^v \, dt \{1+o(1)\}$. Thus letting $\lambda \downarrow 0$ one concludes (9.82). □

Proposition 9.8.5 *Let $1 < \alpha < 2$. If $0 \leq v < \alpha$, then*

$$\lim_{n \to \infty} \frac{1}{c_n^v} \sum_{x=-\infty}^{\infty} p^n(x)|x|^v = \int_{-\infty}^{\infty} \mathfrak{p}_1(\xi)|\xi|^v \, d\xi.$$

Proof Suppose $\alpha \neq 2$, the case $\alpha = 2$ being similar. First, observe that the integral in (9.82) restricted to $|\theta| > M/c_n$ is dominated by $C(c_n/M)^v$, so that we have only to evaluate the integral over $|\theta| < M/c_n$. From (4.20) it follows that as $\theta \to \pm 0$

$$(1 - \psi(\theta))/|\theta|^\alpha L(1/|\theta|) \longrightarrow \zeta^\pm := \Gamma(1-\alpha)[\cos \tfrac{1}{2}\alpha\pi \mp i(p-q)\sin \tfrac{1}{2}\alpha\pi],$$

hence $\psi^n(\theta) = e^{-n[1-\psi(\theta)]\{1+o(1)\}} = e^{-n\zeta^\pm|\theta|^\alpha L(1/\theta)\{1+o(1)\}}$. Substitute this expression into (9.82) and observe that c_n^{-v} times the integral restricted on $\{|\theta| < \varepsilon/c_n\}$ tends to zero as $n \to \infty$ and $\varepsilon \downarrow 0$ in this order, which leads to

$$\lim_{n \to \infty} \frac{1}{c_n^v} \sum_{x=-\infty}^{\infty} p^n(x)|x|^v = \frac{2C(v)}{\pi} \Re \int_0^\infty \frac{1 - \exp\{-\zeta^+ t^\alpha\}}{t^{v+1}} \, dt.$$

Thus, for every $0 \leq \nu < \alpha$, $E_0|Y_n(1)|^\nu$ is bounded, which entails the uniform integrability of $|Y_n(1)|^\nu$ under P_0 and hence the identity of the theorem.   □

As mentioned right after its statement, Proposition 9.8.3 follows from the next

Lemma 9.8.6 *If* $1 < \alpha \leq 2$, $\lim E_0[|Y_n(1)|^\nu \,|\, \sigma_0 > n] = \mathbf{E}^0[|\mathfrak{y}_1|^\nu] < \infty$ *for* $0 \leq \nu < \alpha$.

Proof We use the notation in the proof of Lemma 9.3.5. For $x = 0$, (9.44) reduces to

$$\phi_{n,0}(\theta) = E_0^{n,\{0\}}[e^{i\theta S_n}] = 1 - \sum_{k=0}^{n-1} \frac{r_k^0}{r_n^0}[1 - \psi(\theta)]\psi^{n-k-1}(\theta). \qquad (9.83)$$

Let h be the function defined in the proof of Lemma 9.8.4. Then

$$\int_{-\infty}^{\infty} P_0[S_n = h(t) \,|\, \sigma_0 > n]e^{i\theta t} \, dt = \phi_{n,0}(\theta)\frac{\sin \theta/2}{\theta/2}.$$

As in the proof of Lemma 9.8.4, one therefore obtains

$$\int_{-\infty}^{\infty} |t|^\nu P_0[S_n = h(t) \,|\, \sigma_0 > n] \, dt = C_1 \int_{-\infty}^{\infty} \frac{1 - \phi_{n,0}(\theta)}{|\theta|^{\nu+1}} \cdot \frac{\sin \theta/2}{\theta/2} \, d\theta,$$

and one sees that the integral above restricted to $|\theta| > 1/c_n$ is bounded by a constant multiple of c_n. By Lemma 9.3.3 $r_k^0/r_n^0 \sim (k/n)^{1/\alpha-1}l(k)/l(n)$ (as $k \wedge \to \infty$) for some s.v. l while $|1 - \psi(\theta)| = O(|\theta|^\alpha L(1/|\theta|))$. Hence for a fixed large number M,

$$E_0^{n,\{0\}}[|Y_n(1)|^\nu] \leq C_3 + \frac{C_4}{c_n^\nu} \int_0^{1/c_n} \theta^{\alpha-\nu-1}L(1/\theta) \, d\theta \sum_{k=M}^{n} \frac{k^{1/\alpha-1}l(k)}{n^{1/\alpha-1}l(n)}.$$

It is easy to see that the integral and the sum above are bounded by a constant multiple of c_n^ν/n and n, respectively. This shows that $|Y_n(1)|^\nu$ is uniformly integrable under $P_0^{n,\{0\}}$. Hence the identity of the lemma follows from the convergence of $P_0^{n,\{0\}}$.   □

9.9 Some Related Results

Here we state, without proof except for Lemmas 9.9.1 to 9.9.4, the principal results of Uchiyama [79], [82], Doney [19] and Caravenna and Chaumont [13]. The first two works obtain precise asymptotic forms of the transition probability for the walk killed when it hits the finite set, in the cases when $\sigma^2 < \infty$ and when F is in the domain of attraction of a stable law of exponent $1 < \alpha < 2$, respectively. In [19], the walk killed on entering the negative half-line is studied, and the corresponding result on the transition probability and a local limit theorem for the hitting time of the half-line for the killed walk are obtained. The estimates obtained in these works are uniform in the space variables within the natural space-time region $x = O(c_n)$, where c_n is a norming sequence associated with the walk. In [13] the functional limit theorem for the r.w. conditioned to stay positive forever is obtained.

9.9.1 The r.w. Avoiding a Finite Set in the Case $\sigma^2 < \infty$

This subsection continues Section 9.1.2, and we use the notation introduced there. Let $Q_B^n(x, y) = P_x[S_n = y, \sigma_B \geq n]$ as in Section 9.6, which entails $Q_B^n(x, y) = P_x[S_n = y, \sigma_B = n]$ ($n \geq 1, y \in B, x \in \mathbb{Z}$) as well as $Q_B^0(x, y) = \delta_{x,y}$. Suppose $\sigma^2 < \infty$ and let M be an arbitrarily chosen positive number > 1. The following results (i) to (iii) are shown in [79]:

(i) *Uniformly for* $(x, y) \in [-M, M\sqrt{n}]^2 \cup [-M\sqrt{n}, M]^2$ *with* $K_B(x, y) \neq 0$, *as* $n \to \infty$ *along with* $|x| \wedge |y| = o(\sqrt{n})$,

$$Q_B^n(x, y) = \frac{\sigma^2}{n} K_B(x, y) P_x[S_n = y]\{1 + o(1)\};$$

(ii) *as* $x \wedge y \wedge n \to \infty$ *under* $x \vee y < M\sqrt{n}$ *along with* $P_x[S_n = y] > 0$,

$$Q_B^n(x, y) = \frac{d_*}{\sqrt{2\pi\sigma^2 n}} \left(e^{-(y-x)^2/2\sigma^2 n} - e^{-(x+y)^2/2\sigma^2 n} \right) \{1 + o(1)\},$$

where d_* *denotes the temporal period of the r.w.* S;

(iii) *uniformly for* $|x| < M\sqrt{n}$ *and for* $y_0 \in B$ *with* $K_B(x, y_0) \neq 0$, *as* $n \to \infty$,

$$P_x[\sigma_B = n, S_{\sigma_B} = y_0] = \sigma^2 K_B(x, y_0) \frac{P_x[S_n = y_0]}{n} \{1 + o(1)\}.$$

By (9.12), which may read $K_B(x, y) = u_B(x)u_{-B}(-y) - w_B(x)w_{-B}(-y)$, one finds that the formula of (i) is quite analogous to (2.8) (the corresponding one for the case $B = \{0\}$) valid for fixed x, y for which $P_0[\sigma_0 = n] \sim \sigma^2 P_x[S_n = y]/n$.

On noting $\sum_{y_0 \in B} K_B(x, y_0) = 2u_B(x)$, an elementary computation deduces from (iii)

$$P_x[\sigma_B > n] \sim \sqrt{\frac{2\sigma^2}{\pi}} \frac{u_B(x)}{|x|_+} \int_0^{|x|_+/\sqrt{n}} e^{-y^2/2\sigma^2} \, dy,$$

where $|x|_+ = |x| \vee 1$. Below, combining this and the formula in (i), we derive the asymptotic form of $P_x[S_n = y \mid \sigma_B > n]$.

In the sequel suppose $K_B(x, y) > 0$ for all x, y for simplicity. Since $g_B^-(-y) \sim 2y/\sigma^2$ (resp. $= o(y)$) as $y \to +\infty$ (resp. $-\infty$) and similarly for $g_B^+(-y)$, it follows that

$$\sigma^2 K_B(x, y) = \begin{cases} y g_B^+(x) + o(y) & y \to \infty, \\ -y g_B^-(x) + o(|y|) & y \to -\infty. \end{cases} \quad (9.84)$$

Note that $u_B(x) + \operatorname{sgn}(y)w_B(x)$ equals $g_B^+(x)$ or $g_B^-(x)$ according as $y > 0$ or $y < 0$, and, therefore, is bounded away from zero as $|x| \to \infty$ by virtue of Lemma 9.1.2(c,c').

Then writing $y_n = y/\sqrt{n}$, one sees that uniformly for $|y_n| < M$, as $x/\sqrt{n} \to 0$ and $|y| \to \infty$ along with $P_x[S_n = y] > 0$,

$$Q_B^n(x, y) = \frac{d_*[u_B(x) + \mathrm{sgn}(y)w_B(x)]}{\sqrt{2\pi\sigma^2}} \cdot \frac{|y_n|e^{-y_n^2/2\sigma^2}}{n}\{1 + o(1)\}, \qquad (9.85)$$

and

$$P_x[S_n = y \mid \sigma_B > n] = d_* k_y(x)\frac{|y_n|e^{-y_n^2/2\sigma^2}}{2\sigma^2\sqrt{n}}\{1 + o(1)\}, \qquad (9.86)$$

where

$$k_y(x) := 1 + \frac{\mathrm{sgn}(y)w_B(x)}{u_B(x)}.$$

Note that $k_y(x) \to 2$ or $\sim (\sigma^2 a(x) - |x|)/|x|$ according as $\mathrm{sgn}(y)x \to \infty$ or $-\infty$. Thus (9.85) provides the exact asymptotics of $P_x[S_n = y \mid \sigma_B > n]$ by means of a, indicating how the limit of the conditional 'density' $P_x[S_n/\sqrt{n} \in d\eta \mid \sigma_B > n]/d\eta$ depends on the initial site x. The assertion of Lemma 9.3.2 is valid also in the case $\sigma^2 < \infty$ and accordingly so is Lemma 9.8.6. Thus from (9.86) we deduce its integral version:

$$P_x[S_n/c_n \in I \mid \sigma_B > n] \sim \frac{1}{2\sigma^2} \int_I k_\eta(x)|\eta|e^{-\eta^2/2\sigma^2}\, d\eta,$$

valid as $x/\sqrt{n} \to 0$ for any finite interval I. From this the corresponding functional limit theorem also follows (see the proof of Lemma 9.3.7 for the tightness).

9.9.2 Uniform Estimates of $Q_B^n(x, y)$ in the Case $1 < \alpha < 2$

Below we state the results in [82] on the asymptotic form of $Q_B^n(x, y)$ in the space-time region $|x| \vee |y| < Mc_n$ with an arbitrarily given constant M. It differs in a significant way according as $pq \neq 0$ or $pq = 0$. First, we state the result for the case $pq \neq 0$, which can be formulated in a neat form. We write x_n for x/c_n as before.

Let $pq \neq 0$. Then for each $\varepsilon > 0$ and uniformly for $|x| \vee |y| < Mc_n$, as $n \to \infty$

$$Q_B^n(x, y) \sim \begin{cases} u_B(x)P_0[\sigma_0 = n]u_B(-y) & (|x_n| \vee |y_n| \to 0), \\ \mathfrak{f}^{x_n}(1)u_B(-y)/n & (y_n \to 0, |x_n| > \varepsilon), \\ u_B(x)\mathfrak{f}^{-y_n}(1)/n & (x_n \to 0, |y_n| > \varepsilon), \\ \mathfrak{p}_1^0(x_n, y_n)/c_n & (|x_n| \wedge |y_n| > \varepsilon). \end{cases} \qquad (9.87)$$

One can deduce from (9.87) that uniformly for $|x| \vee |y| < Mc_n$,

$$Q_B^n(x, y) \sim \mathfrak{p}_1^0(x_n, y_n)/c_n = \mathfrak{p}_n^0(x, y) \quad \text{as} \quad n \wedge |x| \wedge |y| \to \infty.[14]$$

---

[14] Use Lemma 9.9.2 if $y_n \to 0$, $|x_n| \in [\varepsilon, M]$ and (2.23) of [82] if $|x_n| \vee |y_n| \to 0$.

Summing over $y \in B$ in the first two formulae of (9.87), one sees that as $n \to \infty$

$$P_x[\sigma_B = n] \sim \begin{cases} \mathsf{u}_B(x)P_0[\sigma_0 = n] & \text{as } x_n \to 0, \\ \mathsf{f}^{x_n}(1)/n & \text{uniformly for } |x_n| \in [\varepsilon, M]. \end{cases} \quad (9.88)$$

(In [82] it is shown $P_0[\sigma_0 = n] \sim (1-\alpha^{-1})\kappa^\sharp c_n/n^2$.) From this, one can easily show $P_x[\sigma_B > n] \sim \mathsf{u}_B(x)P_0[\sigma_0 > n]$ as $x_n \to 0$, and recalling $\kappa^\sharp \mathsf{q}(\eta) = \mathsf{f}^{-\eta}(1)$, one accordingly obtains that if $pq \neq 0$,

$$c_n P_x[S_n = x \mid \sigma_B > n] \longrightarrow \mathsf{q}(\eta) \quad \text{as } x_n \to 0 \text{ and } y_n \to \eta \neq 0. \quad (9.89)$$

In the case $pq = 0$ we have a somewhat delicate situation. Let $q = 0$ so that $\mathsf{p}_n^0(x, y) = \mathsf{p}_n^{(-\infty, 0]}(x, y)$. If $\mathsf{u}_B(x_0) = 0$ for some x_0 (necessarily $x_0 > 0$), then (9.87) holds for x with $\mathsf{u}_B(x) > 0$ and $Q_B^n(x, y) \sim P_x[S_n = y, T > n]$ as $x \wedge y \to \infty$ when $\mathsf{u}_B(x) = 0$ (so, (9.102) below applies). Let $\mathsf{u}_B(x) > 0$ for all x. Then $Q_B^n(x, y)$ behaves quite differently. (9.87) holds at least for (x, y) (with $|x| \vee |y| < Mc_n$) contained in the region

$$(\mathbb{Z} \times (-M, \infty) \setminus D_{n,M}) \cup ((-\infty, M] \times \mathbb{Z} \setminus E_{n,M}), \quad (9.90)$$

where $D_{n,M} = (M, \varepsilon c_n) \times [-M, \infty)$ and $E_{n,M} = (-\infty, M] \times (-\varepsilon c_n, -M)$. In a part of the region $D_{n,M} \cup E_{n,M}$, (9.87) fails, while (9.87) is still valid in a large subregion of it that depends on the behaviour of $a(x)$ as $x \to \infty$. By duality, we only consider $D_{n,M}$. We make the decomposition

$$Q_B^n(x, y) = Q_\Omega^n(x, y) + R_B^n(x, y). \quad (9.91)$$

The second term denotes the remainder, i.e., the probability of the event $\{S_n = y, T < n < \sigma_B\}$. It can be shown that $R_B^n(x, y) \sim \kappa^\sharp \mathsf{u}_B(x)\mathsf{q}(\eta)/n$ as $x_n \to 0$, $y_n \to \eta \neq 0$ (as one may expect from (9.87) together with $\lim_{x_n \downarrow 0} P_x[T < n < \sigma_{[c_n, \infty)}] = 1$). Put

$$\lambda(n, x) = \frac{1}{\kappa_\alpha^\sharp \Gamma(1 - 1/\alpha)|\Gamma(1 - \alpha)|^{1/\alpha}} \cdot \frac{V_\mathrm{d}(x)/\mathsf{u}_B(x)}{V_\mathrm{d}(c_n)c_n/n} \quad (x > 0).$$

Then $Q_\Omega^n(x, y) \sim \kappa_\alpha^\sharp \mathsf{u}_B(x)n^{-1}\lambda(n, x)\mathsf{q}^{\mathrm{mnd}}(\eta)$, where $\mathsf{q}^{\mathrm{mnd}}$ denotes the stable meander (see (9.102) and (A.32)), so that as $x_n \downarrow 0$, $y_n \to \eta > 0$,

$$Q_B^n(x, y) \sim \kappa_\alpha^\sharp \mathsf{u}_B(x)n^{-1}[\mathsf{q}(\eta) + \lambda(n, x)\mathsf{q}^{\mathrm{mnd}}(\eta)]. \quad (9.92)$$

Hence the two terms on the RHS in (9.91) are comparable to each other under $\lambda(n, x) \asymp 1$. If $x_n \asymp 1$, the first term of it is negligible compared to the second since by $a(x)/a(-x) \to 0 \, (x \to \infty)$ we have $\lambda(n, c_n) \to 0$. One can verify the equivalence $\lambda(n, x) \ll 1 \Leftrightarrow x \ll a^\dagger(x)c_n^2/n$. These in particular show that in (9.90) M can be replaced by c_n^2/n if $a(x) \to \infty \, (x \to \infty)$; while if $a(x)$ is bounded on $x > 0$, then $E|\hat{Z}| < \infty$, $V_\mathrm{d}(x) \sim v^\circ x/E|\hat{Z}|$, so that $\lambda(n, x) \sim Cxn/c_n^2$. In any case from (9.92) one obtains

$$P_x[\sigma_B > n] \sim u_B(x)\kappa_\alpha^\# c_n n^{-1}[1 + \lambda(n,x)] \qquad (x_n \downarrow 0),$$

and hence

$$P_x[S_n = y \mid \sigma_B > n] \sim \frac{q(\eta) + \lambda(n,x)q^{\mathrm{mnd}}(\eta)}{1 + \lambda(n,x)} \qquad (x_n \downarrow 0, y_n \to \eta).$$

We know that $q^{\mathrm{mnd}}(\xi) \sim C'\xi \ (\xi \downarrow 0)$, $C''\xi^{-\alpha} \ (\xi \to \infty)$ ([17, Remark 5 and Eq(33)]), which together with (9.96, 9.97) yields $q(\xi)/q^{\mathrm{mnd}}(\xi) \sim C_1\xi^{\alpha-2} \ (\xi \downarrow 0)$, $C_2\xi^{-1} \ (\xi \to \infty)$. In particular, this shows that (9.89) fails if $q = 0$.

9.9.3 Asymptotic Properties of $\mathfrak{p}_t^0(\xi,\eta)$ and $\mathfrak{q}_t(\xi)$ and Applications

We rewrite the scaling relation (9.23) as

$$\mathfrak{f}^\xi(\lambda t) = \mathfrak{f}^{\xi/t^{1/\alpha}}(\lambda)/t = \mathfrak{f}^{\mathrm{sgn}(\xi)}(\lambda t/|\xi|^\alpha)/|\xi|^\alpha \qquad (\xi \neq 0, t > 0, \lambda > 0). \quad (9.93)$$

We know that if $p = 1$, then V_d varies regularly with index 1 and

$$\mathfrak{f}^\xi(t) = P_\xi^Y[\sigma_{(-\infty,0)} \in dt]/dt = \xi t^{-1}\mathfrak{p}_t(-\xi) \quad (\xi > 0) \tag{9.94}$$

(cf. [4, Corollary 7.3]). In the case $pq = 0$, expansions of $\mathfrak{f}^\xi(t)t = \mathfrak{f}^{\xi/t^{1/\alpha}}(1)$ into power series of $\xi/t^{1/\alpha}$ are known. Indeed, if $p = 1$, owing to (9.94) the series expansion for $\xi > 0$ is obtained from that of $t^{1/\alpha}\mathfrak{p}_t(-\xi)$, which is found in [31], while for $\xi < 0$, the series expansion is derived by Peskir [57]. In the recent paper [50, Theorem 3.14] Kuznetsov et al. obtain a similar series expansion for all cases.

Here we let $1 < \alpha < 2$ and we give the leading term for $\mathfrak{f}^1(t)$ and an error estimate (as $t \to \infty$ and $t \downarrow 0$) that are deduced from the series expansions of $\mathfrak{f}^\xi(t)$ mentioned above. To have neat expressions we put

$$\kappa_\alpha^{a,\pm} = [1 \mp (p - q)]/\kappa_\alpha^\circ, \tag{9.95}$$

so that $a(x) \sim \kappa_\alpha^{a,\pm}x^{\alpha-1}/L(x)$ as $x \to \pm\infty$; see Proposition 4.2.1, also (9.76) for κ_α°. (In (9.95) both upper or both lower signs should be chosen in the double signs.)

In the rest of this subsection let $c_\# = 1$ as in Section 9.8. Then, as $t \to \infty$,

$$\mathfrak{f}^{\pm 1}(t) = \begin{cases} \kappa_\alpha^* t^{-1-1/\alpha}\{1 + O(t^{-1/\alpha})\} & \text{if } \kappa_\alpha^{a,\pm} = 0, \\ \kappa_\alpha^{\mathfrak{f},\pm} t^{-2+1/\alpha}\{1 + O(t^{1-2/\alpha})\} & \text{if } \kappa_\alpha^{a,\pm} > 0, \end{cases} \tag{9.96}$$

where $\kappa_\alpha^{\mathfrak{f},\pm} = (1-\alpha^{-1})\kappa_\alpha^\#\kappa_\alpha^{a,\pm}$ and $\kappa_\alpha^* = |\Gamma(1-\alpha)|^{1/\alpha}/|\Gamma(-1/\alpha)|$. Note that $\kappa_\alpha^{a,+} > 0$ ($\kappa_\alpha^{a,-} > 0$) if and only if $q > 0$ ($p > 0$). The first formula above follows from [31, Lemma XVII.6.1] in view of (9.94). In [50, Theorem 3.14] and [82, Lemma 8.2] an asymptotic expansion as $t \to 0$ is obtained which entails that

$$f^1(t) = \frac{C_\Phi \alpha^2 \Gamma(\alpha) \kappa_\alpha^\# \sin \alpha \rho \pi}{\pi \cos(\hat{\rho} - \rho)\alpha \pi} t^{1/\alpha} + O(t^{1+1/\alpha}) \qquad (t \downarrow 0), \qquad (9.97)$$

if $q > 0$, and $f^1(t) = o(t^M)$ for all $M > 1$ if $q = 0$.

We note that the leading term in the second formula of $f^1(t)$ in (9.96) is deduced from (9.88) together with $P_0[\sigma_0 > n] \sim \alpha^{-1}(\alpha - 1)\kappa_\alpha^\# c_n/n$;

$$\lim_{\xi \downarrow 0} \frac{f^\xi(1)}{\xi^{\alpha-1}} = \lim_{\xi \downarrow 0} \lim_{x_n \to \xi} \frac{n f^x(n)}{\xi^{\alpha-1}} = \lim_{\xi \downarrow 0} \lim_{x_n \to \xi} \frac{(\alpha-1)\kappa_\alpha^\# a(x)}{\alpha \xi^{\alpha-1} n/c_n} = \frac{\alpha-1}{\alpha} \kappa_\alpha^{a,+} \kappa_\alpha^\# = \kappa_\alpha^{f,+},$$

giving what we wanted because of the scaling relation (9.93).

Lemma 9.9.1 (i) $\alpha \int_0^\infty [\mathfrak{p}_1(0) - \mathfrak{p}_1(\pm u)] u^{-\alpha} \, du = \kappa_\alpha^{a,\mp}$.
(ii) *If $\varphi(t)$ is a continuous function on $t \geq 0$, then for $t_0 > 0$*

$$\lim_{\eta \to \pm 0} \int_0^{t_0} \frac{\mathfrak{p}_t(0) - \mathfrak{p}_t(\eta)}{|\eta|^{\alpha-1}} \varphi(t) \, dt = \kappa_\alpha^{a,\mp} \varphi(0).$$

Proof Since $\mathfrak{p}_1$ is bounded and differentiable, the existence of the integral in (i) is obvious. We must identify it as $\kappa_\alpha^{a,\mp}$, which we do in the next lemma.

Using the scaling relation as above and changing the variable $t = (|\eta|/s)^\alpha$ we obtain

$$\int_0^{t_0} \frac{\mathfrak{p}_t(0) - \mathfrak{p}_t(\eta)}{|\eta|^{\alpha-1}} \varphi(t) \, dt = \alpha \int_{|\eta|/t_0^{1/\alpha}}^\infty \frac{\mathfrak{p}_1(0) - \mathfrak{p}_1(\mathrm{sgn}(\eta)u)}{u^\alpha} \varphi(|\eta|^\alpha/u^\alpha) \, du.$$

Dominated convergence shows the formula of (ii) by virtue of (i). □

Lemma 9.9.2 *For any $t_0 > 0$, uniformly for $|\xi| > 0$ and $t > t_0$, as $\eta \to \pm 0$,*

$$\mathfrak{p}_t^0(\xi, \eta)/|\eta|^{\alpha-1} \longrightarrow \kappa_\alpha^{a,\mp} f^\xi(t).$$

Proof Although the result follows from (9.87), we use the latter only for the identification of the constants $b_{\alpha,\gamma}^\pm = \alpha \int_0^\infty \frac{\mathfrak{p}_1(\pm u) - \mathfrak{p}_1(0)}{|u|^\alpha} \, du$ in this proof, which is based on

$$\mathfrak{p}_t^0(\xi, \eta) = \mathfrak{p}_t(\eta - \xi) - \int_0^t f^\xi(t - s)\mathfrak{p}_s(\eta) \, ds.$$

Subtracting from this equality the same one, except with $\eta = 0$, for which the LHS vanishes, and then dividing by $|\eta|^{\alpha-1}$, one obtains

$$\frac{\mathfrak{p}_t^0(\xi, \eta)}{|\eta|^{\alpha-1}} = \frac{\mathfrak{p}_t(\eta - \xi) - \mathfrak{p}_t(-\xi)}{|\eta|^{\alpha-1}} + \int_0^t \frac{\mathfrak{p}_s(0) - \mathfrak{p}_s(\eta)}{|\eta|^{\alpha-1}} f^\xi(t - s) \, ds. \qquad (9.98)$$

As $\eta \to 0$, the first term on the RHS tends to zero, and Lemma 9.9.1 applied to the second term yields the equality of the lemma. The uniformity of the convergence is checked by noting that the above integral restricted to $s > t/2$ is negligible.

To show $b^{\pm}_{\alpha,\gamma} = \kappa^{a,\mp}_{\alpha}$, suppose $\kappa^{a,\mp}_{\alpha} \neq 0$; if $\kappa^{a,\mp}_{\alpha} = 0$, the required identity follows from what we have just verified. By the second case of (9.87), for any $\varepsilon > 0$, there exists $0 < \delta < \varepsilon$ such that for $|y_n| < \delta$, $\limsup |Q^n_{\{0\}}(c_n, y)n[\mathfrak{f}^1(1)a(-y)]^{-1} - 1| < \varepsilon$. Hence, by the fourth case of (9.87), for $\varepsilon\delta < |y_n| < \delta$,

$$\limsup \left| \frac{\mathfrak{p}^0_1(1, y_n)n/c_n}{\mathfrak{f}^1(1)a(-y)} - 1 \right| = \limsup \left| \frac{\mathfrak{p}^0_1(1, y_n)}{Q^n_{\{0\}}(c_n, y)c_n} \cdot \frac{Q^n_{\{0\}}(c_n, y)n}{\mathfrak{f}^1(1)a(-y)} - 1 \right| \leq \varepsilon.$$

Letting $\varepsilon \downarrow 0$, this entails that as $y \to \pm\infty$ under $y_n \to \pm 0$, so that $L(c_n)/L(|y|) \to 1$,

$$1 \sim \frac{\mathfrak{p}^0_1(1, y_n)n/c_n}{\mathfrak{f}^1(1)a(-y)} \sim \frac{\mathfrak{p}^0_1(1, y_n)c_n^{\alpha-1}/L(c_n)}{\kappa^{a,\mp}_{\alpha}\mathfrak{f}^1(1)|y|^{\alpha-1}/L(|y|)} \sim \frac{\mathfrak{p}^0_1(1, y_n)}{\kappa^{a,\mp}_{\alpha}\mathfrak{f}^1(1)|y_n|^{\alpha-1}}.$$

Thus $b^{\pm}_{\alpha,\gamma} = [\mathfrak{f}^1(1)]^{-1}\lim_{\eta\to\pm 0}\mathfrak{p}^0_1(1, \eta)/|\eta|^{\alpha-1} = \kappa^{a,\mp}_{\alpha}$, finishing the proof. $\square$

The following lemma shows that $\mathfrak{a}$ is the 'unique' non-negative harmonic function for the transition function $\mathfrak{p}^0_t(\xi, \eta)$ in view of the theory of Martin boundaries.

Lemma 9.9.3 *Put* $\mathfrak{g}_0(\xi, \eta) = \int_0^\infty \mathfrak{p}^0_t(\xi, \eta)dt$. *Then*

(i) $\displaystyle \lim_{\eta\to\pm\infty} \mathfrak{g}_0(\xi, \eta) = \frac{\alpha[1 - \mathrm{sgn}(\xi)(p - q)]}{2\kappa^\circ_\alpha} \int_0^\infty \mathfrak{f}^{\mp r}(1)\frac{dr}{r}|\xi|^{\alpha-1}$;

(ii) $\displaystyle \lim_{|\eta|\to\infty} \mathfrak{g}_0(\xi, \eta)/\mathfrak{g}_0(\pm 1, \eta) = \left(\kappa^{\#}_\alpha\kappa^{a,\pm}_\alpha\right)^{-1}\mathfrak{a}(\xi)$,

where either sign applies if $pq > 0$ *whereas the upper (resp. lower) sign is allowed only if* $q > 0$ *(resp.* $p > 0$*) so that* $\mathfrak{g}_0(\pm 1, \eta) > 0$ *for all* $\eta \neq 0$.

Proof We show (i) in its dual form, which reads

$$(*) \qquad \lim_{\xi\to\pm\infty} \mathfrak{g}_0(\xi, \eta) = \frac{\alpha[1 + \mathrm{sgn}(\eta)(p - q)]}{2\kappa^\circ_\alpha} \int_0^\infty \mathfrak{f}^{\pm r}(1)\frac{dr}{r}|\eta|^{\alpha-1}.$$

By the second scaling relation in (9.26) and the change of variable $t = (|\xi|/r)^\alpha$,

$$\mathfrak{g}_0(\xi, \eta) = \alpha|\xi|^{\alpha-1} \int_0^\infty \mathfrak{p}^0_1(\mathrm{sgn}(\xi)r, |\xi|^{-1}\eta r)\frac{dr}{r^\alpha}. \tag{9.99}$$

A formal application of Lemma 9.9.2 gives the formula of (i), which must be justified because of the singularity of the integrand at zero. For the justification, resuming the procedure leading to Lemma 9.9.2, we apply the decomposition of $\mathfrak{p}^0_t(\xi, \eta)$ in (9.98). Let I_ξ and II_ξ denote the contribution to, respectively, the RHS above from the part corresponding to the first and second terms of the RHS of (9.98). Let $\xi > 0$ and $\delta = \xi^{-1}\eta$. First, we dispose of I_ξ. Using $|(r + \delta r)^{-\alpha} - r^{-\alpha}| < C|\delta|r^{-\alpha}$, we observe that

$$\left| \int_0^\infty \frac{\mathfrak{p}_1(\delta r - r) - \mathfrak{p}_1(-r)}{r^\alpha} \, dr \right| \le C_1 |\delta| \left[\int_0^1 r^{1-\alpha} \, dr + \int_{1-|\delta|}^\infty r^{-\alpha} \, dr \right] \le C_2 |\eta|/\xi,$$

provided $|\delta| < 1/2$. Thus $I_\xi \to 0$.

To evaluate II_ξ we write it as

$$II_\xi = \alpha |\eta|^{\alpha-1} \int_0^\infty \frac{dr}{r} \int_0^1 \frac{\mathfrak{p}_s(0) - \mathfrak{p}_s(\delta r)}{(|\delta|r)^{\alpha-1}} \mathfrak{f}^r(1-s) \, ds.$$

Since $\mathfrak{p}_1$ is continuously differentiable, for every $\varepsilon > 0$, this repeated integral restricted to $r > \varepsilon$ and $1/2 < s \le 1$ is $O(\delta^{2-\alpha})$, hence tends to zero as $\xi \to \infty$. In the remaining range of integration $\mathfrak{f}^r(1-s)$ is bounded by a constant multiple of $r^{\alpha-1}$, hence the repeated integral over that converges to $\kappa_\alpha^{a,\mp} \int_\varepsilon^\infty \mathfrak{f}^r(1) r^{-1} \, dr$ by virtue of Lemma 9.9.1(ii), where $\pm$ accords with the sign of η. Letting $\varepsilon \downarrow 0$ we see that $(*)$ follows when $\xi \to \infty$. The case $\xi \to -\infty$ is similar.

The second assertion (ii) is immediate from (i) and the definition of $\mathfrak{a}$. □

Lemma 9.9.4 (i) $\lim_{|\eta| \to 0} \mathfrak{p}_t^0(\xi, \eta)/\mathfrak{a}(-\eta) = \mathfrak{q}(-\xi)$, where the limit is restricted to η such that $\mathfrak{a}(-\eta) > 0$ when $pq = 0$.
(ii) For every $t > 0$, $\int_{-\infty}^\infty \mathfrak{a}(\xi) \mathfrak{q}_t(\xi) \, d\xi = 1$.

Proof Recalling the definitions in (9.75) and (9.95), we have $\mathfrak{a}(\eta) = \kappa_\alpha^{a,\pm} \kappa_\alpha^\sharp |\eta|^{\alpha-1}$ ($\pm$ accords with the sign of η) and find that (i) follows from Lemma 9.9.2 because $\mathfrak{f}^\xi(t) = \kappa_\alpha^\sharp \mathfrak{q}_t(-\xi)$. To verify (ii), write down the dual of what we have just obtained in the form

$$\lim_{|\eta| \to 0} \mathfrak{p}_t^0(\eta, \xi)/\mathfrak{a}(\eta) = \mathfrak{q}_t(\xi),$$

multiply both sides by $\mathfrak{a}(\xi)$ and integrate over $\xi \in \mathbb{R}$. Then a formal application of (9.78) gives the result, where the limit procedure can be justified as in the proof of the preceding lemma. □

9.9.4 The r.w. Killed Upon Entering the Negative Half-Line

Concerning the conditional law of the r.w. given $T(= \sigma_\Omega) > n$, there are many works that have studied the functional limit theorem [10], [11], [13], [17], [19], [38], [68] and the local limit theorems [1], [11], [12], [17], [88] in various frameworks. Below we state the main results from [11], [13], [17] and [88] in our setting of the arithmetic r.w.

Suppose that $|1 - \alpha| \wedge \rho\hat{\rho} > 0$ and F is strongly aperiodic. We know that $P_0[T > n] \sim t^{-\hat{\rho}} \hat{\ell}(t)$ for some $\hat{\ell}$ s.v. at infinity (Theorem A.3.1). Let $\mathfrak{q}^{\mathrm{mnd}}(\eta)$ and $\hat{\mathfrak{q}}^{\mathrm{mnd}}(\eta)$, $\eta \ge 0$ denote the densities of the stable meander of length 1 at time 1 associated with Y and $-Y$, respectively (see [4] for the definition), and $Q_\Omega^n(x, y) := P_x[S_n = y, \sigma_\Omega > n]$. Vatutin and Wachtel [88] proved that

$$Q_\Omega^n(0,y) \sim \begin{cases} V_d(0)U_a(y)\mathfrak{p}_1(0)/(nc_n) & (y_n \downarrow 0), \\ P[T > n]q^{mnd}(y_n)/c_n & (y_n > \varepsilon), \end{cases} \tag{9.100}$$

and based on it, derived

$$P_0[T = n] \sim \hat{\rho}n^{-1-\hat{\rho}}\hat{\ell}(n). \tag{9.101}$$

Partially based on the analysis made in [88], Doney [17] extended the above result to the general initial point. The result reads (given in the dual setting in [17]): for each $\varepsilon > 0$,

$$Q_\Omega^n(x,y) \sim \begin{cases} V_d(x)U_a(y)\mathfrak{p}_1(0)/(nc_n) & (x_n \downarrow 0, y_n \downarrow 0), \\ U_a(y)P_0[\sigma_{[1,+\infty)} > n]\hat{q}^{mnd}(x_n)/c_n & (y_n \downarrow 0, x_n > \varepsilon), \\ V_d(x)P_0[\sigma_{(-\infty,0]} > n]q^{mnd}(y_n)/c_n & (x_n \downarrow 0, y_n > \varepsilon), \\ \mathfrak{p}_1^{(-\infty,0]}(x_n,y_n)/c_n & (x_n \wedge y_n \geq \varepsilon) \end{cases} \tag{9.102}$$

uniformly for $x \vee y < c_n/\varepsilon$. Based on this result, the following was derived

$$P_x[T = n] \sim \begin{cases} \rho V_d(x)P_0[T = n]/v^\circ & \text{as} \quad x_n \downarrow 0, \\ n^{-1}\mathfrak{f}_{(-\infty,0]}^{x_n}(1) & \text{uniformly for} \quad x_n \in [\varepsilon, \varepsilon^{-1}]. \end{cases} \tag{9.103}$$

(The derivations of (9.101) from (9.100), and (9.103) from (9.102), constitute essential parts of [88] and [17], respectively.)

The conditional limit theorem for the r.w. given $T = \infty$ was studied by Bryn-Jones and Doney [11] and Caravenna and Chaumont [13]. The conditional process is an h-transform of S by the renewal function V_d, which is harmonic on $\mathbb{Z} \setminus \Omega$. Define the probability laws on $D_{[0,\infty)}$ by

$$\mathbf{P}_x^{n,V}(\Gamma) = P_x[V_d(S_{\lfloor nt \rfloor}); Y_n \in \Gamma, T > nt]/V_d(x) \quad x = 0, 1, 2, \ldots,$$

$$\mathbf{P}_\xi(\Gamma) = \begin{cases} E_\xi^Y[(Y(t))^{\alpha\hat{\rho}}; Y \in \Gamma, \sigma_{(-\infty,0]}^Y > t]/\xi^{\alpha\hat{\rho}} & \xi > 0, \\ E_t^{mnd}[\mathfrak{y}_t^{\alpha\hat{\rho}}; \Gamma]/E_t^{mnd}[\mathfrak{y}_t^{\alpha\hat{\rho}}] & \xi = 0. \end{cases}$$

Here for $t > 0$, $\Gamma \in \mathcal{B}_t$, and E_t^{mnd} stands for the expectation with respect to the law of the stable meander of length t. Bryn-Jones and Doney [11] established that for $0 < \rho < 1$, $\mathbf{P}_x^{n,V} \Rightarrow \mathbf{P}_\xi$ as $x/c_n \to \xi \geq 0$ in the case $\alpha = 2$ by showing the local limit theorem, and Caravenna and Chaumont [13] extended it to the case $0 < \alpha \leq 2$ by employing the earlier result (cf. [10], [19], [38], [68]) that the r.w. conditioned on $P_0[T > n]$ converges in law to the stable meander.

Appendix

A.1 Relations Involving Regularly Varying Functions

A.1.1 Results on s.v. Functions I

Here we present some results derived from standard facts about the s.v. functions. Lemmas A.1.1 and A.1.2 below are used for the proof of Theorem 5.4.1 as well as for the deduction of Corollary 5.4.2(i) from Theorem 5.4.1.

Lemma A.1.1 *Let $L(x) > 0$ be an s.v. function at infinity. The following are equivalent*

(a) $m(x) \sim L(x)/2$.
(b) $x\eta(x) = o\,(m(x)), \quad c(x) := \int_0^x t\mu(t)\,\mathrm{d}t \sim m(t)$.
(c) $c(x) \sim L(x)/2$.
(d) $x^2\mu(x)/\int_{-x}^x t^2\,\mathrm{d}F(t) \to 0$.
(e) $\int_{-x}^x t^2\,\mathrm{d}F(t) \sim L(x)$.

Proof The implication (a) $\Rightarrow$ (b) follows from $\frac{1}{2}x\eta(x) \le m(x) - m(\frac{1}{2}x)$ and its converse from $m(x) = m(1)e^{\int_1^x \varepsilon(t)\,\mathrm{d}t/t}$, where $\varepsilon(t) = t\eta(t)/m(t)$. Combining (a) and (b) shows (c) in view of $m(x) = c(x) + x\eta(x)$; conversely, (c) entails $x^2\mu(x) = o(L(x))$, by which one deduces $x\eta(x) \ll x\int_x^\infty L(t)t^{-2}\,\mathrm{d}t \sim L(x)$. Thus (b) $\Leftrightarrow$ (c). The equivalence of (d) and (e) follows from Theorem VIII.9.2 of [31] and shows the equivalence of (e) and (c) since $\int_{-x-}^{x+} t^2\,\mathrm{d}F(t) = -x^2\mu(x) + 2c(x)$. $\qquad\square$

Lemma A.1.2 *Let $m_+/m \to 0$. Then by $\int_0^x t^2\,\mathrm{d}F(t) \le 2c_+(x) = o(m_-(x))$ the slow variation of $m_-(x)$ entails $\int_{-x}^x t^2\,\mathrm{d}F(t) \sim \int_{-x}^0 t^2\,\mathrm{d}F(t) \sim 2m_-(x)$ because of the equivalence (a) $\Leftrightarrow$ (c) in (1).*

A.1.2 Results on s.v. Functions II

We prove three lemmas involving s.v. functions, which are used in the proofs of Lemmas 6.3.1, 6.3.2 and 6.4.2, respectively.

Lemma A.1.3 *Let α be a positive constant, $\ell(t)$ $(t \geq 0)$ an s.v. function, and $u(t)$ and $G(t)$ positive measurable functions of $t \geq 0$. Suppose that $\int_0^x u(t)\, dt \sim x^\alpha / \alpha \ell(x)$ $(x \to \infty)$, $G(t)$ is non-increasing and both u and $1/\ell$ are locally bounded, and put*

$$\tilde{u}(t) = \frac{t^{\alpha-1}}{\ell(t)}, \quad h_\sharp(x) = \int_x^\infty \tilde{u}(t)G(t)\, dt,$$

$$h(x) = \int_0^\infty u(t)G(x+t)\, dt, \quad and \quad \tilde{h}(x) = \int_0^\infty \tilde{u}(t)G(x+t)\, dt.$$

If one of $h(1)$, $\tilde{h}(1)$ and $h_\sharp(1)$ is finite, then so are the other two and

(i) *if either h or $\tilde{h}$ is regularly varying with index > -1, then $h(x) \sim \tilde{h}(x)$;*
(ii) *if either $\tilde{h}$ or $h_\sharp$ is s.v., then $h(x) \sim h_\sharp(x)$.*

Proof We may suppose ℓ is positive and continuous. Integrating by parts verifies that $\int_0^x \tilde{u}(t)G(t) \asymp \int_0^x u(t)G(t)\, dt$, so that $h(1)$, $\tilde{h}(1)$ and $h_\sharp(1)$ are finite if either of them is finite.

Note that $\tilde{h}(1) < \infty$ entails $G(t) = o(t^{-\alpha}\ell(t))$, so that $G(t)\int_0^t u(s)\, ds \to 0$ $(t \to \infty)$. Put $U(x) = \int_0^t u(s)\, ds$ and $\tilde{U}(t) = \int_0^t \tilde{u}(s)\, ds$. Then, interchanging the order of integration and integrating by parts, in turn, one deduces

$$\int_0^x h(t)\, dt = \int_0^\infty u(t)\, dt \int_t^{x+t} G(s)\, ds = \int_0^\infty [G(t) - G(x+t)]U(t)\, dt,$$

and similarly

$$\int_0^x \tilde{h}(t)\, dt = \int_0^\infty [G(t) - G(x+t)]\tilde{U}(t)\, dt.$$

Since G is monotone and $U(t) \sim \tilde{U}(t)$, these identities together show

$$\int_0^x h(s)\, ds \sim \int_0^x \tilde{h}(s)\, ds,$$

provided $\int_0^x h(s)\, ds$ or $\int_0^x \tilde{h}(s)\, ds$ diverges to infinity as $x \to \infty$. It therefore follows that if one of h or $\tilde{h}$ is regularly varying with index > -1, then so is the other and $\tilde{h}(x) \sim h(x)$ owing to the monotone density theorem [8]. Thus (i) is verified.

For the proof of (ii), suppose either $\tilde{h}$ or $h_\sharp$ is s.v. Take $M > 2$ large. Then we see

$$\tilde{h}(x) \geq \int_0^{(M-1)x} \tilde{u}(t)G(x+t)\, dt = (M-1)^\alpha \frac{x^\alpha}{\alpha\ell(x)}G(Mx)\{1 + o(1)\}$$

and

$$\tilde{h}(Mx) = \left(\int_0^x + \int_x^\infty \right) \tilde{u}(t) G(Mx+t)\, dt \le \frac{x^\alpha}{\alpha \ell(x)} G(Mx) + h_\sharp(x). \tag{A.1}$$

If $\tilde{h}$ is s.v., so that $\tilde{h}(Mx) \sim \tilde{h}(x)$, these inequalities lead to

$$h_\sharp(x) \ge \tilde{h}(x) \left[1 + O(M^{-\alpha})\{1 + o(1)\} \right]. \tag{A.2}$$

On the other hand, if $h_\sharp$ is s.v., then

$$h_\sharp(x) \sim h_\sharp(x/2) \ge \int_{x/2}^x \tilde{u}(t) G(t)\, dt \ge (1 - 2^{-\alpha}) \frac{G(x)x^\alpha}{\alpha \ell(x)},$$

so that by (A.1) applied with x/M in place of x, one obtains

$$\tilde{h}(x) \le \left[1 + O(M^{-\alpha})\{1 + o(1)\} \right] h_\sharp(x),$$

hence (A.2). Since M can be made arbitrarily large, one concludes $\tilde{h}(x) \le h_\sharp(x)\{1 + o(1)\}$ in either case. The reverse inequality is verified as follows. If $\alpha \ge 1$, then taking $\varepsilon \in (0, 1)$ so small that for $0 < s < 1$, $(1 - \varepsilon s)^{\alpha - 1} > 1 - \alpha \varepsilon$, one deduces

$$\tilde{h}(\varepsilon x) \ge \int_x^\infty \frac{(t - \varepsilon x)^{\alpha - 1}}{\ell(t - \varepsilon x)} G(t)\, dt \ge (1 - \alpha \varepsilon) h_\sharp(x)\{1 + o(1)\},$$

which shows $\tilde{h}(x) \ge h_\sharp(x)\{1 + o(1)\}$. If $\alpha < 1$, one may suppose ℓ to be normalised so that for all sufficiently large x, $\tilde{u}(t) \ge \tilde{u}(x + t)$ $(t \ge 0)$, hence $\tilde{h}(x) \ge h_\sharp(x)$ as well. Now we can conclude that $\tilde{h}(x) \sim h_\sharp(x)$, and hence $h(x) \sim h_\sharp(x)$ by (i). □

Lemma A.1.4 *Let $V(x)$ and $\mu(x)$ be positive functions on $x \ge 0$. If μ is non-increasing, $\limsup \mu(\lambda x)/\mu(x) < 1$ for some $\lambda > 1$ and V is non-decreasing and s.v., then $\int_1^\infty \mu(t)\, dV(t) < \infty$ and*

$$\int_x^\infty \mu(t)\, dV(t) = o\left(V(x)\mu(x) \right) \quad (x \to \infty).$$

Proof We may suppose that V is continuous and there exists constants $\delta < 1$ and x_0 such that $\mu(\lambda x) < \delta \mu(x)$ for $x \ge x_0$. Let $x > x_0$. It then follows that

$$\int_x^\infty \mu(t)\, dV(t) = \sum_{n=0}^\infty \int_{\lambda^n x}^{\lambda^{n+1} x} \mu(t)\, dV(t) \le \mu(x) \sum_{n=0}^\infty \delta^n [V(\lambda^{n+1} x) - V(\lambda^n x)]. \tag{A.3}$$

Take $\varepsilon > 0$ so that $(1+\varepsilon)\delta < 1$. Then for all sufficiently large x, $V(\lambda x)/V(x) \le 1+\varepsilon$, so that $V(\lambda^n x)/V(x) < (1 + \varepsilon)^n$, hence $\sum_{n=0}^\infty \delta^n V(\lambda^n x) \le C V(x)$, which shows that the last expression in (A.3) is $o(\mu(x)V(x))$, for $V(\lambda^{n+1}x) - V(\lambda^2 x) = o((V(\lambda^n x))$ uniformly in n. □

Lemma A.1.5 *Let $v(t)$ and $G(t)$ be non-negative measurable functions on $t \geq 0$ such that $V(t) := \int_0^t v(s)\,ds$ is s.v. at infinity, $\int_1^\infty v(t)G(t)\,dt < \infty$ and G is non-increasing. Then as $t \to \infty$*

$$\frac{\int_t^\infty v(s)G(s)\,ds}{V(t)G(t) + \int_t^\infty V(s)G(s)\,ds/s} \longrightarrow 0.$$

Proof Let $\tilde{V}$ be a normalised version of V: $\tilde{V}(t) = e^{\int_0^t \varepsilon(s)\,ds}$ with $\varepsilon(t) \to 0$, so that $\tilde{v}(t) := \tilde{V}'(t) = o(V(t)/t)$. We show that

$$\int_t^\infty v(s)G(s)\,ds = \int_t^\infty \tilde{v}(s)G(s)\,ds\{1 + o(1)\} + o\left(V(t)G(t)\right), \qquad (A.4)$$

which implies the asserted result. Put

$$\varDelta(t) = \int_t^\infty v(s)G(s)\,ds, \quad E(t) = -\int_t^\infty V(s)\,dG(s).$$

Then $E(t) = V(t)G(t) + \varDelta(t)$. Since $E(t) \sim -\int_t^\infty \tilde{V}(s)\,dG(s)$, this entails

$$\varDelta(t) = -\tilde{V}(t)G(t)\{1 + o(1)\} - \int_t^\infty \tilde{V}(s)\,dG(s) + o(E(t))$$

$$= \int_t^\infty \tilde{v}(s)G(s)\,ds + o\left(V(t)G(t)\right); +o(E(t)).$$

Since $o(E(t)) = o(V(t)G(t)) + o(\varDelta(t))$, we obtain (A.4).   □

A.2 Renewal Processes, Strong Renewal Theorems and the Overshoot Distribution

Let $T_0 = 0$ and $(T_n)_{n=0}^\infty$ be a r.w. on $\{0, 1, 2, \ldots\}$ with i.i.d. increments. Put

$$u(x) = \sum_{n=0}^\infty P[T_n = x] \quad (x = 1, 2, \ldots)$$

(renewal mass function) and $U(x) = u(0) + \cdots + u(x)$. Suppose that T_1 is strongly aperiodic so that $u(x)$ is positive for all sufficiently large x. We shall consider asymptotic forms of u and/or U and the overshoot distribution. Most results (some with auxiliary assumptions) can be extended to non-arithmetic cases.

A.2.1 Strong Renewal Theorems

We give some results on the asymptotics of $u(x)$ in two extremal cases, one for r.s. variables and the other for those having s.v. tails, which are dealt with in Sections A.2.1.1 and A.2.1.2, respectively.

A.2.1.1 Case: T_1 is r.s.

First, consider the case when T_1 is r.s. Put

$$L(t) = \int_0^t P[T_1 > s]\, ds \quad (t \geq 0).$$

Erickson [28, §2(ii)] shows that $\lim u(x)L(x) = 1$ if $tP[T_1 > t]$ is s.v. at infinity. This restriction on T_1 is relaxed as in the following lemma, which is used to obtain (5.28). (As mentioned before, the result is extended to general r.s. variables in [83].)

Lemma A.2.1 *If L is s.v. at infinity, then $u(x) \sim 1/L(x)$ as $x \to \infty$.*

Proof We follow the argument made by Erickson [28]. Let $\phi(\theta) = E \exp\{i\theta T_1\}$. Unlike [28], we take up the sine series of coefficients $u(x)$ that represents the imaginary part of $1/(1 - \phi(\theta))$. Suppose $ET_1 = \infty$; otherwise, the result is the well-known renewal theorem. Fourier inversion yields

$$u(x) = \frac{2}{\pi} \int_0^\pi S(\theta) \sin x\theta\, d\theta, \quad \text{where} \quad S(\theta) = \Im\left(\frac{1}{1 - \phi(\theta)}\right), \tag{A.5}$$

where the integral is absolutely convergent, as is verified shortly. The proof of this representation of $u(x)$ will be given after we derive the assertion of the lemma by taking it for granted. The assumed slow variation of L implies – in fact, is equivalent to – each of

$$\int_0^t sP[T_1 \in ds] \sim L(t) \quad \text{and} \quad tP[T_1 > t]/L(t) \to 0 \quad (t \to \infty)$$

(cf. [31, Theorem VIII.9.2], [8, Corollary 8.1.7]). Using this we observe that as $\theta \downarrow 0$, $\int_0^{\varepsilon/\theta} tP[T_1 \in dt] \sim L(1/\theta)$ for each $\varepsilon > 0$ and hence

$$1 - \phi(\theta) = \left[\int_0^{1/\theta} + \int_{1/\theta}^\infty\right](1 - e^{i\theta t})P[T_1 \in dt] \tag{A.6}$$

$$= -i\theta L(1/\theta)\{1 + o(1)\}.$$

In particular, $S(\theta) \sin x\theta$ is summable on $(0, \pi)$, as mentioned above.

Decomposing $u(x) = \frac{2}{\pi} \int_0^{B/x} + \frac{2}{\pi} \int_{B/x}^{\pi} = J_1 + J_2$, we deduce from (A.6) that

$$J_1 = \frac{2}{\pi} \int_0^B \frac{\sin u \, du}{uL(x/u)} \{1 + o(1)\} \tag{A.7}$$

with $o(1) \to 0$ as $x \to \infty$ for each $B > 1$. On the other hand

$$\pi J_2 = \int_{B/x}^{\pi} \left[S(\theta) - S\left(\theta + \frac{\pi}{x}\right) \right] \sin x\theta \, d\theta + \left(\int_{B/x}^{(B+\pi)/x} - \int_{\pi}^{\pi+\pi/x} \right) S(\theta) \sin x\theta \, d\theta$$

$$= J_2' + J_2'' \quad \text{(say)}.$$

With the help of the bound $|\phi(\theta) - \phi(\theta')| \le 2|\theta - \theta'|L(1/|\theta - \theta'|)$ $(\theta \ne \theta')$ (Lemma 5 of [28]) the same proof as given in [28, (5.15)] yields the bound $J_2' \le C'/BL(x)$. By the strong aperiodicity of T_1 and (A.6), $|S(\theta)| \le C/\theta L(1/\theta)$ $(0 < \theta \le \pi)$, and it is easy to see that $|J_2''| \le C''[B/x + 1/BL(x/B)]$. Thus $\lim_{x\to\infty} L(x)J_2 \le C'/B$. Since (A.7) implies that $L(x)J_1 \to 1$ as $x \to \infty$ and $B \to \infty$ in this order, we can conclude $L(x)u(x) \to 1$ as desired.   □

Proof (of (A.5)) The 'cosine formula', namely $u(n) = \frac{2}{\pi} \int_0^{\pi} C(\theta) \cos x\theta \, d\theta$, where $C(\theta) := \Re(1 - \phi(\theta))^{-1}$, is applied in [33] and [28] without proof. In [33] is cited the article [37], which proves the Herglotz representation theorem (named after its author) of the analytic functions in the unit open disc with non-negative real parts. Noting $\Re(1 - \phi(\theta))^{-1} \ge 0$ $(|\theta| < 1)$ one can easily obtain the cosine formula by deducing the summability of $C(\theta)$ from the representation theorem. The same argument does not go through for (A.5), $S(\theta)$ not always being summable. An elementary proof of the cosine formula is given in [71, p.98-99] and easily modified to obtain (A.5) as given below.

Put

$$w_r(\theta) = \frac{1}{2\pi} \sum_{n=0}^{\infty} r^n \phi^n(\theta) = \frac{1}{2\pi(1 - r\phi(\theta))},$$

where $1/2 < r < 1$. Noting $P[T_n = x] = (2\pi)^{-1} \int_{-\pi}^{\pi} \phi^n(\theta) e^{-ix\theta} \, d\theta$ and $u(x) = \lim_{r \uparrow 1} \sum_{n=0}^{\infty} r^n P[T_n = x]$ for all $x \in \mathbb{Z}$ as well as $u(-x) = 0$ for $x \ge 1$ one deduces that

$$u(x) = u(x) - u(-x) = \lim_{r \uparrow 1} \int_{-\pi}^{\pi} w_r(\theta)(-i2 \sin x\theta) \, d\theta.$$

Since $\Re w_r(\theta)$ is even and $\Im w_r(\theta)$ is odd we see that the limit above equals

$$4 \lim_{r \uparrow 1} \int_0^{\pi} \Im w_r(\theta) \sin x\theta \, d\theta = \frac{2}{\pi} \int_0^{\pi} S(\theta) \sin x\theta \, d\theta.$$

Here the equality is justified by the bound $|\Im w_r(\theta)| \le 1/ [2\pi r|\Im \phi(\theta)|] \le 1/\theta L(1/\theta)$ for $\theta > 0$ small enough, which follows from (A.6). Thus yielding (A.5).   □

A.2.1.2 Case: $\ell(t)$ is s.v.

Next consider the case when the tail

$$\ell(t) := P[T_1 > t]$$

is s.v., or what is the same thing, the renewal function $U(x)$ is s.v. Nagaev [54] shows that if $xP[T_1 = x]$ is s.v., then $u(x) \sim P[T_1 = x]/[\ell(x)]^2$. For the proof of Theorem 7.1.6 (cf. (7.97)) we needed the estimate $u(x) = o\,(U(x)/x)$. The next lemma, an extension of [54], gives a better bound on

$$q(x) := P[T_1 = x],$$

under a condition.

Lemma A.2.2 *Suppose $P[T_1 > t]$ is s.v. If*

$$C := \lim_{\delta \uparrow 1} \limsup_{x \to \infty} \frac{1}{q(x)} \sup_{\delta x < y \le x} q(y) < \infty, \tag{A.8}$$

then $u(x) \le q(x)[\ell(x)]^{-2}\{C + o(1)\}$;   and if

$$c := \lim_{\delta \uparrow 1} \liminf_{x \to \infty} \frac{1}{q(x)} \inf_{\delta x < y \le x} q(y) > 0, \tag{A.9}$$

then $u(x) \ge q(x)[\ell(x)]^{-2}\{c + o(1)\}$. In particular, if both (A.8) and (A.9) hold with $C = c = 1$, then

$$u(x) \sim \frac{q(x)}{[\ell(x)]^2}.$$

Proof The proof elaborates on that of [54]. Put

$$u^{(2)}(x) = \sum_{y=0}^{x} u(x - y)u(y).$$

Then it holds [54, Lemma 2.6] that

$$xu(x) = \sum_{y=0}^{x-1}(x - y)q(x - y)u^{(2)}(y). \tag{A.10}$$

This is derived by means of the generating functions. Indeed if $f(s) = \sum q(x)s^x$ and $h(s) = \sum u(x)s^x$ ($|s| < 1$) (the generating functions of $q(\cdot)$ and $u(\cdot)$), then $h(s) = 1/[1 - f(s)]$. The identity (A.10) follows by comparing the identities

$$\sum(x + 1)q(x + 1)s^x = f'(s), \quad \sum u^{(2)}(x)s^x = \frac{1}{[1 - f(s)]^2}$$

and $\sum(x + 1)u(x + 1)s^x = h'(s)$ ($|s| < 1$).

The slow variation of ℓ entails

$$U(x) \sim 1/\ell(x), \quad u^{(2)}(x) \leq \frac{2 + o(1)}{\ell(x)} \max_{\frac{1}{2}x \leq y \leq x} u(y), \text{ and} \qquad (A.11)$$

$$\sum_{y=0}^{x} u^{(2)}(y) = \sum_{y=0}^{x} u(y) \sum_{z=y}^{x} u(z-y) = [U(\tfrac{1}{2}x)]^2 \{1 + o(1)\} + \sum_{y=x/2}^{x} U(x-y)u(y).$$

The last sum being less than $U(x) \sum_{y=x/2}^{x} u(y) = o([U(x)]^2)$, it follows that

$$\sum_{y=0}^{x} u^{(2)}(y) \sim \frac{1}{[\ell(x)]^2}. \qquad (A.12)$$

Let $3/4 < \delta < 1$ and $\varepsilon = 1 - \delta$, put

$$C_\delta = \delta^{-1} \limsup_{x \to \infty} \sup_{\delta x \leq y \leq x} q(y)/q(x)$$

and split the range of the sum on the RHS of (A.10) according to $y \leq 2\varepsilon x$ or $y > 2\varepsilon x$. Then using (A.12), (A.8) and (A.11) one sees that for all sufficiently large x,

$$xu(x) \leq C_\delta \frac{xq(x)}{[\ell(x)]^2} + \frac{3 \sum_{y=0}^{\delta x} yq(y)}{\ell(x)} \max_{\varepsilon x \leq y \leq x} u(y). \qquad (A.13)$$

Since $\sum_{y=0}^{x} yq(y) = -x\ell(x) + \sum_{y=0}^{x-1} \ell(y) = o(x\ell(x))$, on writing

$$N(x) = \frac{q(x)}{[\ell(x)]^2}, \quad M_\delta(x) = \max_{\varepsilon x \leq y \leq x} u(y),$$

(A.13) yields that for x large enough

$$u(x) \leq C_\delta N(x) + o(M_\delta(x)).$$

We claim that there exist positive constants r_0 and α such that $q(x) > x^{-\alpha}$ for $x \geq r_0$. For the proof, let $\bar{q}(t)$ denote the continuous function of $t \geq 1$ that is obtained from $q(x)$ by linearly interpolating between successive positive integers. By (A.8), one can choose constants $r_0 \geq 2$, $0 < \delta < 1$ and $M > 1$ so that $\max_{\delta r \leq t \leq r} \bar{q}(t) \leq M\bar{q}(r)$ for $r \geq r_0$. One may suppose $\min_{\delta r_0 \leq s \leq r_0} \bar{q}(s) > 0$, for if $q(x_n) = 0$ for some sequence $x_n \uparrow \infty$, then q vanishes for some point on, contradicting the slow variation of ℓ. Put $\lambda = [\log M]/\log \delta^{-1}$. Then for $\delta r_0 \leq s \leq r_0$ and $t = \delta^{-n}s$, $(n = 1, 2, \ldots)$,

$$\bar{q}(t) \geq M^{-n}\bar{q}(s) = \delta^{\lambda n}\bar{q}(s) \geq \left[(\delta r_0)^\lambda \min_{\frac{1}{2}r_0 \leq s \leq r_0} \bar{q}(s) \right] t^{-\lambda}.$$

Hence the claim is verified.

Take $\eta = \eta_\delta > 0$ such that

$$\eta C_\delta < \varepsilon \quad \text{and} \quad \log \eta^{-1} > 2\alpha \log \varepsilon^{-1},$$

and choose $r \geq r_0$ so that for $x > r$,

$$u(x) \leq C_\delta N(x) + \eta M_\delta(x) \quad \text{and} \quad \sup_{\delta x \leq y < x} N(y) < C_\delta N(x).$$

Let $x_1 \in [\varepsilon x, x]$ be such that $u(x_1) = M_\delta(x)$, so that

$$u(x) \leq C_\delta N(x) + \eta u(x_1).$$

If $x \geq r/\varepsilon$, $u(x_1) \leq C_\delta N(x_1) + \eta M_\delta(x_1)$, hence

$$u(x) \leq C_\delta \left(N(x) + \eta N(x_1) \right) + \eta^2 M_\delta(x_1).$$

For $x \geq r/\varepsilon^2$, $x_2 \ (\in [\varepsilon x_1, x_1])$ such that $u(x_2) = M_\delta(x_1)$ is not less than r, assuring $u(x_2) \leq C_\delta N(x_2) + \eta M_\delta(x_2)$, hence $u(x) \leq C_\delta \left[N(x) + \eta N(x_1) + \eta^2 N(x_2) \right] + \eta^3 M_\delta(x_2)$. If $x \geq \varepsilon^{-k} r$ one can repeat this procedure k times to obtain

$$u(x) \leq C_\delta \left[N(x) + \eta N(x_1) + \cdots + \eta^k N(x_k) \right] + \eta^{k+1} M_\delta(x_k).$$

Let $n(x)$ be the integer n such that $\varepsilon^{-n} r \leq x < \varepsilon^{-n-1} r$. Since $N(x_j) \leq C_\delta^j N(x)$, recalling $\eta C_\delta < \varepsilon$ one infers that

$$u(x) \leq \delta^{-1} C_\delta N(x) + \eta^{n(x)+1} M_\delta(x_{n(x)}).$$

By the choice of η, $[n(x) + 1] \log \eta < -2\alpha \log x + O(1)$, so that $\eta^{n(x)}/N(x) = o\left(x^{-2\alpha}/q(x) \right) \to 0$. Hence $\limsup u(x)/N(x) \leq \delta^{-1} C_\delta$. This concludes the asserted upper bound, for $\delta^{-1} C_\delta$ can be made arbitrarily close to C.

The lower bound is easily deduced from (A.10). Indeed, the restriction to $y \leq (1 - \delta)x$ of the sum on its RHS is larger than $\delta x \inf_{\delta x \leq y \leq x} q(y) \sum_{y=0}^{(1-\delta)x} u^{(2)}(y)$, hence (A.9) together with (A.12) yields the asserted lower bound of $u(x)$. □

A.2.2 Renewal Function and Over- and Undershoot Distributions

Let γ be a constant with $0 \leq \gamma \leq 1$. Put

$$N_R = \sup\{n \geq 0 : T_n \leq R\} \quad \text{and}$$
$$Z_R = T_{N_R+1} - R.$$

Theorem A.2.3 *Given a number* $\gamma \in [0,1]$, *the following (A.14) to (A.17) are equivalent:*

$$\begin{cases} \int_0^x P[T_1 > y]\,dy \sim \ell(x) & \gamma = 1, \\ P[T_1 > x] \sim \ell(x)/x^\gamma & 0 \le \gamma < 1, \end{cases} \quad \text{for an s.v. function } \ell; \quad \text{(A.14)}$$

$$U(x) \sim c_\gamma x^\gamma / \ell(x) \quad (0 \le \gamma \le 1) \quad \text{for an s.v. function } \ell, \quad \text{(A.15)}$$

where $c_\gamma = (\gamma\pi)^{-1} \sin \gamma\pi$ *for* $0 < \gamma < 1$ *and* $c_\gamma = 1$ *for* $\gamma \in \{0,1\}$;

$$\begin{cases} Z_R/R \longrightarrow_P 0 & \gamma = 1, \\ P[Z_R/R < \xi] \longrightarrow F_\gamma^{(Z)}(\xi) & 0 < \gamma < 1, \quad \text{as } R \to \infty, \\ Z_R/R \longrightarrow_P \infty & \gamma = 0, \end{cases} \quad \text{(A.16)}$$

where

$$F_\gamma^{(Z)}(\xi) = \frac{\sin \gamma\pi}{\pi} \int_0^\xi \frac{dt}{t^\gamma(1+t)} \quad (\xi \ge 0) \quad \text{for} \quad 0 < \gamma < 1;$$

$$\begin{cases} T_{N_R}/R \longrightarrow_P 1 & \gamma = 1, \\ P[T_{N_R}/R \le \xi] \longrightarrow F_\gamma^{(N)}(\xi) & 0 < \gamma < 1, \quad \text{as } R \to \infty, \\ T_{N_R}/R \longrightarrow_P 0 & \gamma = 0, \end{cases} \quad \text{(A.17)}$$

where

$$F_\gamma^{(N)}(\xi) = \frac{\sin \gamma\pi}{\pi} \int_0^\xi \frac{dt}{t^{1-\gamma}(1-t)^\gamma} \quad (0 \le \xi \le 1, \ 0 < \gamma < 1).$$

Moreover, each of the above relations implies

$$U(x)P[T_1 > x] \longrightarrow \begin{cases} 0 & \gamma = 1, \\ c_\gamma & 0 \le \gamma < 1; \end{cases} \quad \text{(A.18)}$$

if $\gamma = 1$, *then the converse is true; in particular (A.18) (with* $\gamma = 1$) *implies that* $U(x)\int_0^x P[T_1 > t]\,dt/x \to 1$.

The equivalence of (A.14) to (A.17) is shown by Dynkin [26] for $0 < \gamma < 1$ and by Rogozin [63] for $\gamma \in \{0,1\}$, where the proofs are made fully analytically by using the Laplace transform $\varphi(\lambda) := E[e^{-\lambda T_1}]$ and are somewhat involved. Note that the equivalence of (A.14) and (A.15) is immediate from the identity

$$\widehat{u}(\lambda) := \sum_{x=0}^\infty u(x)e^{-\lambda x} = \frac{1}{1 - \varphi(\lambda)} \quad \text{(A.19)}$$

in view of the Tauberian theorem that says (A.14) is equivalent to

$$1 - \varphi(\lambda) \sim \Gamma(1-\gamma)\lambda^\gamma \ell(1/\lambda) \quad (\lambda \downarrow 0), \quad \text{(A.20)}$$

and similarly for U, the Laplace transform of which equals $\sum \varphi^n = (1 - \varphi)^{-1}$. For $0 < \gamma < 1$, Feller [31, §XIV.3] provides a rather simple derivation of (A.16) and (A.17) from the conjunction of (A.14) and (A.18) based on the identities

$$P[Z_R > x] = \sum_{y=0}^{R} u(y)P[T_1 > R - y + x] \qquad (A.21)$$

and

$$P\left[T_{N_R} = x\right] = u(x)P[T_1 > R - x], \qquad (A.22)$$

respectively.[1] On taking $x = 0$, (A.21) in particular yields

$$U(R)P[T_1 > R] \leq 1. \qquad (A.23)$$

In the sequel we let $\gamma \in \{0, 1\}$ and we give a direct proof of the assertion of the theorem based on (A.21) and (A.22), except for the implication (A.18) $\Rightarrow$ (A.14) in the case $\gamma = 1$, for which we resort to an analytic method. Therein we shall also show

$$Z_R/R \longrightarrow_P 0 \iff U(R)P[T_1 > R] \to 0 \iff \exists k > 0, \ P[Z_R > kR] \to 0. \qquad (A.24)$$

For simplicity we write $\mu_T(x)$ for $P[T_1 > x]$.

Let $\gamma = 0$. First we show that (A.16) implies (A.14). Using (A.23), one infers that for any $\lambda \geq 1$, $P[Z_R > \lambda R] \leq U(R)\mu_T(\lambda R)$. By virtue of (A.23), this inequality leads to $U(R)\mu_T(\lambda R) \to 1$ under (A.16), showing that μ_T is s.v. (as well as (A.18)). Similarly one verifies the implication (A.17) $\Rightarrow$ (A.14) by using the identity $P\left[T_{N_R}/R < \xi\right] = \sum_{y \leq \xi R} u(y)\mu_T(R - y)$, which follows from (A.22). Now it is easy to show the equivalence of (A.16) and (A.17), and that each of them follows from (A.14).

Let $\gamma = 1$. Since then (A.14) implies $x\mu_T(x) = o(\ell(x))$, (A.18) follows from (A.14), which is equivalent to (A.15), as mentioned at (A.19). We show (A.24). By using (A.21) and the sub-additivity of U in turn we see for any constant $k > 0$

$$P[Z_R > kR] \geq U(R)\mu_T(R + kR) \geq (k + 2)^{-1}U(R + kR)\mu_T(R + kR).$$

The convergence to zero of the right-most member entails that of $U(R)\mu_T(R)$, as is readily verified; hence $\lim P[Z_R > kR] = 0$ implies $\lim U(R)\mu_T(R) = 0$. Similarly, if $\lim U(R)\mu_T(R) = 0$,

$$P[Z_R > \varepsilon R] \leq U(R)\mu_T(\varepsilon R) \leq (\varepsilon^{-1} + 1)U(\varepsilon R)\mu_T(\varepsilon R) \to 0.$$

Thus (A.24) is verified. In particular, (A.18) is equivalent to (A.16). The implication (A.16) $\Rightarrow$ (A.14) seems hard to prove by a simple method as above, and the proof by the use of Laplace transform, as in [63], must be natural. The proof of Theorem

---

[1] In [31], the necessity of (A.14) for each of (A.16) and (A.17) is stated without proof. One can prove it by using a double transform, as given in (A.25) (see [8, §8.6.2]).

A.2.3 for $\gamma = 1$ is complete if we can show that the following are equivalent

$$(a)\ Z_R/R \longrightarrow_P 0, \quad (b)\ T_{N_R}/R \longrightarrow_P 1, \quad (c)\ 1 - \varphi(\lambda) \sim \lambda\ell(1/\lambda)\ (\lambda \downarrow 0),$$

where ℓ is some s.v. function in (c). (a) follows from $U(x)\mu_T(x) \to 0$ in view of (A.24), hence from (c) by what we have observed above. Thus it suffices to show (a) $\Rightarrow$ (b) $\Rightarrow$ (c). The following proof is standard.

Since $P\left[T_{N_R}/R < \delta\right] = \sum_{y \leq \delta R} u(y)\mu_T(R - y) \leq U(\delta R)\mu_T((1 - \delta)R)$ for $0 < \delta < 1$, by (A.24) it follows that (a) $\Rightarrow$ (b). To verify (b) $\Rightarrow$ (c) we first deduce

$$(1 - e^{-\lambda}) \sum_{r=0}^{\infty} E\left[e^{-sT_{N_R}}\right] e^{-\lambda r} = \frac{1 - \varphi(\lambda)}{1 - \varphi(s + \lambda)}. \tag{A.25}$$

The sum on the LHS equals $\sum_{r=0}^{\infty} e^{-\lambda r}\ dr\ \sum_{y=0}^{r} e^{-sy}\mu_T(r - y)u(y)$, which, interchanging the order of summation and integration, transforms to

$$\sum_{y=0}^{\infty} e^{-sy}u(y) \sum_{r=y}^{\infty} e^{-\lambda r}\mu_T(r - y) = \sum_{y=0}^{\infty} e^{-(s+\lambda)y}u(y)\widehat{\mu_T}(\lambda) = \widehat{u}(s + \lambda)\widehat{\mu_T}(\lambda).$$

Here $\widehat{\mu_T}(\lambda)$ denotes $\sum_{r=0}^{\infty}\mu_T(r)e^{-\lambda r}$ and equals $(1 - \varphi(\lambda))/(1 - e^{-\lambda})$. Hence (A.25) follows from (A.19) (valid with no extra assumption). Now using (A.25), we deduce (c) from (b) as follows. On the LHS of (A.25), for any constant $M > 1$, the sum over $r < M$ is less than M and negligible as $\lambda \to 0$. On the other hand, (b) entails that $e^{-s(r+1)} \leq E\left[e^{-sT_{N(r)}}\right]\{1 + o(1)\} \leq e^{-s(r-1)}$ as $r \to \infty$. Hence it follows that the LHS of (A.25) can be written as $\lambda/(s+\lambda)\{1+o(1)\}$ as $s+\lambda \to 0$ under $0 < s < \lambda$, so that putting $s = \delta\lambda$ with $0 < \delta < 1$ yields $(1 - \varphi(\lambda))/(1 - \varphi((1+\delta)\lambda)) \to 1/(1+\delta)$. Thus we have (c). This finishes the proof of Theorem A.2.3 for $\gamma \in \{0, 1\}$.

A.3 The First Ladder Epoch and Asymptotics of U_a

This section concerns the strict ascending ladder process (H_n, τ_n), $n = 0, 1, 2, \ldots$, associated with the r.w. S studied in this treatise. Here $\tau_0 = Z_0 = 0$, $\tau_1 = \sigma_{[1,\infty)}$, $H_1 = S_{\tau_1}(= Z)$ and for $n > 1$

$$\tau_{n+1} = \inf\{k > \tau_n : S_k > H_n\} \quad \text{and} \quad H_{n+1} = S_{\tau(n+1)}.$$

We shall write P for P_0 (the r.w. S always starting at zero) in the sequel and adhere to our primary hypothesis of irreducibility and oscillation of arithmetic r.w.'s, although most of them can be extended to non-arithmetic r.w.'s. We shall apply Baxter's formula ([71, Prop.17.5], [31, XVIII, (3.7)], [16]), given in the form

$$1 - E[s^{\tau_1}e^{-\lambda Z}] = e^{-\sum_1^{\infty} n^{-1}s^n E[e^{-\lambda S_n};S_n>0]} \quad (\lambda \geq 0, 0 < s \leq 1). \tag{A.26}$$

A.3.1 Stability of τ_1 and Spitzer's Condition

The following result is due to Rogozin [63]. We write τ for τ_1.

Theorem A.3.1 (i) *There exists* $\rho := \lim P[S_n > 0]$ *if and only if*

$$P[\tau > t] \sim t^{-\rho} \ell_\tau(t) \text{ if } 0 \le \rho < 1 \text{ and } \int_0^t P[\tau > s] \, ds \sim \ell_\tau(t) \text{ if } \rho = 1,$$

with an s.v. function ℓ_τ, *in other words,* τ *is in the domain of attraction of a stable law of exponent* ρ *if* $0 < \rho < 1$; *r.s. if* $\rho = 1$ *and has a distribution function that is s.v. if* $\rho = 0$. *If this is the case and* $\hat\tau = \sigma_{(-\infty,0]}$, *then*

$$tP[\tau > t]P[\hat\tau > t] \longrightarrow \pi^{-1} \sin \rho\pi \quad \text{as } t \to \infty. \tag{A.27}$$

(ii) *In order that the law of* $\tau_n n^{-1/\rho}$ *converges to a (non-degenerate) stable law as* $n \to \infty$ *for* $0 < \rho < 1$ *and that* $E\tau_1 < \infty$ *for* $\rho = 1$, *it is necessary and sufficient that the series* $\sum_{n=1}^\infty n^{-1}(P_0[S_n > 0] - \rho)$ *converges.*

The proof of the first half of Theorem A.3.1(i) is performed by combining Sparre Andersen's identity ([31, Theorem XII.8.2], [16, 8.5.2]), Spitzer's arcsine law for occupation times of the positive half line (cf. [69]) and Theorem A.2.3 as follows.[2] Put $K_n = \min\{0 \le k \le n : S_k = M_n\}$ (the first time ($\le n$) the maximum $M_n := \max\{S_0, \ldots, S_n\}$ is attained). Then the first and second of these theorems yield that the distribution of K_n/n converges to $F_\rho^{(N)}$ (given in (A.17) with obvious interpretation for $\rho \in \{0, 1\}$) if and only if $P[S_n > 0] \to \rho$ (see [71, Problems 7, 9 in p233]). Let $N_n^{(\tau)}$ denote the value taken by the renewal process of epochs $(\tau_k)_{k=0}^\infty$ just before it enters the half line $\{n+1, n+2, n+3, \ldots\}$. After a little reflection, one sees

$$K_n = N_n^{(\tau)}.$$

By Theorem A.2.3, this identity, the gist of the proof, shows that τ satisfies the properties asserted in the theorem if and only if the law of $N_n^{(\tau)}/n$ converges to $F_\rho^{(N)}$, hence the equivalences asserted in the first half of (i).

As for the second half of (i), we apply (A.26), which, together with its dual, yields

$$1 - E[s^\tau] = e^{-\sum_{n=1}^\infty s^n P[S_n > 0]/n} \quad \text{and} \quad 1 - E[s^{\hat\tau}] = e^{-\sum_{n=1}^\infty s^n P[S_n \le 0]/n}. \tag{A.28}$$

Suppose $\rho = \lim P[S_n > 0] \in [0, 1)$. Then, on putting $\theta_n = \rho - P[S_n > 0]$ and $h(s) = e^{\sum_{n=1}^\infty s^n \theta_n/n}$, the RHSs of (A.28) are written, respectively, as

$$(1 - s)^\rho h(s) \quad \text{and} \quad (1 - s)^{1-\rho}/h(s).$$

If $\rho < 1$, from the first half of (i) we obtain $h(s) \sim \Gamma(1-\rho)\ell_\tau(1/(1-s))$ as $s \uparrow 1$ (see (A.20)) and, on noting $\sum_{n=0}^\infty s^n P[\hat\tau > n] = (1 - s)^{-1}(1 - E[s^{\hat\tau}]) = (1 - s)^{-\rho}/h(s)$,

---

[2] In [63] Spitzer's arcsine law is directly derived from Theorem A.2.3 by using Sparre Andersen's result and the formula (A.28).

an application of the Tauberian theorem leads to (A.27). For $\rho = 1$, use the identities $\sum_{n=0}^{\infty} s^n \sum_{k=0}^{n} P[\tau > k] = 1 - E[s^\tau] = (1-s)h(s)$.

If $0 < \rho < 1$, for the law of $\tau_n n^{-1/\rho}$ to converge to a stable law, it is necessary and sufficient that the sum appearing as the exponent in the definition of $h(s)$ approaches a finite number as $s \uparrow 1$, which condition is equivalent to the convergence of the series $\sum \theta_n/n$ owing to Littlewood's Tauberian theorem. One can dispose of the case $\rho = 1$ similarly. The proof of Theorem A.3.1 is finished.

A.3.2 Asymptotics of U_a Under (AS)

Theorem A.3.2 *If (AS) holds, then $U_a(x) \sim x^{\alpha\rho}/\ell(x)$ for some s.v. function ℓ.*

This is Theorem 9 of [63] except in the extreme cases $\rho = 0$ and $\alpha = \rho = 1$. If $\alpha\rho = 1$, we know Z is r.s., so that $U_a(x) \sim x/\ell^*(x)$ (see Section 5.2); we also know U_a is s.v. if $\rho = 0$ (see the table in Section 2.6).

Let $0 < \alpha\rho < 1$. For the proof of Theorem A.3.2, it suffices to show

$$P[Z(c_n)/c_n > \xi] \to 1 - F_{\alpha\rho}^{(Z)}(\xi), \tag{A.29}$$

because of the equivalence of (A.16) and (A.15). Since the law of the scaled process $S_{\lfloor nt \rfloor}/c_n$, $t \geq 0$, converges in Skorokhod's topology to a stable process, Y, of exponent α with $P[Y(1) > 0] = \rho$, it is not hard to show that $Z(c_n)/c_n$ converges in law to $\zeta := Y(\sigma_{[1,\infty)}^Y) - 1$. It is proved in [4, Lemma VII.3] that ζ is subject to the stable law of exponent $\alpha\rho$, which immediately gives (A.29) in view of Theorem A.2.3. Although this approach is elegant, the result of [4] mentioned above is based on a sophisticated theory about the local time of Lévy processes. A direct proof, as given in [63], is also of interest. We give it below, with some simplifications.

In view of the invariance principle mentioned above, (A.29) holds if it does for the stable r.w. whose step distribution is given by $F_{\alpha,\rho}(x) = P^Y[Y(1) \leq x]$. Let $Z^\sharp$ be the first ascending ladder height for this stable r.w. Since the overshoot across a level R of an r.w. is the same as the corresponding overshoot of the ladder height process of the r.w., by Theorem A.2.3 (see (A.20)), (A.29) follows if we can verify

$$1 - E^Y e^{-\lambda Z^\sharp} \sim C\lambda^{\alpha\rho} \qquad (\lambda \downarrow 0) \tag{A.30}$$

with some positive constant C. Because of (A.26) the Laplace transform on the LHS is represented by using the distributions of $Y(n)$, $n = 1, 2, \ldots$ while by the scaling property of the stable law we have $P^Y[Y(n) \leq x] = P^Y[Y(1) \leq x/n^{1/\alpha}]$. It accordingly follows that

$$\log \frac{1}{1 - E^Y[e^{-\lambda Z^\sharp}]} = \sum_{n=1}^{\infty} n^{-1} E^Y[e^{-\lambda Y(n)}; Y(n) > 0] = \rho \sum_{n=1}^{\infty} \frac{1}{n} \varphi_+^\sharp(n^{1/\alpha}\lambda), \tag{A.31}$$

where $\varphi_+^\sharp(\lambda) = E^Y[e^{-\lambda Y(1)} \,|\, Y(1) > 0]$. We write the right-most expression as

$$\rho \sum_{1 \le n \le \lambda^{-\alpha}} \frac{1}{n} + \rho \sum_{n=1}^\infty \frac{1}{n} \left[\varphi_+^\sharp((n\lambda^\alpha)^{1/\alpha}) - 1_{[0,1]}(n\lambda^\alpha) \right].$$

As $\lambda \downarrow 0$, the first term may be written as $\rho[\log \lambda^{-\alpha} + C^*] + o(1)$ (C^* is Euler's constant) while the second term converges to $C_1 := \rho \int_0^\infty t^{-1} \left[\varphi_+^\sharp(t^{1/\alpha}) - 1_{[0,1]}(t) \right] dt \in \mathbb{R}$. Now returning to (A.31) one finds (A.30) to hold with $C = e^{-\rho C^* - C_1}$.

Remark A.3.3 Suppose (AS) holds with $0 < \rho < 1$. According to [88, Lemma 12] there then exist positive constants C, C_0 and C_0' such that

$$P[\tau > n] \sim \begin{cases} CP[Z > c_n] & \text{if } \alpha\rho < 1, \\ Cc_n^{-1} \int_0^{c_n} P[Z > x]\,dx & \text{if } \alpha\rho = 1, \end{cases}$$

and

$$U_d(c_n) \sim C_0/P[\tau > n] \quad \text{and} \quad V_d(c_n) \sim C_0'/P[\hat{\tau} > n]$$

in either case. (Recall $\tau = \sigma_{[1,\infty)}$, $\hat{\tau} = \sigma_{(-\infty,0]}$.) By the last relation ℓ is related to ℓ_τ in Theorem A.3.1 by $\ell_\tau(n) \sim C_0 \ell(c_n)[L(c_n)]^{-\rho}$. In Theorem 5.4.1 ($\rho = 1$ is not excluded), we have observed that under $m_+/m \to 0$, Spitzer's condition holds if and only if $m(x)$ varies regularly with index $2 - 1/\rho$: $m(x) \sim c_\rho x^{-1/\rho+2} L(x)$ for $\frac{1}{2} \le \rho \le 1$ (with a certain $c_\rho > 0$). It is shown in [18], [77] that if this is the case and $EZ < \infty$, then ℓ_τ is given by a positive multiple of the de Bruijn ρ^{-1}-conjugate of L, $L_{1/\rho}^*$ say, which is determined by $L_{1/\rho}^*(x^{1/\rho}L(x)) \sim [L(x)]^\rho$, so that $\ell_\tau(x^{1/\rho}L(x)) \sim \text{const.}[L(x)]^{-\rho}$, hence ℓ_τ is directly related to L. This fails if $EZ = \infty$.

We also point out that if $E|\hat{Z}| < \infty$ and $1 < \alpha \le 2$, then $\ell_\tau(x) \sim C_0'[L(c_n)]^\rho$, where C_0' is equal to $v^\circ C_0/[2E|\hat{Z}|]$ if $\alpha = 2$ and to $v^\circ \kappa_\alpha C_0/[2(\alpha-1)(2-\alpha)E|\hat{Z}|]$ if $1 < \alpha < 2$, and $\ell_\tau(x) = o([L(c_n)]^\rho)$ otherwise. This follows from the dual of (2.15) with the help of $a(-x) \sim 2\kappa_\alpha^{-1} x/m(x)$.

In the extreme case $pq = 0$ of $1 < \alpha < 2$ and the case $\alpha = 2$ we can derive explicit expressions of the constants C_0 and C_0' above. If $1 < \alpha \le 2$ and $\alpha\hat{\rho} = 1$, then by the third and fourth cases of (9.102) it follows that for $\varepsilon^2 < x_n < \varepsilon$, as $y_n \to \eta > 0$,

$$\frac{\mathfrak{p}_1^{\{0\}}(x_n, y_n)}{x_n} = \frac{C_0' V_d(x) \mathfrak{q}^{\text{mnd}}(y_n)}{x_n V_d(c_n)} \{1 + o_\varepsilon(1)\} = C_0' \mathfrak{q}^{\text{mnd}}(y_n)\{1 + o_\varepsilon(1)\},$$

where $o_\varepsilon(1) \to 0$ as $n \to \infty$ and $\varepsilon \downarrow 0$ in this order. One can show that $\mathfrak{p}_1^{\{0\}}(\xi, \eta)/\xi \to \alpha\mathfrak{p}_1(0)\mathfrak{q}^{\text{mnd}}(\eta)$ ($\xi \downarrow 0$) and $\alpha\mathfrak{p}_1(0)$ equals $\sqrt{2/[\pi c_\sharp]}$ if $\alpha = 2$ and $|\Gamma(1-\alpha)c_\sharp|^{-1/\alpha}$ if $0 < \alpha < 2$. This together with (A.27) and Lemma 6.5.1 leads to the following:

(i) if $\alpha = 2$, then $P[\tau > n]U_a(c_n) \sim P[\hat{\tau} > n]V_d(c_n) \longrightarrow \sqrt{2/\pi c_\sharp}$;
(ii) if $q = 0$ and $1 < \alpha < 2$, then

$$P[\tau > n]U_a(c_n) \longrightarrow |\Gamma(1 - \alpha)c_\sharp|^{1/\alpha - 1} [\Gamma(\alpha)\Gamma(1/\alpha)]^{-1},$$
$$P[\hat{\tau} > n]V_d(c_n) \longrightarrow [\Gamma(1 - 1/\alpha)]^{-1} |\Gamma(1 - \alpha)c_\sharp|^{-1/\alpha}. \tag{A.32}$$

A.4 Positive Relative Stability and Condition (C3)

By specialising the central limit theorem applied to the triangular array $X_n, k = X_k/B_n$ ($k = 1, \ldots, n$) to the degenerate case (cf. 1° (with $(\alpha, \sigma) = (1, 0)$) in §22.5.1 of [52]), or [31, Theorem of §XVII.7]) we have the following criterion for S to be p.r.s. Put

$$v(x) = E[X; |X| \leq x] \quad x > 0.$$

Given a positive sequence (B_n), *in order that* $S_n/B_n \to 1$ *in probability, it is*

necessary and sufficient that for any constant $\delta > 0$, [3]

$$v(\delta B_n) \sim B_n/n, \quad \mu(\delta B_n) = o(1/n), \quad E[X^2; |X| \leq \delta B_n] = o(B_n^2/n). \tag{A.33}$$

Note that the first condition above prescribes how one may choose B_n.
Here we show that (A.33) holds with $B_n > 0$ for large n if and only if

$$(*) \qquad v(x)/[x\mu(x)] \longrightarrow \infty \quad (x \to \infty).$$

[Since $v(x) = x[\mu_-(x) - \mu_+(x)] + A(x - 1) \sim A(x)$ under $(*)$, this equivalence shows what is mentioned at (2.20) of Section 2.4.] If (A.33) holds with $B_n > 0$, then its second relation implies that $B_n \to \infty$ and by $S_n/B_n \to_P 1$ we have $B_{n+1}/B_n \to 1$, which shows that v is s.v. because of the first two relations of (A.33) (cf. [8, Theorem 1.9.2]). Hence (A.33) implies $(*)$. For the converse, we have only to verify the third condition of (A.33) under $(*)$, since $(*)$ implies $v(x)$ is s.v., and hence the first two relations of (A.33) hold for some sequence (B_n). To this end we show that $xv(x)/E[X^2; |X| \leq x] \to \infty$, which shows that the third relation follows from the first, hence from $(*)$. But the slow variation of $v(x)$ entails $E[X^2; |X| \leq x] = -(x+1)^2\mu(x+1) + 2\sum_0^x y\mu(y) = o(xv(x))$, which concludes the required estimate.

Theorem A.4.1 *The following are equivalent (under our basic framework):*

(a) *(C3) holds, namely* Z *is r.s. and* V_d *is s.v.;*
(b) $\lim P_x(\Lambda_{2x}) = 1$ *and* Z *is r.s.;*
(c) *the r.w.* S *is p.r.s.;*
(d) *there exists a positive increasing sequence* B_n *such that for each* $M \geq 1$ *and* $\varepsilon > 0$, $P_0\left[\sup_{0 \leq t \leq M} |B_n^{-1}S_{\lfloor nt \rfloor} - t| < \varepsilon\right] \to 0$ (B_n *is necessarily unbounded);*
(e) $\lim P_x(\Lambda_{2x}) = \lim P[S_n > 0] = 1$, *and* $Z(R)/R \to_P 0$.

---

[3] The sufficiency part is readily verified by the elementary inequality Eq(VII.7.5) of [31].

Proof (of Theorem A.4.1) That (a) implies (b) follows immediately from Theorem 6.1.1. The implication (d) $\Rightarrow$ (e) is obvious; (a) follows from (e) since by (6.12), V_d is s.v. if $P_x(\Lambda_{2x}) \to 1$, and Z is r.s. if $Z(R)/R \to_P 0$. Thus we have only to show (b) $\Rightarrow$ (c) and (c) $\Rightarrow$ (d). First, we show the latter.

PROOF OF (c) $\Rightarrow$ (d) (adapted from [63]). Recall (c) implies that $\mu(x) = o(A(x)/x)$, $E[X; |X| \le x] \sim A(x)$, $A(x)$ is s.v. (normalised), and $S_n/B_n \to_P 1$ with B_n defined by $B_n = nA(B_n)$ for large n. In particular, it follows that for each $t \ge 0$, $S_{\lfloor nt \rfloor}/B_n \to t$ in probability and it suffices to show the tightness (with respect to the uniform topology of paths) of the law of the continuous process on $[0, M]$ obtained by the linear interpolation of the points $(n^{-1}k, B_n^{-1}S_k) \in [0, M] \times \mathbb{R}$, $0 \le k \le Mn$. To this end, it suffices to show that for any $\varepsilon > 0$, one can choose $\delta > 0$ so that

$$P_0 \left[\max_{1 \le k \le \delta n} |S_k| > \varepsilon B_n \right] \longrightarrow 0 \quad (n \to \infty) \tag{A.34}$$

(cf. [7, Theorem 2.8.3]). For the proof of (A.34), we introduce the random variables

$$\theta_{n,\delta}(k) = \mathbf{1}(|X_k| \le \delta B_n), \quad S_k' = \sum_{j=1}^{k} \left(X_j \theta_{n,\delta}(j) - E[X\theta_{n,\delta}] \right),$$

where $\theta_{n,\delta}$ stands for $\theta_{n,\delta}(1)$. Then the condition (A.33) is written as

$$E[X\theta_{n,\delta}] \sim B_n/n, \quad E[1 - \theta_{n,\delta}] = o(1/n), \quad E[X^2\theta_{n,\delta}] = o(B_n^2/n). \tag{A.35}$$

Writing $S_k = S_k' + kE[X\theta_{n,\delta}] + \sum_{j=1}^{k} X_j[1 - \theta_{n,\delta}(j)]$, from the first estimate of (A.35) one infers that if $\delta < \varepsilon/3$, for all n large enough

$$P \left[\max_{1 \le k \le \delta n} |S_k| > \varepsilon B_n \right] \le P \left[1 \le^\exists k \le \delta n, \ \theta_{n,\delta}(k) = 0 \right] + P \left[\max_{1 \le k \le \delta n} |S_k'| > \frac{\varepsilon}{2} B_n \right].$$

The second (resp. third) estimate of (A.35) implies that the first (resp. second) probability on the RHS tends to zero. Thus (A.34) has been verified.

PROOF OF (b) $\Rightarrow$ (c). Here we need to apply the following implications

$$(**) \qquad \lim P_x(\Lambda_{2x}) = 1 \implies \lim P[S_n > 0] = 1 \implies \tau_1 \text{ is r.s.},$$

the first implication following from (2.22) and the second from Theorem A.3.1. Below we denote by $(\tau_n, H_n)_{n=0,1,2,\ldots}$ the strict ascending ladder process associated with the r.w. S ($\tau_0 = H_0 = 0$, $\tau_1 = \sigma_{[1,\infty)}$, $H_1 = Z$; see the beginning of Section A.3 for $n > 1$). We shall write $\tau(n)$ for τ_n when it appears as a subscript like $H_n = S_{\tau(n)}$.

Suppose (b) holds. Then τ_1 is r.s. by $(**)$. Hence both $\ell(t) := \int_0^t P[\tau_1 > s] \, ds$ and $\ell^*(t) = \int_0^t P[Z > s] \, ds$ are normalised s.v. functions. Let $s(t)$ and $w(t)$ be continuous increasing functions of $t \ge 0$ such that $s(t) = t\ell(s(t))$ and $w(t) = t\ell^*(w(t))$ for all t large enough, and put $B(t) = w(s^{-1}(t))$. Then it follows that s, w and B are all regularly varying at infinity with index one, and that

$$\tau_n/s_n \longrightarrow_P 1, \quad H_n/w_n \longrightarrow_P 1 \quad \text{and} \quad H_n/B_{\tau(n)} = [H_n/w_n] \cdot [B_{s(n)}/B_{\tau(n)}],$$

where we write s_n, w_n and B_n for $s(n)$, $w(n)$ and $B(n)$. Hence

$$H_n/B_{\tau(n)} \longrightarrow 1 \quad \text{in probability.} \tag{A.36}$$

We are going to show that for any $\varepsilon > 0$ there exists a constant $\delta \in (0, 1)$ such that, as $n \to \infty$,

$$P\left[|S_k/B_k - 1| > \varepsilon \quad \text{for} \quad s_n \le k \le s_{n/\delta}\right] \longrightarrow 0, \tag{A.37}$$

which is stronger than (c). We write τ_r or $\tau(r)$ for $\tau_{\lfloor r \rfloor}$ for positive real number r and $\xi_n \prec \eta_n$ when two sequences of random variables (ξ_n) and (η_n) satisfy $P[\xi_n < \eta_n] \to 1$. Given $\delta \in (0, 1)$ one observes the following obvious relations

$$\tau_{\delta^2 n} \prec \tau_{\delta n} \prec s_n \prec s_{n/\delta} \prec \tau_{n/\delta^2} \quad \text{and} \quad (\sharp) \begin{cases} S_{\tau(\delta^2 n)} \prec S_{\tau(\delta n)} - \varepsilon_1 w_n, \\ S_{\tau(n/\delta^2)} \prec S_{\tau(\delta n)} + \varepsilon_2 w_n, \end{cases}$$

where $\varepsilon_1 = 2^{-1}(\delta - \delta^2)$, $\varepsilon_2 = 2(\delta^{-2} - \delta)$. By the first condition of (b) the probability of the r.w. $\bar{S}_k := S_{\tau(\delta n)+k} - S_{\tau(\delta n)}$ ($k = 0, 1, \ldots$) exiting the interval $[-\varepsilon_1 w_n, \varepsilon_2 w_n]$ on the upper side approaches unity as $n \to \infty$. Combined with the relation ($\sharp$) above and the trivial fact that $S_k < S_{\tau(n/\delta^2)}$ for $k < \tau_{n/\delta^2}$, this yields

$$P[S_{\tau(\delta^2 n)} < S_k < S_{\tau(n/\delta^2)} \quad \text{for} \quad \tau_{\delta n} \le k < \tau_{n/\delta^2}] \longrightarrow 1.$$

Here one can replace the range of k by $s_n \le k \le s_{n/\delta}$, for which

$$\frac{S_{\tau(\delta^2 n)}}{B_k} \ge \frac{S_{\tau(\delta^2 n)}}{w_{\delta^2 n}} \cdot \frac{w_{\delta^2 n}}{B_{s(n/\delta)}} \xrightarrow{P} \delta^3 \quad \text{and} \quad \frac{S_{\tau(n/\delta^2)}}{B_k} \le \frac{S_{\tau(n/\delta^2)}}{w_{n/\delta^2}} \cdot \frac{w_{n/\delta^2}}{B_{s_n}} \xrightarrow{P} \frac{1}{\delta^2}.$$

Consequently, taking δ so close to 1 that $\delta^{-2} - \delta^3 < \varepsilon$, one finds (A.37) to be true.
□

A.5 Some Elementary Facts

A.5.1 An Upper Bound of the Tail of a Trigonometric Integral

Let $G(x)$ be a right continuous function of bounded variation on $x \ge 0$ such that $G(x) \to 0$ ($x \to \infty$). Then

$$\left| \int_x^\infty G(y) \begin{Bmatrix} \cos \theta y \\ \sin \theta y \end{Bmatrix} dy \right| \le \frac{1}{\theta} \left[|G(x)| + \int_x^\infty |dG(y)| \right];$$

in particular if $E|X| < \infty$,

$$\left| \int_x^\infty y\mu(y) \begin{Bmatrix} \cos\theta y \\ \sin\theta y \end{Bmatrix} dy \right| \le \frac{2}{\theta}[x\mu(x)+\eta(x)] \le \frac{4m(x)}{\theta x}. \tag{A.38}$$

We consider the bound of the sine transforms only; the other one is similar. The first inequality follows from the identity

$$\int_x^\infty G(y)\sin\theta y\, dy = \frac{1}{\theta}G(x)\cos\theta x + \frac{1}{\theta}\int_x^\infty \cos\theta y\, dG(y).$$

Applying it with $G(x) = x\mu(x)$ and observing $\int_x^\infty |d(y\mu(y))| \le \eta(x)+\int_x^\infty y\, d(-\mu(y))$ $= 2\eta(x)+x\mu(x)$ lead to the first inequality of (A.38). The second one follows from $x^2\mu(x) \le 2c(x)$.

A.5.2 A Bound of an Integral

If $h(t), t \ge 0$, is continuous and increasing, vanishes at $t = 0$, and satisfies $h(t)\, dt \le Ct\, dh(t)$, then for any $s > 0$ and $\varepsilon > 0$,

$$\int_0^s \frac{\varepsilon h(t)}{\varepsilon^2 + h^2(t)}\frac{dt}{t} \le C\arctan(h(s)/\varepsilon),$$

since by the assumption the integral above is not larger than $C\int_0^{h(s)/\varepsilon} dh/[1+h^2]$.

A.5.3 On the Green Function of a Transient Walk

Let F be transient so that we have the Green function

$$G(x) = \sum_{n=0}^\infty P[S_n = x] < \infty.$$

For $y \ge 0, x \in \mathbb{Z}$ it holds that

$$G(y-x) - g_\Omega(x,y) = \sum_{w=1}^\infty P_x[S_T = -w]G(y+w).$$

According to the Feller–Orey renewal theorem [31, Section XI.9], $\lim_{|x|\to\infty} G(x) = 0$ (under $E|X| = \infty$), showing that the RHS above tends to zero as $y \to \infty$ uniformly in $x \in \mathbb{Z}$; in particular $g_\Omega(x,x) \to G(0) = 1/P_0[\sigma_0 = \infty]$ as $x \to \infty$. It also follows that $P_x[\sigma_0 < \infty] = G(-x)/G(0) \to 0$ ($|x| \to \infty$).

References

1. L. Alili and R. Doney, Wiener–Hopf factorization revisited and some applications, *Stochastics Stochastics Rep.* **66**, 87–102 (1999).
2. B. Belkin, A limit theorem for conditioned recurrent random walk attracted to a stable law, *Ann. Math. Statist.*, **42**, 146–163 (1970).
3. B. Belkin, An invariance principle for conditioned recurrent random walk attracted to a stable law, *Zeit. Wharsch. Verw. Gebiete* **21**, 45–64 (1972).
4. J. Bertoin, *Lévy Processes*, Cambridge Univ. Press, Cambridge (1996).
5. J. Bertoin and R.A. Doney, On conditioning a random walk to stay nonnegative, *Ann. Probab.* **22**, no. 4, 2152–2167 (1994).
6. J. Bertoin and R.A. Doney, Spitzer's condition for random walks and Lévy processes, *Ann. Inst. Henri Poincaré* **33**, 167–178 (1997).
7. P. Billingsley, *Convergence of Probability Measures*, John Wiley and Sons, Inc. NY. (1968).
8. N.H. Bingham, G.M. Goldie and J.L. Teugels, *Regular Variation*, Cambridge Univ. Press (1989).
9. R. Blumenthal, R. Getoor and D. Ray, On the distribution of first exits for symmetric stable processes, *Trans. Amer. Math. Soc.* **99**, 540–554 (1961).
10. E. Bolthausen, On a functional central limit theorem for random walks conditioned to stay positive. *Ann. Probab.* **4**, 480–485 (1976).
11. A. Bryn-Jones and R.A. Doney, A functional central limit theorem for random walk conditioned to stay non-negative, *J. London Math. Soc.* (2) **74** 244–258 (2006).
12. F. Caravenna, A local limit theorem for random walks conditioned to stay positive, *Probab. Theory Related Fields* **133**, 508–530 (2005).
13. F. Caravenna and L. Chaumont, Invariance principles for random walks conditioned to stay positive, *Ann. Inst. Henri Poincaré- Probab et Statist.* **44**, 170–190 (2008).
14. F. Caravenna and R.A. Doney, Local large deviations and the strong renewal theorem, arXiv:1612.07635v1 [math.PR] (2016).
15. Y.S. Chow, On the moments of ladder height variables, *Adv. Appl. Math.* **7**, 46–54 (1986).
16. K.L. Chung, *A Course in Probability Theory*, 2nd ed. Academic Press (1970).
17. R.A. Doney, A note on a condition satisfied by certain random walks, *J. Appl. Probab.* **14**, 843–849 (1977).
18. R.A. Doney, On the exact asymptotic behaviour of the distribution of ladder epochs, *Stoch. Proc. Appl.* **12**, 203–214 (1982).
19. R.A. Doney, Conditional limit theorems for asymptotically stable random walks, *Z. Wahrsch. Verw. Gebiete* **70**, 351–360 (1985).
20. R.A. Doney, Spitzer's condition and ladder variables for random walks, *Probab. Theor. Rel. Fields* **101**, 577–580 (1995).
21. R.A. Doney, One-sided large deviation and renewal theorems in the case of infinite mean, *Probab. Theor. Rel. Fields* **107**, 451–465 (1997).
22. R.A. Doney, *Fluctuation Theory for Lévy Processes*, Lecture Notes in Math. 1897, Springer, Berlin (2007).
23. R.A. Doney, Local behaviour of first passage probabilities, *Probab. Theor. Rel. Fields* **152**, 559–588 (2012).

24. R.A. Doney and R.A. Maller, The relative stability of the overshoot for Lévy processes, *Ann. Probab.* **30**, 188–212 (2002).
25. R. Durrett, Conditioned limit theorems for some null recurrent Markov processes, *Ann. Probab.* **6**, 798–828 (1978).
26. E.B. Dynkin, Some limit theorems of independent random variables with infinite mathematical expectation. *Select. Transl. Math. Statist. Prob.* **1**, 171–178 (1961).
27. D.J. Emery, On a condition satisfied by certain random walks, *Zeit. Wharsch. Verw. Gebiete* **31**, 125–139 (1975).
28. K.B. Erickson, Strong renewal theorems with infinite mean, *Trans. Amer. Math. Soc.* **151**, 263–291 (1970).
29. K.B. Erickson, The strong law of large numbers when the mean is undefined, *Trans. Amer. Math. Soc.* **185**, 371–381 (1973).
30. S. Ethier and T. Kurtz, *Markov Processes: Characterization and Convergence*, John Wiley and Sons. Inc. Hoboken, NJ. (1986).
31. W. Feller, *An Introduction to Probability Theory and Its Applications*, vol. 2, 2nd ed., John Wiley and Sons, Inc. NY. (1971).
32. J.B.G. Frenk, The behavior of the renewal sequence in case the tail of the time distribution is regularly varying with index −1, *Advances in Applied Probability* **14**, 870–884 (1982).
33. A. Garsia and J. Lamperti, A discrete renewal theorem with infinite mean, *Comment. Math. Helv.* **37**, 221–234 (1962/3).
34. P.E. Greenwood, E. Omey and J.I. Teugels, Harmonic renewal measures, *Z. Wahrsch. verw. Geb.* **59**, 391–409 (1982).
35. P. Griffin and T. McConnell, Gambler's ruin and the first exit position of random walk from large spheres, *Ann. Probab.* **22**, 1429–1472 (1994).
36. V. Gnedenko and A.N. Kolmogorov, *Limit Distributions for Sums of Independent Random Variables*, Addison-Wesley, Reading, Mass. (1954) [Russian original 1949].
37. G. Herglotz, Über Potenzreihen mit positivem reellen Teil im Einheitskreis, *Ber. Verh. Sächs. Akad. Wiss. Leipzig. Math. Nat. Kl.* **63**, 501–511 (1911).
38. D.L. Iglehart, Functional central limit theorems for random walks conditioned to stay positive. *Ann. Probab.* **2**, 608–619 (1974).
39. H. Kesten, On a theorem of Spitzer and Stone and random walks with absorbing barriers, *Illinois J. Math.* **5**, 246–266 (1961).
40. H. Kesten, Random walks with absorbing barriers and Toeplitz forms, *Illinois J. Math.* **5**, 267–290 (1961).
41. H. Kesten, Ratio limit theorems II, *Journal d'Analyse Math.* **11**, 323–379 (1963).
42. H. Kesten, Hitting probabilities of single points for processes with stationary independent increments. *Mem. Amer. Math. Soc.* No. 93, Providence (1969).
43. H. Kesten, Problem 5716, *Amer. Math. Monthly* **77**, 197 (1970).
44. H. Kesten, The limit points of a normalized random walk, *Ann. Math. Statist.* **41**, 1173–1205 (1970).
45. H. Kesten and R.A. Maller, Ratios of trimmed sums and other statistics, *Ann. Probab.* **20**, 1805–1842 (1992).
46. H. Kesten and R.A. Maller, Infinite limits and infinite limit points of random walks and trimmed sums, *Ann. Probab.* **22**, 1473–1513 (1994).
47. H. Kesten and R.A. Maller, Divergence of a random walk through deterministic and random subsequences, *J. Theor. Probab.* **10**, 395–427 (1997).
48. H. Kesten and R.A. Maller, Stability and other limit laws for exit times of random walks from a strip or a half line, *Ann. Inst. Henri Poincaré* **35**, 685–734 (1999).
49. H. Kesten and F. Spitzer, Ratio limit theorems I, *Journal d'Analyse Math.* **11**, 285–322 (1963).
50. A. Kuznetsov, A.E. Kyprianou, J.C. Pardo and A.R. Watson, The hitting time for a stable process, *Electron. J. Probab.* **19** no. 30, 1–26 (2014).
51. A.E. Kyprianou, V.M. Rivero and W. Satitkanitkul, Conditioned real self-similar Markov processes. arXiv:1510.01781v1 (2015).
52. M. Loéve, *Probability Theory*, 3rd ed., Van Nostrand, New York (1963).
53. R.A. Maller, Relative stability, characteristic functions and stochastic compactness, *J. Austral. Math. Soc. (Series A)* **28**, 499–509 (1979).
54. S.V. Nagaev, The renewal theorem in the absence of power moments, *Theory Probab. Appl.* **56** no. 1, 166–175 (2012).
55. D. Ornstein, Random walks I and II. *Trans. Amer. Math. Soc.* **138**, 1–43 and 45–60 (1969).
56. H. Pantí, On Lévy processes conditioned to avoid zero, preprint: arXiv:1304.3191v1 [math.PR] (2013).

57. G. Peskir, The law of the hitting times to points by a stable Lévy process with no negative jumps, *Electronic Commun. Probab.* **13**, 653–659 (2008).
58. E.J.G. Pitman, On the behaviour of the characteristic function of a probability distribution in the neighbourhood of the origin, *J. Amer. Math. Soc. (A)* **8**, 422–43 (1968).
59. S.C. Port and C.J. Stone, Hitting time and hitting places for non-lattice recurrent random walks, *Journ. Math. Mech.* **17**, no. 1, 35–57 (1967).
60. S.C. Port and C.J. Stone, Infinitely divisible processes and their potential theory I & II, *Ann. Inst. Fourier* **21**, Fasc. 2, 157–275 & Fasc 4. 179–265 (1971).
61. S.C. Port, The exit distribution of an interval for completely asymmetric stable processes, *Ann. Math. Statist.* **41**, 39–43 (1970).
62. B.A. Rogozin, Local behavior of processes with independent increments, *Theory Probab. Appl.* **13**, 482–486 (1968).
63. B.A. Rogozin, On the distribution of the first ladder moment and height and fluctuations of a random walk, *Theory Probab. Appl.* **16**, 575–595 (1971).
64. B.A. Rogozin. The distribution of the first hit for stable and asymptotically stable walks on an interval, *Theory Probab. Appl.* **17**, 332–338 (1972).
65. B.A. Rogozin, Relatively stable walks, *Theory Probab. Appl.* **21**, 375–379 (1976).
66. L.C.G. Rogers and D. Williams, *Diffusions, Markov Processes and Martingales*, vol. 1, 2nd ed. Cambridge University Press (2000).
67. K. Sato, *Lévy Processes and Infinitely Divisible Distributions*, Cambridge Univ. Press, Cambridge (1999).
68. M. Shimura, A class of conditional limit theorems related to ruin problem, *Ann. Probab.* **11**, no. 1, 40–45 (1983).
69. F. Spitzer, A combinatorial lemma and its application to probability theory, *Trans. Amer. Math. Soc.* **82**, 323–339 (1956).
70. F. Spitzer, Hitting probabilities, *J. of Math. and Mech.* **11**, 593–614 (1962).
71. F. Spitzer, *Principles of Random Walks*, 2nd ed., Springer-Verlag, NY. (1976).
72. F. Spitzer and C.J. Stone, A class of Toeplits forms and their applications to probability theory, *Illinois J. Math.* **4**, 253–277 (1960).
73. C.J. Stone, Weak convergence of stochastic processes on semi-infinite time intervals, *Proc. Amer. Math. Soc.* **14**, 694–696 (1963).
74. C.J. Stone, On the potential operator for one-dimensional recurrent random walks, *Trans. Amer. Math. Soc.* **136**, 413–426 (1969).
75. M.L. Silverstein, Classification of coharmonic and coinvariant functions for a Lévy process, *Ann. Probab.* **8**, 539–575 (1980).
76. K. Uchiyama, One dimensional lattice random walks with absorption at a point / on a half line. *J. Math. Soc. Japan* **63**, 675–713 (2011).
77. K. Uchiyama, A note on summability of ladder heights and the distributions of ladder epochs for random walks, *Stoch. Proc. Appl.* **121**, 1938–1961 (2011).
78. K. Uchiyama, The first hitting time of a single point for random walks, *Elect. J. Probab.* vol. 16, no. 71, 1960–2000 (2011).
79. K. Uchiyama, One dimensional random walks killed on a finite set, *Stoch. Proc. Appl.* **127**, 2864–2899 (2017).
80. K. Uchiyama, Asymptotic behaviour of a random walk killed on a finite set, *Potential Anal.* **46**(4), 689–703 (2017).
81. K. Uchiyama, On the ladder heights of random walks attracted to stable laws of exponent 1, *Electron. Commun. Probab.* **23**, no. 23, 1–12 (2018).
82. K. Uchiyama, Asymptotically stable random walks of index $1 < \alpha < 2$ killed on a finite set, *Stoch. Proc. Appl.* **129**, 5151–5199 (2019).
83. K. Uchiyama, A renewal theorem for relatively stable variables, *Bull. London Math. Soc.* **52**, issue 6, 1174–1190, Dec. (2020).
84. K. Uchiyama, The potential function and ladder variables of a recurrent random walk on $\mathbb{Z}$ with infinite variance, *Electron. J. Probab.* **25**, article no. 153, 1–24 (2020).
85. K. Uchiyama, Estimates of potential functions of random walks on $\mathbb{Z}$ with zero mean and infinite variance and their applications, preprint available at: http://arxiv.org/abs/1802.09832 (2020).
86. K. Uchiyama, The two-sided exit problem for a random walk on $\mathbb{Z}$ with infinite variance I, preprint available at: http://arxiv.org/abs/1908.00303 (2021).
87. K. Uchiyama, The two-sided exit problem for a random walk on $\mathbb{Z}$ with infinite variance II, preprint available at: http://arxiv.org/abs/2102.04102 (2021).

88. V.A. Vatutin and V. Wachtel, Local probabilities for random walks conditioned to stay positive, *Probab. Theory Rel. Fields* **143**, 177–217 (2009).

89. S. Watanabe, On stable processes with boundary conditions, *J. Math. Soc. Japan* **14** no. 2, 170–198 (1962).

90. H. Widom, Stable processes and integral equations, *Trans. Amer. Math. Soc.* **98**, 430–449 (1961).

91. A. Zygmund, *Trigonometric Series*, vol. 2, 2nd ed., Cambridge Univ. Press (1959).

92. V.M. Zolotarev, *One-Dimensional Stable Distributions*, Translations of mathematical monographs, vol. 65, AMS, Providence Rhode Island (1986).

Notation Index

© The Author(s), under exclusive license to Springer Nature Switzerland AG 2023
K. Uchiyama, *Potential Functions of Random Walks in $\mathbb{Z}$ with Infinite Variance*,
Lecture Notes in Mathematics 2338, https://doi.org/10.1007/978-3-031-41020-8

Subject Index

almost decreasing/increasing, 10
asymptotic properties of $p_t^0(\xi, \eta)$ and
    $q_t(\xi)$, 244
asymptotics of hitting probabilities
    $P_x[\sigma_R < \sigma_0]$, 76, 77
    $P_x[\sigma_R < \sigma_{B(R)}]$, 188
    $P_x[\sigma_R < T]$, 139, 156,
    180–182
    $P_x[\sigma_R < T | \Lambda_R]$, 181

comparison between σ_R and $\sigma_{[R,\infty)}$,
    94

dominated variation, 10, 149
duality, — in Feller's sense, 10

escape probability
    one-sided, 78, 94
    two-sided, 95, 96, 99, 227, 230

Feller's duality, 10

Green's function $g_B(x, y)$, 8
    for a finite set, 208, 209
    for a single point, 8
    for the complement of an
    interval, 102, 141, 186, 187
    for the negative half-line, 9, 85,
    86, 136, 141, 176, 177
    scaling limit, 141, 189

Green's kernel, $G(x)$, for a transient
    walk, 136, 157, 211

harmonic function, 8, 76, 111, 209

ladder epoch, 260
ladder height, 9
    distribution, 97, 111, 114, 118,
    122–124
    strictly ascending, 9
    strictly descending, 9
left/right-continuous random walk,
    75

normalised s.v. function, 10, 107
n.r.s. (negatively relatively stable),
    14, 135

oscillatory random walk, 7
overshoot, 15–17, 79, 257
    distribution, 84, 131, 166, 203,
    258

partial sums of g_Ω, 111
potential function, 8
    asymptotics, 51, 57, 62
    asymptotics of the increments,
    68, 72
    for a finite set, 208, 209
    irregular behaviour, 45
    known results, 11
    upper and lower bounds, 20–22

K. Uchiyama, *Potential Functions of Random Walks in* $\mathbb{Z}$ *with Infinite Variance*,
Lecture Notes in Mathematics 2338, https://doi.org/10.1007/978-3-031-41020-8

Printed in the United States
by Baker & Taylor Publisher Services